New Challenges in the Cosmetics Industry

New Challenges in the Cosmetics Industry

Topic Editors

Ana Catarina Silva
Hugo Almeida
Ana Barros

MDPI

Basel • Beijing • Wuhan • Barcelona • Belgrade • Novi Sad • Cluj • Manchester

Topic Editors
Ana Catarina Silva
Faculty of Health Sciences
University Fernando Pessoa
Porto
Portugal

Hugo Almeida
Mesosystem Investigação & Investimentos by Spinpark
Guimarães
Portugal

Ana Barros
Centre for the Research and Technology of Agro-Environmental and Biological Sciences (CITAB)
University of Trás-os-Montes and Alto Douro (UTAD)
Vila Real
Portugal

Editorial Office
MDPI AG
Grosspeteranlage 5
4052 Basel, Switzerland

This is a reprint of the Topic, published open access by the journals *Pharmaceuticals* (ISSN 1424-8247), *Pharmaceutics* (ISSN 1999-4923), *Antioxidants* (ISSN 2076-3921), *Journal of Functional Biomaterials* (ISSN 2079-4983) and *Cosmetics* (ISSN 2079-9284), freely accessible at: https://www.mdpi.com/topics/1RAC67W74N.

For citation purposes, cite each article independently as indicated on the article page online and as indicated below:

Lastname, A.A.; Lastname, B.B. Article Title. *Journal Name* **Year**, *Volume Number*, Page Range.

ISBN 978-3-7258-5523-0 (Hbk)
ISBN 978-3-7258-5524-7 (PDF)
https://doi.org/10.3390/books978-3-7258-5524-7

Contents

About the Editors

Ana Catarina Silva

Ana Catarina Silva is an associate professor of pharmaceutical technology and pharmaceutical biotechnology and a senior researcher at UCIBIO, Faculty of Pharmacy, University of Porto, Portugal, as well as a researcher at FP-I3ID, FP-BHS, Faculty of Health Sciences, University Fernando Pessoa, Porto, Portugal. She is also a senior researcher at UCIBIO, REQUIMTE, Faculty of Pharmacy, University of Porto, Portugal. Her research focuses on the development of lipid-based nanosystems for enhanced drug delivery, particularly for nose-to-brain and cutaneous administration. Professor Silva has published over 50 articles, 6 book chapters, 60 posters, and 23 oral communications, and has a Scopus h-index of 28. She has edited three books and has served as an invited editor on 12 special issues. She was included in Stanford University's World's Top 2% Scientists list from 2021 to 2024 and has served as a referee for multiple international funding agencies. She is a member of the editorial boards of the Springer book series "Advances in Biochemical Engineering/Biotechnology" and the journals *Dermatological Drug Delivery* (Frontiers in Drug Delivery), *Chinese Chemical Letters*, *Current Drug Metabolism*, and *Biochemistry & Pharmacology: Open Access.*

Hugo Almeida

Hugo Almeida received his degree in pharmaceutical sciences from the Faculty of Pharmacy, University of Coimbra, Portugal, in 2006, and his master's degree in quality control (focusing on drug substances and medicinal plants) from the Faculty of Pharmacy, University of Porto, Portugal, in 2009. In 2016, he received his PhD degree in pharmaceutical sciences (focusing on pharmaceutical technology) from the Faculty of Pharmacy, University of Porto, Portugal. He has more than 20 years of experience in the medical device industry, as a technical director and head of the Quality Control Department in a medical device company, as well as one year of experience in the pharmaceutical industry. At present, he is a production manager in a medical device and cosmetics company. He is also a researcher at UCIBIO, Laboratory of Pharmaceutical Technology, Faculty of Pharmacy, University of Porto, Portugal. He is developing his research into the application of stimuli-responsive polymers in controlled and self-regulated drug delivery systems and also in developing lipid-based nanosystems to improve drug delivery. He is the author of several scientific articles, the editor of one book, the author of one book chapter, an invited editor on two special issues in the journals *Pharmaceutics* and *Current Pharmaceutical Design*, and an invited editor of two Topics for MDPI.

Ana Barros

Ana Barros is an associate professor with habilitation from the University of Trás-os-Montes and Alto Douro. She has been the director of the Centre for the Research and Technology of Agro-Environmental and Biological Sciences (CITAB) for 5 years (May 2017–July 2022) and of the Institute for Innovation, Capacity building and Sustainability of Agri-Food production (Inov4Agro) for a year and half (January 2021–July 2022). She is responsible for the Phytochemicals Laboratory, and Coordinator of the Agri-Food Quality Group. Her research targets are mainly in the identification, separation, and recovery of functional molecules from different natural products, as well as their implementation as ingredients and bioactive compounds in the food, pharma, and cosmetics industries, with the ultimate goal of extracting high added value molecules and reuse them in the circular economy and industrial symbiosis concepts. In recent years, she has been the recipient of several awards and distinctions. She has published more than 140 scientific documents, including research articles, reviews, and editorials, within the highest impact factor journals in the food science and technology field (h-index 38, SCOPUS, September 2025). She has registered 10 national and international patents. She has international collaborations and is a PI and member of several research projects with different typologies (last 5 years). She is the coordinator from the European project WASTELESS—TOPIC ID: HORIZON-CL6-2022-FARM2FORK-01-08. Her scientific work has raised interest and has led to the supervision of post-docs (3 concluded, 1 ongoing), PhDs (12 concluded and 7 ongoing), master's studies (56 concluded and 7 ongoing), and bachelor (48 concluded) students. She has been a member of the organizing committees of scientific conferences, presented invited lectures at international and national conferences, and has produced over 400 oral/poster communications, including papers in conference proceedings.

Preface

Currently, cosmetics manufacturers face several challenges in developing innovative products, with the main objective of meeting the different needs of consumers. In this context, the increasingly widespread use of natural compounds as a source of bioactive molecules, associated with the reuse of industrial and natural waste, in a circular economy perspective, brings economic, environmental and performance advantages to this new product. These bioactive molecules have been showing different activities, such as antioxidant, anti-inflammatory, and antimicrobial. In addition, the advent of the biotechnology industry has brought new ingredients, including growth factors, enzymes, peptides, and stem cells, among others. However, to make the use of these bioactive molecules feasible, optimization of their extraction, purification, and formulation is required. In the past 20 years, we have been witnessing the introduction of nanotechnology systems in cosmetic and medical device products. Their use brings several advantages, such as enabling targeted delivery and enhancing the absorption of molecules through the skin and external mucous membranes, protecting compounds from degradation, and allowing prolonged release, among others. Nevertheless, applying this type of technology in cosmetic products raises regulatory challenges that manufacturers must carefully consider and that deserve thorough reflection. With this topic, the editors aim to discuss the future perspectives of cosmetic products, considering not only the incorporation of bioactive molecules derived from residues and biotechnological processes, but also the application of nanotechnology to enhance product efficacy.

We are pleased to share the latest research and review articles addressing the new challenges facing the cosmetic industry. Readers will have the opportunity to read four insightful review articles which focus on different aspects of the cosmetics industry. The first article highlights the latest advances in the encapsulation of marine bioactive compounds in lipid nanoparticles (SLN and NLC) for health applications. The second article covers the potential of coffee and olive by products as ingredients in cosmetic formulations, including their extraction techniques. The third article discusses recent developments of multifunctional surgical sutures designed to promote wound regeneration through antibacterial properties, drug elution, and biodegradability. Finally, the fourth review provides a systematic analysis of the main outcomes of using lipid mixtures as Stratum Corneum (SC) lipid matrix models for pharmaceutical and cosmetic applications. On the other hand, the first research article investigates the bioactive potential of molecules extracted from a blend of grape stems (GS blend extract) and their use in cosmetic products with antioxidant, anti-aging and depigmenting properties. The second research article concluded that the combination of bipolar radiofrequency and non-crosslinked hyaluronic acid mesotherapy improved the skin appearance and the epidermal barrier function. The third research article evaluated the systemic safety of Althaea rosea flower extract (ARFE) as a cosmetic ingredient. The fourth research article demonstrates the modulatory influence of polyflavonoids extracted from Citrus aurantifolia, lemon peel extract (LPE-polyflavonoids) on attenuating diabetes mellitus (DM) and its complications. The fifth research article describes the design and optimization of PLGA and PCL nanoparticles for delivering baricitinib (BTB) for topical ocular administration.

We would like to thank all the authors for their outstanding contributions through their valuable literature reviews and scientific research on the latest advances and future challenges in the cosmetics industry.

Ana Catarina Silva, Hugo Almeida, and Ana Barros
Topic Editors

pharmaceutics

MDPI

Review

Advances in Encapsulating Marine Bioactive Compounds Using Nanostructured Lipid Carriers (NLCs) and Solid Lipid Nanoparticles (SLNs) for Health Applications

Rita Favas [1,2], Hugo Almeida [1,2,3], Andreia F. Peixoto [4], Domingos Ferreira [1,2] and Ana C. Silva [1,2,5,*]

1 UCIBIO (Applied Molecular Biosciences Unit), Laboratory of Pharmaceutical Technology, Faculty of Pharmacy, University of Porto, 4050-313 Porto, Portugal
2 Associate Laboratory i4HB Institute for Health and Bioeconomy, Faculty of Pharmacy, University of Porto, 4050-313 Porto, Portugal
3 Mesosystem Investigação & Investimentos by Spinpark, 4805-017 Guimarães, Portugal
4 LAQV-REQUIMTE (Associated Laboratory for Green Chemistry of the Network of Chemistry and Technology), Departamento de Química e Bioquímica, Faculdade de Ciências, Universidade do Porto, 4169-007 Porto, Portugal
5 FP-BHS (Biomedical and Health Sciences Research Unit), FP-I3ID (Instituto de Investigação, Inovação e Desenvolvimento), Faculty of Health Sciences, University Fernando Pessoa, 4249-004 Porto, Portugal
* Correspondence: anacatsil@gmail.com

Abstract: As life expectancy rises and modern lifestyles improve, there is an increasing focus on health, disease prevention, and enhancing physical appearance. Consumers are more aware of the benefits of natural ingredients in healthcare products while also being mindful of sustainability challenges. Consequently, marine bioactive compounds have gained popularity as ingredients in cosmetics and food supplements due to their diverse beneficial properties. Nonetheless, the use of some of these compounds is restricted by their low stability and poor aqueous solubility, necessitating solutions to overcome these limitations. In this context, lipid nanoparticles, such as solid lipid nanoparticles (SLNs) and nanostructured lipid carriers (NLCs), have been investigated for their potential to protect and improve the absorption of molecules through various routes, including oral and cutaneous. Numerous studies have shown that nanoencapsulating these compounds and incorporating them into cosmetics and food supplements can be effective. However, this application remains unregulated at the global level and is not currently addressed by existing legislation. Additional in vivo studies in both animals and humans are necessary to fully assess safety concerns.

Keywords: healthcare; cosmetics; food supplements; marine bioactive compounds; lipid nanoparticles

1. Introduction

As life expectancy rises and modern lifestyles continue to evolve, there is a growing focus on health, particularly in disease prevention and enhancing physical appearance. Consumers are increasingly recognizing the advantages of incorporating natural compounds into health products [1].

The marine ecosystem provides a rich source of active ingredients that can be incorporated into cosmetics and food supplements. Among marine organisms, algae, crustaceans, bacteria, and fungi stand out as promising sources of bioactive compounds with different properties, such as antimicrobial and antioxidant, which are promising at different stages of product development, from processing to formulation and storage [2]. This vast source of bioactive compounds includes fatty acids, carotenoids, and vitamins, each of which has a wide range of health benefits.

Fatty acids are essential components of fats and oils, playing a crucial role in regulating various physiological processes [3]. They are categorized as monounsaturated fatty acids (MUFAs) or polyunsaturated fatty acids (PUFAs) based on their length, degree of unsaturation, and the number of chemical bonds. PUFAs, particularly omega-3, are the predominant type of fatty acids found in the marine environment and can be sourced from algae (especially red macroalgae), marine by-products, yeasts, bacteria, and fungi, serving as alternative options to traditional seafood for human consumption [3,4].

Carotenoids are isoprenoid pigments in the yellow-to-red spectrum and play a crucial role in various physiological activities [5]. They are produced by all photosynthetic organisms and obtained through diet by other organisms who cannot synthesize them [6]. Carotenoids are categorized into two main groups based on their structure and chemical composition. The first group includes carotenes and hydrocarbons without oxygen, such as β-carotene and lycopene. The second group includes xanthophylls, which are oxygenated derivatives of carotenes, such as astaxanthin, fucoxanthin, and zeaxanthin [7].

Vitamins are essential for proper metabolism and have an impact on overall health. As the body cannot produce enough vitamins, humans must get them from food or supplements. Vitamins are classified based on their properties, whether they are water-soluble or fat-soluble. While fat-soluble (i.e., lipophilic) vitamins, such as A, D, E, and K, are easily stored upon absorption, water-soluble (i.e., hydrophilic) vitamins, such as C and B, are eliminated and not stored in the body [8,9]. Common sources of these vitamins include fish, algae, bacteria, fungi, corals, sea cucumbers, and sponges [10–13].

Various methods and carriers have been used in the cosmetic and food sectors, such as microparticles (i.e., microcapsules or microspheres) and nanocarriers (e.g., liposomes, nanoemulsions, and lipid nanoparticles) [14]. Due to the lipophilic nature of some bioactive compounds, they cannot be incorporated into aqueous formulations. Encapsulation techniques have, therefore, been explored to increase the solubility, stability, and activity of these compounds [15]. They are used in the food industry to preserve components from environmental conditions, mask unpleasant tastes, or extend product shelf life [16]. In cosmetics, they can improve skin hydration and epidermal delivery, increasing local compound concentration with consequent less/no systemic absorption [17]. Although microencapsulation has been extensively studied, nanoencapsulation techniques have been found to resolve some limitations, leading to the production of more stable carriers that improve the absorption of bioactive molecules [16]. Among these, lipid nanoparticles are gaining attention for their effectiveness in encapsulating, protecting, and promoting the absorption of lipophilic molecules [15,18,19]. Furthermore, hydrophilic bioactive compounds, such as polysaccharides and proteins, are being used to modify the surface of lipid nanoparticles. Polysaccharides and proteins are important macromolecules in various living organisms and are critical in promoting overall body function and balance. They serve as energy sources and are essential for molecular recognition, defense, tissue formation, energy storage, and cellular transport [20–22]. Polysaccharides derived from different marine sources can be categorized into marine animal polysaccharides (i.e., chitin and chitosan), marine plant polysaccharides (i.e., alginate), and marine microbial polysaccharides (i.e., glucan) [20,23,24]. Fish and algae are the major sources of marine proteins, with the most interesting and useful being collagen and gelatin. These polymers have been applied to increase the stability and compatibility of lipid nanoparticles in different products and processes [25].

This review aims to provide an overview of the State of the Art in the use of marine bioactive compounds encapsulated in lipid nanoparticles, namely, nanostructured lipid carriers (NLCs) and solid lipid nanoparticles (SLNs), in cosmetics and food supplements. Regulatory and safety aspects of the use of this type of nanoparticles and the future directions of this area of research are also discussed.

2. Lipid Nanoparticles

Lipid nanoparticles are spherical particles containing a solid matrix composed of one or more lipids, where the encapsulated molecules are solubilized or dispersed, which is stabilized by one or two emulsifying agents. Depending on their internal structure and composition, these nanoparticles can be classified as solid lipid nanoparticles (SLNs) or nanostructured lipid carriers (NLCs) [17,26,27].

SLNs (Figure 1) were the first described and are composed of one solid lipid (5–40%) that contains dissolved or dispersed encapsulated molecules, one or two emulsifiers (0.5–5%), and water (q.s. 100%). The solid lipid forms a protective matrix that improves physical–chemical stability and slows the release of encapsulated molecules. The lipid matrix is solid at body temperature, protecting the molecules from degradation caused by environmental factors, such as light, heat, pH, and enzymes. Being composed of physiological lipids (known as Generally Recognized as Safe—GRAS), this type of lipid nanoparticle is biocompatible, exhibits good permeation through various body tissues, and can be easily incorporated in different pharmaceutical dosage forms for topical, oral, and parenteral administration [19,26,27]. Additionally, researchers have investigated the potential of targeting molecules encapsulated within surface-modified lipid nanoparticles for a range of therapeutic applications, such as neurological disorders, infections, and cancer. The findings so far have shown promising outcomes, highlighting the versatility and effectiveness of these nanoparticles in delivering treatments to specific sites within the body [28–32].

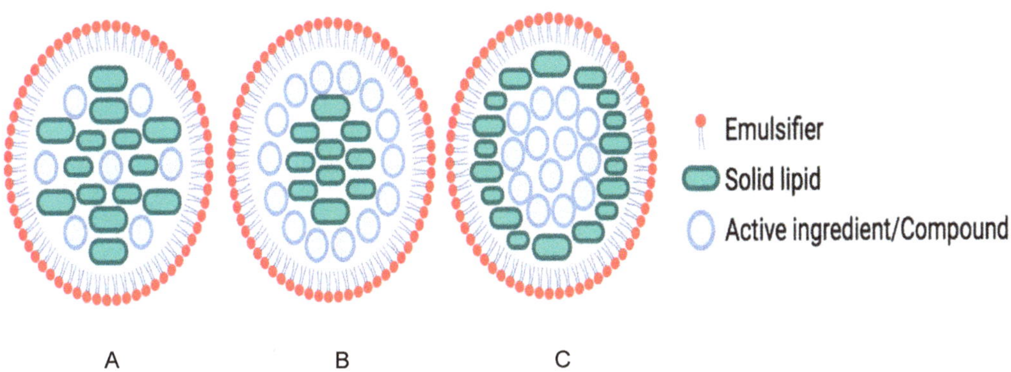

Figure 1. Different types of SLNs: (**A**) Homogenous matrix, (**B**) drug-enriched shell, and (**C**) drug-enriched core.

Depending on the production method and the location of the encapsulated molecules, SLNs are classified as different types or models: homogenous matrix (Figure 1A), drug-enriched shell (Figure 1B), or drug-enriched core (Figure 1C). However, SLNs have limitations, mainly related to their low capacity to encapsulate molecules in the lipid matrix and excessive crystallization of the lipid during storage, leading to leakage of the encapsulated molecules. Thereby, more advanced nanocarriers derived from SLNs have been developed, nanostructured lipid carriers (NLCs), which have a solid matrix resulting from the combination of a solid lipid and a liquid lipid, the former in a higher percentage (Figure 2). Based on the location of the encapsulated molecules, there are different types of NLCs: Type 1, or the imperfect type (Figure 2A), occurs when the NLCs have an imperfect solid matrix structure and solid and liquid lipids are mixed in different nanostructures. The second type is the amorphous type (Figure 2B) and is obtained by mixing solid and liquid lipids in such a way as to prevent the formation of a solid crystalline structure, avoiding molecules from leakage during storage. The third type (Figure 2C) is the multiple type. Since lipophilic molecules are more soluble in liquid lipids than in solid lipids, in this type of NLC, there are oil moieties blended into the solid lipid, forming nano-compartments where the molecules are confined. Compared to SLNs, NLCs have been shown to have a higher loading capacity and a greater ability to prevent the release of molecules during storage by inhibiting or reducing the crystallization of lipids. These advantages have been attributed to the presence of the liquid lipid, which results in a more disorganized lipid matrix with more space to accommodate molecules [17,26,27,33–35].

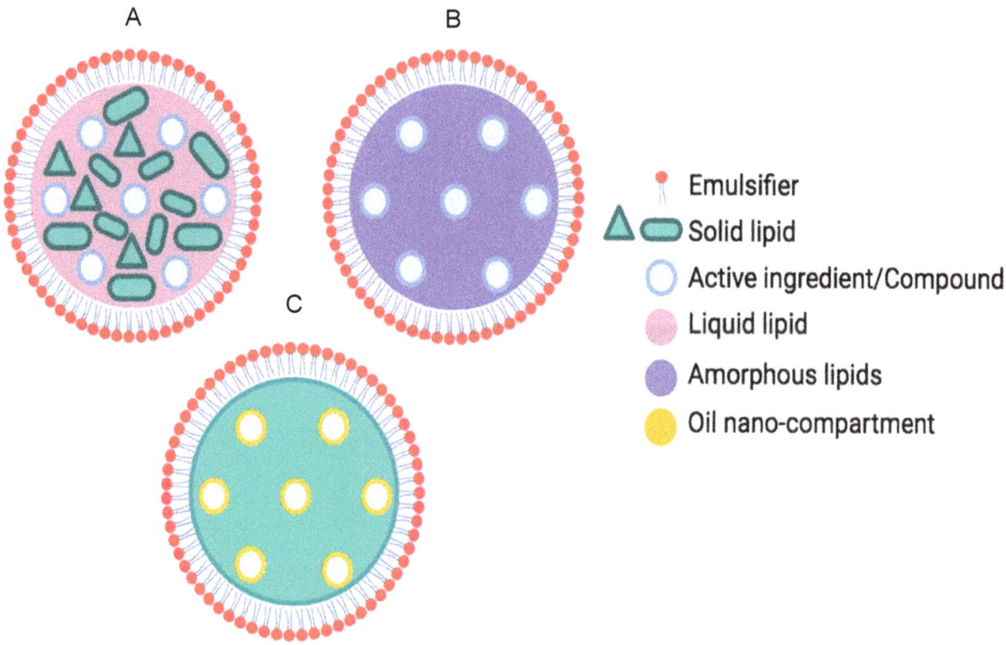

Figure 2. Different types of NLCs: (**A**) Imperfect, (**B**) amorphous, and (**C**) multiple.

Interested readers can consult the references provided for very comprehensive reviews of the characteristics of SLNs and NLCs.

2.1. Relevant Studies with Bioactive Marine Compounds Encapsulated in SLNs and NLCs

Given the various applications of marine bioactive compounds in healthcare and the possibility of encapsulating them in SLNs and NLCs to overcome their limitations, an analysis of published studies will provide good prospects for their potential use. Although most of the research focuses on the encapsulation of these compounds, some studies also present information on their potential health applications.

Additionally, marine bioactive compounds, such as chitosan, alginate, and gelatin, have been tested as coating polymers to modify the surface of SLNs and NLCs. They can increase their stability, performance, delivery efficiency, absorption rate, and biocompatibility [25]. The summary of key findings from relevant studies reveals a notable trend over the years. Starting from a single publication in 2006, there was sporadic research until a gradual increase beginning in 2017. The number of studies then surged, with the highest concentration occurring in 2020 and 2023, each with five published works. Table 1 provides details of the studies conducted between 2006 and 2023, with a total of 24 studies.

Table 1. Relevant studies and applications of the main bioactive marine compounds encapsulated in SLNs and NLCs.

Marine Bioactive Compound	Type of Lipid Nanoparticle	Type of Study	Relevant Results	Healthcare Application	References
Docosahexaenoic and α-linolenic acid	SLN	In vitro	⇒ Encapsulation in SLNs improved the docosahexaenoic acid (277.2%) and α-linolenic acid (222.7%) uptake by colorectal cancer cells. ⇒ Docosahexaenoic acid-loaded SLNs showed greater inhibition of cell growth (68.6% and 80% in colorectal adenocarcinoma—HT29 and human colon cancer cells—and HCT116 cells, respectively) when compared to free docosahexaenoic acid at the same concentration (29.7% and 55.3% in HT-29 and HCT116 cells, respectively). ⇒ In HT-29 cells, α-linolenic acid-loaded SLNs were significantly more effective ($p < 0.02$) in inhibiting tumor cell growth than free α-linolenic acid at all the concentrations tested (5 μM, 10.7% vs. 34.6%; 10 μM, 12% vs. 36.3%; 50 μM, 2.3% vs. 38.2%). ⇒ In HCT116 cells, α-linolenic acid-loaded SLNs were significantly more effective in inhibiting tumor cell growth than free α-linolenic acid at concentrations of 10 and 50 μM (22.5% vs. 29%, $p < 0.05$ and 29% vs. 79.1%, $p < 0.001$, respectively).	Food supplement	[36]
Docosahexaenoic acid	NLC	In vitro and in vivo	⇒ NLC-loaded docosahexaenoic acid showed improved anti-inflammatory and antioxidant properties when compared to free docosahexaenoic acid, with DPPH of 0.57 ± 0.03 and 0.17 ± 0.003, respectively. ⇒ In vivo studies show that NLC-loaded docosahexaenoic acid had the most effective suppression of gingival inflammation compared to the pure free compound.	Food supplement	[37]
β–carotene	NLC	In vivo	⇒ In vivo permeation studies in humans show that β-carotene-loaded NLC improved skin penetration when compared to pure β-carotene: 34% of the encapsulated compound reached the deeper layers of the skin, while 45% of pure β-carotene was retained in the outer layers of the stratum corneum.	Cosmetic	[38]
		In vitro	⇒ β-carotene-loaded NLC demonstrated higher bioavailability (60.7%) and antioxidant properties (ABTS: 91.47 ± 1.9%, DPPH: 24.72 ± 0.38%, and IC_{50}: 7.0 μg/mL) when compared to a β-carotene emulsion, a β-carotene–Tween® 80 phosphate-buffered solution, and a β-carotene–phosphate-buffered solution.	Food supplement	[39]

Table 1. *Cont.*

Marine Bioactive Compound	Type of Lipid Nanoparticle	Type of Study	Relevant Results	Healthcare Application	References
Astaxanthin	NLC	In vitro and ex vivo	⇒ Astaxanthin-loaded NLC demonstrated improved stability, sustained release ($83.0 \pm 3.4\%$ at 48 h), skin permeability (174.10 ± 4.38 µg/cm^2), and skin retention (18.60 ± 1.62 µg/cm^2) properties when compared to pure astaxanthin.	Cosmetic	[40]
		In vitro and in vivo	⇒ In vitro studies in human fibroblasts show an ROS reduction of 81.6%, DNA damage reduction of 41.6%, and cell death reduction of 62.69%. ⇒ In vivo studies show protection of all tested mice from skin damage.	Cosmetic	[41]
		In vitro	⇒ The antioxidant activity of astaxanthin was maintained after encapsulation in NLC: the results of DPPH were 95% for the free astaxanthin and 100% for astaxanthin-loaded NLC ($p \leq 0.05$). ⇒ In vitro release studies show that compared to the free astaxanthin (90% in 12 h), the formulation exhibited a more sustained release profile (88% in 24 h).	Food supplement	[42]
	SLN and NLC	In vitro	⇒ Both astaxanthin-loaded NLC and astaxanthin-loaded SLNs were safe for nasal (RPMI 2650) and neuronal cells (SH-SY5Y) at concentrations up to 100 µg/mL, showing ability to inhibit neurodegenerative pathways, particularly oxidative stress.	Food supplement	[43]
	NLC	In vitro and in vivo	⇒ In vitro studies show a significant decrease in oxidative stress, neuroinflammation, amyloidogenic pathway, and apoptosis with astaxanthin-loaded NLC when compared to a free astaxanthin solution. ⇒ In vivo studies in mice show that astaxanthin-loaded NLC improved cholinergic neurotransmission when compared to a free astaxanthin solution.		[44]
Lycopene	NLC	In vitro	⇒ Lycopene-loaded NLC exhibited higher antioxidant capacity (36.6 ± 0.4 mM/mg and IC_{50}: 14.1 ± 0.6mg/mL) when compared to placebo NLC (26.6 ± 0.1 mM/mg and IC_{50}: 17.7 ± 0.4mg/mL).	Cosmetic	[45]

Table 1. *Cont.*

Marine Bioactive Compound	Type of Lipid Nanoparticle	Type of Study	Relevant Results	Healthcare Application	References
Fucoxanthin	NLC	In vitro	⇒ Fucoxanthin-loaded NLC was safe for concentrations up to 5 mM in human fibroblasts and significantly decreased the expression of psoriatic markers by approximately 40%.	Cosmetic	[46]
	SLN		⇒ The sun protection factor of fucoxanthin-loaded SLNs was significantly higher (1.85 times) than that of the other formulations tested (3 different chemical sunscreens).		[47]
Vitamin A	SLN	In vitro and ex vivo	⇒ 3400 ng of vitamin A was detected in the stratum corneum and upper epidermis when using vitamin A-loaded SLN formulation. After 24 h, the concentration increased in the deeper skin layers (from 0 to 250 ng).	Cosmetic	[48]
		In vitro, ex vivo and in vivo	⇒ In vitro studies show 54.38% of vitamin A release from SLNs in 24 h. ⇒ Ex vivo studies show that the concentration of vitamin A in the skin after applying vitamin A-loaded SLNs was almost 2 times higher when compared to a vitamin A hydrogel. ⇒ In vivo studies in mice showed an increased thickness of the stratum corneum and a primary irritation index of 0.00, after applying vitamin A-loaded SLNs.		[49]
	SLN and NLC	In vitro	⇒ Both vitamin A-loaded SLNs and vitamin A-loaded NLC demonstrated excellent stability under food processing conditions and gastric conditions, and it was observed that approximately 80% of the encapsulated vitamin A remained intact.	Food supplement	[50]
Vitamin D	NLC	In vitro	⇒ Vitamin D3-loaded NLC was more effective in inhibiting breast cancer cell proliferation (37 ± 5.1%) and increasing the percentage of cells in the apoptotic phase (40 ± 3.45%) when compared to a free vitamin D3 solution.	Food supplement	[51]
		In vivo	⇒ In vivo studies in mice show that the colon concentration of vitamin D3 was three times higher than the basal level and persisted for 12 h after the administration of vitamin D3-loaded NLC. Progressive mitigation of the symptoms was observed from day 3.		[52]

Table 1. *Cont.*

Marine Bioactive Compound	Type of Lipid Nanoparticle	Type of Study	Relevant Results	Healthcare Application	References
Vitamin E	NLC	In vitro	⇒ Vitamin E-loaded NLC was biocompatible with human skin keratinocytes (HaCAT cells) up to concentration of 10 μg/mL and showed an irritation score of 0.00.	Cosmetic	[53]
		In vitro and in vivo	⇒ In vitro studies show that vitamin E-loaded NLC demonstrated superior occlusive properties (80%) when compared to a vitamin E nanoemulsion. ⇒ In vivo studies in humans show that vitamin E-loaded NLC (12 g/H·m^2) reduced the TEWL when compared to a vitamin E nanoemulsion (14 g/h·m^2).		[54]
			⇒ In vivo studies in humans show no signs of erythema after 48 h of applying the vitamin E-loaded NLC. In addition, skin moisture increased by 80%, and skin elasticity improved significantly ($p = 0.0319$) over 12 weeks.		[55]
	SLN	In vitro	⇒ Hemolysis was minimal at all tested concentrations of vitamin E-loaded SLNs. ⇒ Vitamin E-loaded SLNs significantly boosted lymphocyte proliferation by approximately 150% and reduced DNA damage.	Food supplement	[56]
Vitamin K	SLN	In vitro	⇒ Vitamin K1-loaded SLNs showed stability in simulated gastrointestinal conditions and during storage.	Food supplement	[57]
Chitosan	NLC surface coating	In vitro and ex vivo	⇒ Chitosan-coated NLC showed excellent retention in the skin (581 ng after 4 h), biocompatibility, and efficacy in photodynamic therapy (50% reduction in melanoma cell viability after irradiation).	Cosmetic	[58]
Alginate	NLC surface coating	In vitro and ex vivo	⇒ Ex vivo studies in pig skin show that alginate-coated NLC originated high skin penetration of vitamin E (1.9-fold) and quercetin (2.3-fold) in damaged skin. ⇒ TEWL was reduced to −7 g/h·m^2 when compared to the water control (3 g/h·m^2).	Cosmetic	[59]
Gelatin	NLC surface coating	In vitro	⇒ Gelatin-coated NLC showed improved antioxidant and antimicrobial activities (0.1 ± 0.00 mg/mL) of the encapsulated compound for 90 days.	Food supplement	[60]

Abbreviations: ABTS: 2,2′-azino-bis(3-ethylbenzothiazoline-6-sulfonic acid) assay; DPPH: 2,2-diphenyl-1-pyridylohydrazinyl assay; NLCs: nanostructured lipid carriers; SLNs: solid lipid nanoparticles. TEWL: transepidermal water loss.

2.1.1. Omega 3

Research on SLNs and NLCs loaded with omega-3 mostly focuses on their use as a component of the oil phase rather than as a bioactive compound. However, a few studies have explored the latest application. For instance, Serini et al. encapsulated α-linolenic and docosahexaenoic acid in SLNs. In the experiments, the researchers evaluated the physiochemical properties of SLNs and their effect on human colorectal cancer cells (HT-29 and HCT116 cell lines). Regarding docosahexaenoic acid and α-linolenic acid encapsulation, the particle size was 100 ± 1.8 nm and 842.2 ± 1.3 nm, the polydispersity index (PDI) was 0.220 ± 0.020 and 0.126 ± 0.017, and the encapsulation efficiency (EE) was 100% and 77%, respectively. The results show that after encapsulation, there was a significant increase in the uptake of α-linolenic acid (222.7%, $p < 0.02$) in cancer cells after 24 h of incubation. The incorporation of docosahexaenoic acid was also notably higher at 24 h when encapsulated in SLNs (277.2%, $p < 0.009$). At all tested concentrations, free docosahexaenoic acid and docosahexaenoic acid-loaded SLNs demonstrated a time-dependent inhibition of colorectal cancer cell growth. After 48 and 72 h, the 50 μM docosahexaenoic acid-loaded SLNs demonstrated a significantly greater inhibition of cell growth (68.6%, $p < 0.01$ and 80%, $p < 0.001$ in HT29 and HCT116 cells, respectively) compared to free docosahexaenoic acid at the same concentration (29.7% in HT-29 and 55.3% in HCT116). In addition, α-linolenic acid exhibited a significantly inhibitory effect on the growth of HT-29 and HCT116 cells after 48 h, with a more pronounced effect observed after 72 h. Notably, in HT-29 cells, α-linolenic acid-loaded SLNs were significantly more effective ($p < 0.02$) in inhibiting tumor cell growth than free α-linolenic acid at all the concentrations tested (5 μM, 10.7% vs. 34.6%; 10 μM, 12% vs. 36.3%; 50 μM, 2.3% vs. 38.2%). In HCT116 cells, α-linolenic acid-loaded SLNs demonstrated significantly enhanced efficacy than free α-linolenic acid at concentrations of 10 and 50 μM (22.5% vs. 29%, $p < 0.05$ and 29% vs. 79.1%, $p < 0.001$, respectively) [36].

The potential of using docosahexaenoic acid encapsulated in NLCs for peri-implantitis treatment was studied. The docosahexaenoic acid-loaded NLC had a particle size of 163.7 ± 2.0 nm, PDI of 0.118 ± 0.01, zeta potential (ZP) of 40.1 ± 1.3 mV, and EE of 78.13% ± 1.85%. The release of docosahexaenoic acid from the NLC showed a gradual and steady pattern over 144 h. According to the 2,2-diphenyl-1-pyridylohydrazinyl (DPPH) assay in vitro studies, the free-radical scavenging rate of the docosahexaenoic acid-loaded NLC was significantly higher (0.57 ± 0.03) than that of pure docosahexaenoic acid (0.17 ± 0.003, $p < 0.001$). Moreover, the docosahexaenoic acid-loaded NLC exhibited a superior inhibitory effect on the expression of cellular inflammatory factors compared to the pure form. In vivo studies on rats also showed that the docosahexaenoic acid-loaded NLC group displayed the most effective suppression of gingival inflammation after a 2-week reagent treatment [37].

2.1.2. β–Carotene

Maretti et al. evaluated the skin penetration of a β-carotene-loaded NLC in humans ($n = 4$) by measuring the concentration of β-carotene in the stratum corneum after application to the skin and using pure β-carotene as a control. The results show that pure β-carotene was mainly retained in the outer layers of the stratum corneum (45%). In contrast, the β-carotene-loaded NLC demonstrated improved penetration into deeper skin layers, with 34% of β-carotene detected. The characterization studies reveal that the β-carotene-loaded NLC was spherical and showed a particle size of 222.8 ± 87.3 nm, PDI of 0.666, ZP of −43.46 ± 1.74 mV, and EE of 23.96 ± 3.13% [38].

The application of β-carotene in the food industry was also investigated by Rohmah et al. The study aimed to evaluate the bioavailability and antioxidant activity of a β-carotene-loaded NLC. For the experiments, an in vitro gastrointestinal tract model was used to simulate the normal course of ingested food through the mouth, stomach, and intestines. The particle size of the β-carotene-loaded NLC was 166.3 ± 0.19 nm, the PDI was 0.35 ± 0.1, the ZP was −26.9 ± 0.17 mV, and the EE was 91.2 ± 0.15%. The results show that the NLC has a superior capacity to encapsulate β-carotene compared to the other tested systems, such as a β-carotene emulsion, a β-carotene–Tween 80 phosphate-buffered solution, and a β-carotene–phosphate-buffered solution ($p < 0.05$). After 4 h of incubation, the release of β-carotene in the small intestine was significantly higher for the β-carotene-loaded NLC (233 μg/mL) compared to the other formulations, indicating that the NLC effectively released the β-carotene in the small intestine, reducing stomach degradation. Furthermore, the bioavailability of the β-carotene-loaded NLC (60.7%) in the small intestine exceeded the ones obtained from the emulsion (34.1%), the β-carotene–Tween 80 solution (23.4%), and the β-carotene–phosphate-buffered solution (8.7%). In terms of antioxidant activity, the β-carotene-loaded NLC exhibited moderate to strong antioxidant properties, with 2,2′-azino-bis(3-ethylbenzothiazoline-6-sulfonic acid) (ABTS) and DPPH assay values of 91.47 ± 1.9 and 24.72 ± 0.38%, respectively. Additionally, the IC_{50} of the β-carotene-loaded NLC was 7.0 μg/mL, while the pure β-carotene solution was 10.0 μg/mL, indicating enhanced antioxidant activity for the encapsulated β-carotene. In this study, the radical scavenging activity during the digestive process was also evaluated, which consistently demonstrated that the β-carotene-loaded NLC exhibited higher radical scavenging activity compared to the other tested formulations. These results suggest that NLCs may be a promising vehicle for application in functional food and beverages [39].

2.1.3. Astaxanthin

Geng et al. optimized and assessed the stability, skin retention, and permeability of astaxanthin-loaded NLCs. The optimization process involved adjusting different independent variables, such as the solid lipid/liquid lipid ratio, total amount of lipids, astaxanthin concentration, and type and amount of emulsifiers, while the dependent variables were particle size, PDI, and EE. For the experiments, 21 formulations were prepared, and the optimized formulation showed stable nanoparticles with a spherical surface, a particle size of 67.4 ± 2.1 nm, a PDI of 0.26, and an EE of 94.3 ± 0.5%. This formulation was also non-irritating, homogeneous, and exhibited excellent stability. The in vitro release studies indicate that the cumulative release rate of astaxanthin from the NLC was 83.0 ± 3.4% for 48 h, while pure astaxanthin dissolved completely within 4 h. Moreover, the antioxidant and anti-linoleic lipid peroxidation activities of astaxanthin were effectively preserved following encapsulation in the NLC. The results of the skin penetration studies indicate that the cumulative permeability within 24 h was 174.10 ± 4.38 μg/cm^2, and the retention was 18.60 ± 1.62 μg/cm^2 for the astaxanthin-loaded NLC, while pure astaxanthin demonstrated permeability and retention of 295.20 ± 6.04 μg/cm^2 and 8.00 ± 1.62 μg/cm^2, respectively. Although the astaxanthin-loaded NLC showed a reduction in skin permeation, this formulation showed a sustained release profile compared with free astaxanthin, reducing the amount of free astaxanthin that can permeate through the skin. However, the skin retention of astaxanthin was effectively improved by encapsulation in the NLC, maximizing the local astaxanthin concentration in the skin while reducing its systemic absorption. These findings suggest that NLCs have the potential to serve as a carrier for astaxanthin, offering enhanced stability and improved skin retention capabilities [40]. Another study tested the ability of astaxanthin-loaded NLCs to reduce skin damage caused by radiotherapy. In the experiments, the effects of the astaxanthin-loaded NLCs on human fibroblasts and mice after exposure to X-irradiation were tested. The results show that the astaxanthin-loaded NLC had a particle size of 114.4 nm, ZP of −34.1 mV, and EE of 85.67%. Furthermore, it was observed that a concentration of 0.25 μg/mL of astaxanthin-loaded NLC successfully reduced reactive oxygen species (ROS) production by 81.6%, decreased DNA damage by 41.6%, and lowered cell death by 62.69% compared to the control. In the in vivo experiments, all six mice treated with an astaxanthin-loaded NLC were protected from acute skin damage after nine days of X-irradiation. In contrast, five out of six untreated mice exhibited grade 1 skin damage. Furthermore, after 28 days of treatment, the histological images indicate significant skin recovery with minimal differences in collagen fibers and sebaceous glands compared to normal skin [41]. Other studies have suggested that astaxanthin-loaded NLCs could be employed in the production of food supplements [42]. For example, researchers encapsulated astaxanthin in an NLC and analyzed its behavior during in vitro digestion. The particle size was 145.3 nm, the PDI was 0.468 ± 0.036, and the ZP was −30.8 ± 0.3 mV. The EE was 94.8 ± 1.0%, and a long-term stability study reveals that approximately 75% of astaxanthin remained encapsulated after 28 days of storage at 25 °C. In addition, the antioxidant activity test demonstrated that the astaxanthin retained its biological activity after encapsulation in the NLC (the DPPH assay was 95% for free astaxanthin and 100% for the astaxanthin-loaded NLC, $p \leq 0.05$). The in vitro release study shows that free astaxanthin was rapidly released in the initial 6 h, reaching approximately 90% after 12 h, while encapsulated astaxanthin exhibited a more sustained release profile, with 88% released in 24 h. The results of the formulation's digestion test indicate that the particle size and PDI remained stable during oral and stomach digestion. However, during intestine digestion, the particle size increased significantly ($p < 0.05$) to 227.6 ± 1.8 nm. This change was attributed to the presence of anionic components in the micelle mixture, such as bile salts, phospholipids, and free fatty acids. Additionally, the digestion process in the intestine initially occurred rapidly but then slowed down. The researchers concluded that astaxanthin was trapped in the inner oil phase and suggested that the lipid matrix might contribute to increased stability. These results suggest that NLCs can be a promising delivery system for astaxanthin, offering enhanced stability and a prolonged release effect [42].

Recent studies have suggested that encapsulating astaxanthin in NLCs or SLNs can enhance the treatment of neurodegenerative diseases via the nose-to-brain route. For instance, comparative in vitro experiments were carried out to evaluate the compatibility of an astaxanthin-loaded NLC and astaxanthin-loaded SLNs with nasal and neuronal cells. In this study, the antioxidant activity (neuroprotective effect) and the cellular uptake of astaxanthin-loaded SLNs and the astaxanthin-loaded NLC were also assessed. The results reveal that the astaxanthin-loaded SLNs had particle size of 106.967 ± 2.515 nm, PDI of 0.220 ± 0.017, ZP of −24.133 ± 0.379 mV, and EE of 99.99 ± 0.00%, while the astaxanthin-loaded NLC exhibited a particle size of 117.300 ± 2.163 nm, PDI of 0.222 ± 0.016, ZP of 23.267 ± 0.451 mV, and EE of 99.61 ± 0.04%. Additionally, both the astaxanthin-loaded NLC and astaxanthin-loaded SLNs were found to be safe for nasal and neuronal cells at concentrations up to 100 μg/mL. Regarding the neuroprotective effects, both formulations exhibited the ability to inhibit neurodegenerative pathways, such as oxidative stress. Notably, the astaxanthin-loaded NLC demonstrated superior neuroprotective effects against cytotoxicity induced by aggressors in comparison to the astaxanthin-loaded SLNs [43]. In addition, in the nose-to-brain route, a different study involving an astaxanthin-loaded NLC was carried out to test the in vivo effectiveness of this approach in slowing the progression of Alzheimer's disease. The optimized astaxanthin-loaded NLC had a particle size of 142.8 ± 5.02 nm, PDI of 0.247 ± 0.016, ZP of −32.2 ± 7.88 mV, and EE of 94.1 ± 2.46% and remained stable at 4–8 ± 2 °C for six months. When administered to rats with Alzheimer's disease-like symptoms, the astaxanthin-loaded NLC led to a significant decrease in oxidative stress, amyloidogenic pathway, neuroinflammation, and apoptosis while also showing improvement in cholinergic neurotransmission compared to a free astaxanthin solution [44].

2.1.4. Lycopene, Fucoxanthin, and Zeaxanthin

Okonogi et al. developed a lycopene-loaded NLC for skin use, which had a particle size of 166 ± 4 nm, PDI of 0.15 ± 0.05, ZP of −74.6 ± 2.0 mV, and EE of 100 ± 0%. The in vitro release studies indicate a biphasic profile with faster release in the first 6 h, followed by sustained release over the following 18 h. The in vitro occlusion test reveals increased occlusive properties with higher lycopene-loading in the NLC. Additionally, the lycopene-loaded NLC exhibited higher antioxidant capacity (36.6 ± 0.4 mM/mg) compared to the placebo NLC (26.6 ± 0.1 mM/mg). The IC_{50} obtained by the DPPH assay was 14.1 ± 0.6 mg/mL for the lycopene-loaded NLC and 17.7 ± 0.4 mg/mL for the placebo NLC. Physical stability studies show that the particle size, PDI, and ZP of the lycopene-loaded NLC remained stable over 120 days when stored at different temperatures (4, 30, and 40 °C). Chemical stability data suggest that encapsulation in the NLC improved the stability of lycopene, particularly at lower temperatures (4 °C), which was demonstrated by the delay in lycopene degradation (0.4 µg/mL/day) [45].

Two distinct studies explored the potential use of fucoxanthin encapsulated in SLNs and NLCs for cosmetic applications. Cordenonsi et al. investigated the potential of a fucoxanthin-loaded NLC to prevent skin hyperproliferative diseases, particularly psoriasis. The fucoxanthin-loaded NLC had a particle size of 427.3 ± 5.7 nm, PDI of 0.309 ± 0.11, ZP of 21.21 ± 1.23 mV, and EE of 90.12 ± 2.51%. In vitro experiments on human fibroblast (NHDF) cultures show that up to a fucoxanthin concentration of 5 mM, there was no reduction in cell viability. Additionally, the fucoxanthin-loaded NLC decreased the expression of psoriatic markers by approximately 40%, suggesting their potential to manage skin hyperproliferation and inflammation [46]. In another study, Lee et al. assessed the efficacy of fucoxanthin-loaded SLNs to increase the effectiveness of sun protection. The results show that fucoxanthin-loaded SLNs had a particle size of 168 nm and PDI of 0.162. The sun protection factor (SPF) of the fucoxanthin-loaded SLNs was significantly higher (1.85 times) than that of the other formulations tested (three different chemical sunscreens), demonstrating a remarkable sun protection-enhancing effect [47].

The only study found on zeaxanthin involved the optimization of SLNs and an NLC loaded with the compound, which had the respective particle sizes of 179.16 ± 0.94 nm and 130.16 ± 1.58 nm; the PDI were 0.34 ± 0.01 and 0.30 ± 0.01; the ZP were −19.44 ± 1.19 mV and −21.49 ± 1.13 mV; and the EE were 81.14 ± 3.06% and 90.43 ± 2.85%. The researchers suggested that both SLNs and the NLC could have a wide range of applications, particularly as carriers for bioactive compounds in nutraceutical beverages [61].

2.1.5. Vitamin A

The first documented studies on vitamin A-loaded SLNs date back to the early 2000s. Jenning et al. developed vitamin A-loaded SLNs and tested their efficacy in permeating porcine skin compared to a vitamin A nanoemulsion. The results reveal that SLNs significantly enhanced the stability of vitamin A compared to the nanoemulsion. The vitamin A-loaded SLNs had a particle size of 224 nm and a PDI of 0.205. Six hours after application, a substantial concentration of vitamin A (approximately 3400 ng) was detected in the stratum corneum and the upper epidermis. In contrast, the nanoemulsion only transported 2500 ng of vitamin A to the upper skin layers ($p < 0.05$). Surprisingly, after 24 h, a change in the distribution pattern was observed. The concentration of vitamin A in the upper epidermis decreased significantly to 900 ng ($p < 0.05$), while the concentration in the deeper skin layers increased from 0 to 250 ng ($p < 0.05$). In contrast, the distribution pattern of the vitamin A nanoemulsion did not change [48]. In another study, Pople et al. developed vitamin A-loaded SLNs with a particle size of 350 nm and compared their efficacy with the one from a vitamin A hydrogel. The in vitro release studies show that the vitamin A-loaded SLNs had a prolonged release of vitamin A for up to 24 h, with 54.38% of the compound released compared to 70% released from the vitamin A hydrogel. This extended release was attributed to the compound being embedded in the solid lipid core. In the in vitro penetration studies, after 24 h, the vitamin A concentration in the skin was almost 2 times higher with the vitamin A-loaded SLN formulation compared to the one obtained with the vitamin A hydrogel. Only 1.2% of the compound was detected in the receptor compartment of the vitamin A-loaded SLNs, while almost 10 times more was detected in the vitamin A hydrogel, resulting in a significantly higher amount of unabsorbed compound (around 72%) compared to the vitamin A-loaded SLNs (around 67%). The in vivo skin hydration studies in albino rats show that the application of the vitamin A hydrogel results in a slight change in the thickness of the stratum corneum, while the application of the vitamin A-loaded SLNs led to a substantial increase in its thickness: 3-fold compared to the hydrogel and 3.5-fold compared to the untreated skin. Additionally, the developed vitamin A-loaded SLNs showed a primary irritation index of 0.00, with no observed erythema or edema [49].

The potential use of vitamin A-loaded SLNs and vitamin A-loaded NLCs in the food industry has recently been investigated. The study focused on ensuring the stability and oral bioavailability of these nanoparticles under food processing conditions. Six formulations were developed for the experiments, from which two were selected for further investigation. The particle size, PDI, ZP, and EE for the vitamin A-loaded SLNs were 223 ± 10 nm, 0.171 ± 0.008, −25 ± 1 mV, and 97 ± 3%, respectively. For the vitamin A-loaded NLC, the corresponding values were 228 ± 7 nm, 0.146 ± 0.007, −22 ± 1 mV, and 90 ± 1%. The results of the stability tests show that the vitamin A-loaded SLNs and the vitamin A-loaded NLC remained stable for one month after storage at room temperature (25 °C). Furthermore, exposure to typical food processing temperatures (60 and 70 °C) did not significantly change the nanoparticle properties, although these temperatures were above the melting point of the solid lipid used in the nanoparticles (43 °C). The vitamin A-loaded SLNs and the vitamin A-loaded NLC were also tested in different media, including phosphate buffer at pH 5, highly concentrated sucrose solution, and high-ionic-strength sodium chloride solution. Both the vitamin A-loaded SLNs and vitamin A-loaded NLC showed stability in all the tested media, with variations of < 10% compared to fresh formulations not exposed to the various media. Afterward, to simulate gastric conditions, both tested formulations were subjected to a simulated gastric environment containing an acidic pH, salts, and gastric enzymes. Despite an increase of 0.03 in the PDI after two hours, the size of the vitamin A-loaded NLC remained approximately the same. A similar situation was observed for the vitamin A-loaded SLNs. Furthermore, after two hours in a simulated gastric environment, approximately 80% of the encapsulated vitamin A remained intact in the SLNs and NLC, suggesting their potential to deliver the compound to the small intestine where it can be absorbed [50].

2.1.6. Vitamins D, E, and K

The potential of using NLCs to improve the effectiveness of vitamin D3 against breast cancer cells has been investigated. A vitamin D3-loaded NLC showed a particle size of 87 ± 5 nm, PDI of 0.24, and ZP of −12.2 ± 4.86 mV. The 3-(4,5-Dimethylthiazol-2-yl)-2,5-Diphenyltetrazolium Bromide (MTT) assay showed that the vitamin D3-loaded NLC was more successful in inhibiting cancer cell proliferation compared to a free vitamin D3 solution ($p < 0.05$), reducing cell proliferation from 49 ± 7.2% to 37 ± 5.1%, respectively. Furthermore, the vitamin D3-loaded NLC increased the percentage of cells in the apoptotic phase to 40 ± 3.45% ($p < 0.05$). These findings suggest that a vitamin D3-loaded NLC can improve the efficacy of chemotherapeutics in breast cancer patients [51]. Another study suggested using an oral vitamin D3-loaded NLC to improve the management of inflammatory bowel disease. The vitamin D3-loaded NLC had a particle size of 110 ± 4 nm, PDI of 0.23 ± 0.01, and ZP of −17.10 ± 0.30 mV. In the in vivo studies, when the vitamin D3-loaded NLC was orally administered to mice, the concentration of vitamin D3 in the colon was observed to be three times higher than the basal level, persisting for at least 12 h ($p < 0.01$). Subsequently, the researchers assessed the effects of the vitamin D3-loaded NLC on intestinal inflammation. The results demonstrate that although the symptoms worsened in mice treated with the free vitamin D3 solution and the vitamin D3-loaded NLC, the group tested with the vitamin D3-loaded NLC experienced progressive mitigation of clinical symptoms from day 3, and it suppressed a decrease in body weight from day 6. Additionally, the histological examination of colonic tissue showed significantly reduced inflammatory manifestations, immune cell infiltration, and crypt destruction in the group treated with the vitamin D3-loaded NLC ($p < 0.01$) [52].

Eiras et al. developed a vitamin E-loaded NLC and assessed its potential for skin application. The characterization of the NLC loaded with vitamin E showed that the particles had a consistent size in the nanometer range, with 90% having a size of less than 328 nm and 50% having a size of less than 105 nm, even after 7 months of storage. The results of in vitro studies in human keratinocytes (HaCAT) indicated that the vitamin E-loaded NLC is biocompatible (0.1–10 μg/mL) and non-irritant (with a score of 0.00), being suitable for skin application. From these findings, the researchers proposed using a vitamin E-loaded NLC in cosmetic and dermatological formulations to improve skin hydration [53]. Later, the same researchers conducted a more comprehensive study with this formulation, where they found no significant differences in particle size after incorporating a vitamin E-loaded NLC into a hydrogel (D90 = 386 ± 0.00 nm vs. D90 = 397 ± 0.021 nm). The biocompatibility studies were assessed by the MTT assay, where it was observed that the vitamin E-loaded NLC was more cytotoxic to human keratinocytes before incorporation in the hydrogel (IC_{50} = 14.38 μg/mL vs. IC_{50} = 28.74 μg/mL, $p < 0.001$). In vitro and in vivo tests reveal that the vitamin E-loaded NLC demonstrated superior occlusive properties (80% vs. 67%) and significantly reduced (12 g/H·m^2) the skin's transepidermal water loss (TEWL) compared to a vitamin E nanoemulsion (14 g/h·m^2) ($p < 0.05$). In this study, the researchers demonstrated the safety and hydration potential of the hydrogel containing a vitamin E-loaded NLC for skin application. However, the need for more in vivo experiments over extended periods and involving more volunteers to confirm this evidence was emphasized [54]. In another study, the effectiveness of a vitamin E-loaded NLC was examined for its moisturizing and anti-aging properties. The optimized vitamin E-loaded NLC had a particle size of 82 nm, a PDI of 0.261, a ZP of −28.6 mV, and an EE of 95.83 ± 0.02%. The stability study reveals that the nanoparticles maintained their spherical shape and physical stability for 12 weeks after storage at room temperature. The release profile indicated that 30% of the vitamin E was released within the first 5 h, with 70% released after 24 h. The vitamin E-loaded NLC was incorporated into a hydrogel, and in vivo tests were performed on 13 human volunteers over 12 weeks to assess skin moisture content and mechanical properties. In the skin irritation test, all volunteers showed no signs of erythema after 48 h of exposure to the hydrogel containing the vitamin E-loaded NLC. However, 3 out of 13 volunteers experienced mild erythema when exposed to the vitamin E hydrogel alone. This suggests that the free form of vitamin E can be irritating to the skin, while the NLC provides a protective effect against this irritation. Additionally, throughout the 12-week study period, the hydrogel containing the vitamin E-loaded NLC enhanced the skin elasticity capacity ($p = 0.0319$) and consistently increased skin moisture levels by 80%, while a vitamin E hydrogel only resulted in a 20% increase in moisture levels ($p < 0.01$) [55]. In a different approach, the researchers evaluated the effectiveness of encapsulating vitamin E in SLNs to reduce the negative effects of anemia treatment. The characterization of the vitamin E-loaded SLNs revealed a particle size of 228.2 ± 3.5 nm, a PDI of 0.34 ± 0.02, a ZP of −8.92 ± 2.2 mV, and an EE of 99.9 ± 0.1%. The in vitro release studies reveal that approximately 65% of the encapsulated vitamin E was released after 24 h in simulated gastrointestinal media at 37 °C. Furthermore, the vitamin E-loaded SLNs demonstrated good hemocompatibility at various concentrations. After 3 h of incubation, there was minimal hemolysis (less than 0.3%). Even after 24 and 48 h, the level of hemolysis remained low, with less than 3.0% at the highest SLN concentration. It is important to note that biomaterials causing hemolysis of less than 5% are generally considered safe for human use, according to ISO/TR 7406. These results indicate their potential for clinical applications. Notably, the vitamin E-loaded SLNs significantly boosted lymphocyte cell proliferation by approximately 150% and decreased DNA damage caused by iron treatment [56].

Regarding vitamin K, only one significant study involving the encapsulation of vitamin K1 in SLN for oral delivery was found. In the experiments, seventeen different formulations were prepared and the most effective one was chosen for further tests. The particle size, PDI, ZP and EE for the vitamin K-loaded SLN were 132 nm, 0.17 ± 0.02, and −26.83 ± 2.83 mV, 98%, respectively. This formulation was shown to be stable in a 54-h in vitro release study in simulated gastric and intestinal media and after 4 months storage at 25 °C [57].

2.1.7. Polymer Surface Coating

Chitosan

Almeida et al. aimed to evaluate the efficacy of chloroaluminium phthalocyanine-loaded NLCs coated with chitosan in the photodynamic therapy of skin cancer. A chloroaluminium phthalocyanine-loaded NLC coated with chitosan showed a particle size of 231.5 ± 5.8 nm, a PDI of 0.18 ± 0.01, a ZP of +19.96 ± 0.3 mV, and an EE of 96%. It is important to highlight that the chitosan coating resulted in a larger but equally viable NLC. Ex vivo studies show that the phthalocyanine-loaded NLC coated with chitosan significantly retained the compound in the skin after 2 h (5.8 ng) and 4 h (581 ng), with no detection in the bloodstream, indicating limited systemic exposure and no potential adverse events in patients. The biocompatibility test performed on L929 fibroblasts showed that the phthalocyanine-loaded NLC coated with chitosan did not induce cytotoxicity at any of the concentrations tested, indicating that the chitosan coating did not affect the biocompatibility of the NLC. It was also observed that the phthalocyanine-loaded NLC coated with chitosan was localized around the cellular nucleus. The photodynamic therapy tests performed on BF16-F10 melanoma cells showed that the phthalocyanine-loaded NLC coated with chitosan did not have toxic effects when not exposed to irradiation. However, after irradiation, there was a 50% reduction in cell viability ($p < 0.001$). These results highlight that the chitosan coating improved stability and biocompatibility and facilitated drug passage through the skin [58].

Alginate

Costa-Fernandez et al. developed an NLC co-encapsulated with vitamin E and quercetin and coated with alginate to investigate its potential to promote wound healing and evaluate the influence of the alginate coating in enhancing the penetration of the encapsulated compounds into the skin damaged by an incision. The results show that the alginate-coated NLC co-encapsulated with vitamin E and quercetin had a particle size of 321.2 ± 18.3 nm, a PDI of 0.23 ± 0.02, a ZP of −16.0 ± 3.0 mV, and an EE of 85% for quercetin and 92% for vitamin E. The irritation potential of the alginate-coated NLC was assessed by the Hen's Egg Test–chorioallantoic membrane, showing it did not induce bleeding, lysis, or coagulation. Regarding the in vitro release, it was observed that after 16 h, 44.2% of quercetin and 32.3% of vitamin E were released. Furthermore, in ex vivo permeation studies, the experiments were performed on intact and damaged skin. In the intact skin, the permeation of the alginate-coated NLC co-encapsulated with vitamin E and quercetin was higher compared to a conventional formulation containing the same compounds. In the damaged skin, higher penetration of vitamin E and quercetin was observed with both formulations. The delivery of vitamin E to the epidermis and dermis increased by 1.9-fold and 2.3-fold for quercetin compared to the intact skin. The TEWL was also reduced (from 3 $g/H{\cdot}m^2$ to -7 $g/h{\cdot}m^2$) when using the alginate-coated NLC co-encapsulated with vitamin E and quercetin and water as the control. The results of this study highlight the advantages of using polysaccharides, such as alginate, as bioadhesive polymers to promote the penetration of compounds into the upper layers of skin [59].

Gelatin

Gelatin, a natural biopolymer (i.e., a protein composed of long chains of amino acids), is widely used in the food industry because of its advantages, such as non-toxicity, low cost, availability, biocompatibility, and biodegradability. Studies have shown that gelatin can stabilize nanocarriers due to its amine and carboxylic groups. Malekmohammadi et al. conducted a study aiming to synthesize a gelatin-coated NLC to encapsulate sage extract to inhibit microbial growth and lipid oxidation in beef burgers. The optimized sage extract-loaded NLC coated with gelatin had a particle size of 100.4 nm, PDI of 0.36, ZP of −18.4 mV, and EE of 80%. The results of the DPPH assay show that the sage extract-loaded NLC coated with gelatin exhibited significantly higher antioxidant activity than the free extract after 30 days of storage at 25 °C ($p < 0.05$). A higher inhibitory effect against *E. coli* compared to the free extract was also observed in the minimum inhibitory concentration (0.1 ± 0.00 vs. 0.2 ± 0.00 mg/mL) and the minimum bactericidal concentration (0.1 ± 0.00 vs. 0.3 ± 0.00 mg/mL) tests. Furthermore, incorporating the formulation into beef burgers increased their oxidation stability during 90 days of storage at 4 and −18 °C ($p < 0.05$). The sage extract-loaded NLC coated with gelatin also effectively decreased the total counts of various bacteria, yeasts, and molds in treated beef burger samples during storage in comparison to the controls (all $p < 0.05$). The sensory tests showed no significant differences in the color, odor, texture, and flavor attributes of the beef burger samples immediately after preparation. However, over time, the treated samples showed greater acceptability. These improved sensory properties were attributed to reduced proteolytic and lipolytic reactions, as well as reduced microbial activities. Overall, the study suggests that sage extract-loaded NLCs coated with gelatin can be an effective preservative for extending the shelf life of beef burgers [60].

3. Regulatory and Safety Concerns of Lipid Nanoparticles for Healthcare Applications

3.1. Cosmetics

Notwithstanding the regulatory frameworks governing the global market for nanocosmetics, the recognition of nanomaterials as cosmetic ingredients remains inconsistent across jurisdictions [62]. Consequently, each country adheres to its legal system. Given that the European Union (EU) and the United States of America (USA) represent the two most significant markets for cosmetics products, their respective regulatory frameworks are of significant importance. However, there are notable discrepancies between the two. Prior to 2023, the USA did not require cosmetic product registration, while the EU has mandated registration since 2013, especially for products containing nanoparticles. Furthermore, notification to the European Commission (EC), including information on the identification, specifications, toxicological profile, and safety of nanomaterials, is required. In cases of uncertainty regarding the safety of a nanomaterial, the EC can request an opinion from the Scientific Committee on Consumer Safety (SCCS) [63].

Under the stipulations set in the EC Regulation 1223/2009, the term "nanomaterial" is defined to regulate cosmetics products as a deliberately insoluble or bio-persistent manufactured material that has external dimensions (one or more) or an internal structure within the range of 1 to 100 nanometers [63]. On the other hand, in the USA, the Food and Drug Administration (FDA) oversees the use of nanomaterials in cosmetics through the Federal Food, Drug, and Cosmetic Act (FFDCA). However, the FDA does not have a legal definition for nanotechnology, does not approve the ingredients in cosmetic formulations, and does not require manufacturers to disclose the presence of nanomaterials in their products. The FDA has thus far refrained from making a definitive determination regarding the inherent safety or potential risks associated with nanotechnology [64]. As a consequence, the FDA has created the National Nanotechnology Initiative (NNI) and the Nanotechnology Task Force (NTF) intended to assess the restrictions required for nanotechnology products. These entities have published two documents addressing the safety issues of nanotechnology and cosmetics, which merely make recommendations. The first document is concerned with the determination of a material's classification as a nanomaterial based on the size of its particles and their properties/phenomena. In particular, it considers whether the material or final product is (a) intentionally designed to have at least one external dimension, internal dimension, or surface structure within the nanoscale range (approximately 1 nm to 100 nm) and (b) meant to exhibit properties or phenomena, including physical, chemical, or biological effects, attributed to its size, even if these dimensions extend beyond the nanoscale range to one micrometer (1000 nm). It is important to highlight that the properties of nanomaterials relevant to safety, effectiveness, performance, quality evaluation, public health impact, and product regulatory status can be linked to materials with one or more dimensions larger than the 1–100 nm range. In a succeeding document, the FDA advises a thorough assessment of safety by characterizing the nanomaterial itself and examining a wide range of chemical and physical properties. This involves evaluating the toxicity, absorption, distribution, metabolism, and excretion of the particles [62,65].

The scientific community has been debating whether the use of insoluble nanoparticles in cosmetics constitutes a health risk. The results so far have been inconclusive due to inconsistent findings and a lack of long-term toxicological studies. Safety evidence originates mainly from the EU, which has strict regulations [65]. Regarding the use of SLNs and NLCs in cosmetics, these can be classified as nanomaterials according to the EU standards, as they consist of water-insoluble materials, such as solid and liquid lipids. However, these lipids are similar to the body's natural lipids. They can readily adhere to and interact with the outer layer of the skin, causing lipid rearrangement and allowing the encapsulated compounds to penetrate deeper layers of the skin. Furthermore, the nanoscale dimensions of the particles allow for greater adhesion and surface contact area, which, in turn, improves the permeation of the compound through the skin [17]. They are, therefore, considered to be soluble substances [65]. In addition, it has been documented that the use of SLNs and NLCs in topical cosmetics reduces the likelihood that these carriers will cause systemic toxicity. Considering the points outlined above, SLNs and NLCs in cosmetic products are fully compliant with the relevant legislation and do not raise any safety concerns or objections. The current regulatory framework supports their use, confirming that they are safe and appropriate for inclusion in cosmetic formulations [34].

3.2. Food Supplements

The EC Regulation 2015/2283 defines engineered nanomaterials in food as novel foods, categorizing them as intentionally produced materials with dimensions of approximately 100 nm or less [66]. These materials can have discrete functional parts at the surface or internally, as well as structures, aggregates, or agglomerates, which may maintain nanoscale-specific properties despite having a size above 100 nm. Such properties may be related to specific physicochemical properties that differ from those of the non-nano form of the same material and/or to the large specific surface area of the materials considered. Nevertheless, the EU legislation applicable to novel food is the same as all food (Regulation EC 178/2002) [66]. Concerning this legislation, certain safety stipulations give rise to uncertainty, particularly the potential adverse effects of novel food on the health of both the current and future generations [67]. Regarding long-term effects, there is a gap in data to substantiate the safety of nanoparticles. However, the European Food Safety Authority (EFSA) has issued an exceptional document that provides an overview of the risk assessment of nanomaterials in food and the requisite information [68]. The document elucidates that lipid nanoparticles can facilitate the delivery of select food components, particularly those exhibiting pronounced lipophilicity. Moreover, the components of the nanoparticle can be derived from naturally occurring body lipids or approved food additives, such as emulsifiers. The primary safety concerns pertain to the extent of degradation of the encapsulation materials within the gastrointestinal tract. This encompasses not only the active ingredient itself but also the encapsulating material and the entire encapsulate/nanocarrier. In this regard, it has been proposed that specific adaptations in hazard characterization are required, including the assessment of the quantity of the encapsulated compound and the amount present in its free form in the food. Moreover, it may be advisable to examine the pertinent chemical constituents of a nanocarrier system and present evidence on how intestinal cells absorb and transport them [69].

In the USA, the FDA adheres to the same position as the EFSA when monitoring the safety and effectiveness of nanotechnology products. As well as regulating these products within its current regulatory framework, they are customized to the specific standards for each product category under its authority. Thus, the industry is committed to ensuring compliance with all applicable legal requirements and is encouraged to collaborate with the FDA to address any concerns regarding the safety or regulatory status of these products [64].

It is noteworthy that the safety of SLNs and NLCs in food supplements has been reported. Once ingested, these nanoparticles undergo the same physiological processes as the lipids present in food. This process includes digestion in the stomach, absorption primarily in the small intestine, and systemic blood uptake. In the small intestine, SLNs and NLCs can be broken down by lipase enzymes, releasing and facilitating the absorption of the compounds. Additionally, lipid nanoparticles have adhesive properties, allowing them to adhere to the gut wall, specifically the enterocyte surface, leading to the prompt absorption of the compounds within the enterocytes. Additionally, oral SLNs and NLCs are generally larger than 100 nm and biodegradable, indicating that their cellular uptake is not expected. Therefore, it is reasonable to conclude that no significant toxicological concerns are expected [70,71].

4. Future Perspectives

Most current research on encapsulating bioactive marine compounds in SLNs and NLCs is limited to in vitro studies, with minimal guidance on regulatory and safety standards for human application. To advance this field, future research should prioritize standardizing clinical trial protocols to facilitate comparisons across studies from different research groups; performing in vivo studies to better understand the absorption, distribution, and excretion of bioactive compounds and their respective nanocarriers; extending testing to human volunteers to confirm that findings from animal studies are applicable to humans; and publishing "negative results" to reduce redundancy, foster transparency, and accelerate collective learning.

Collaboration between academia and industry is also crucial to overcoming technological and financial challenges. Such partnerships drive innovation, bridge the gap between theory and practice, and help align research with real-world needs. Additionally, keeping regulatory standards in sync with the latest scientific developments is essential to ensure both safety and progress.

5. Conclusions

In recent years, the growing focus on sustainability and well-being has driven research into bioactive compounds from natural sources that are safe for human consumption. Marine bioactive compounds, in particular, have garnered attention for their potential health benefits. Despite their promise, these molecules often face challenges, such as poor stability and low water solubility. To address these issues, lipid nanoparticles, like SLNs and NLCs, have proven to be effective delivery systems, enhancing the stability and bioavailability of these compounds, protecting them from degradation, and improving their solubility in aqueous environments. This makes them more versatile for applications in healthcare, particularly in cosmetics and food supplements. In addition, SLNs and NLCs are often considered safer for human consumption and topical use compared to polymeric delivery systems, as lipids are naturally occurring substances.

Compared to other nanocarriers, such as liposomes, nanoemulsions, and polymeric nanoparticles, SLNs and NLCs offer several advantages, such as enhanced stability, lower toxicity, and simpler formulation processes. For instance, liposomes, while also effective at encapsulating bioactive compounds, can be more expensive and less stable in certain formulations. Nanoemulsions, though offering improved solubility, may have issues with stability. Polymeric nanoparticles can offer targeted delivery, but they are often associated with more complex production processes and potential biocompatibility concerns.

Despite their promising advantages, SLNs and NLCs have limitations that need to be addressed, such as the development of more comprehensive regulatory standards to ensure their safe use in healthcare products, particularly in cosmetics and food supplements. Additionally, further clinical and pre-clinical studies are required to establish their full safety profile, especially concerning long-term use in humans.

In conclusion, while SLNs and NLCs present an exciting solution for improving the delivery of bioactive marine compounds, more research, particularly in vivo studies, is essential to confirm their efficacy, safety, and broader applicability in healthcare products.

Author Contributions: Conceptualization, R.F. and A.C.S.; formal analysis, R.F.; investigation, R.F.; writing—original draft preparation, R.F. and A.C.S.; writing—review and editing, A.C.S., H.A., D.F. and A.F.P.; supervision, A.C.S. All authors have read and agreed to the published version of the manuscript.

Funding: This work was supported by the Fundação para a Ciência e a Tecnologia (FCT) (2023.02170.BDANA) and the Applied Molecular Biosciences Unit (UCIBIO), which is financed by national funds from the FCT (UIDP/04378/2020 and UIDB/04378/2020) and by the LAQV-REQUIMTE through UIDP/50006/2020 and UIDB/50006/2020 (https://doi.org/10.54499/UIDP/50006/2020, https://doi.org/10.54499/UIDB/50006/2020, all accessed on 17 October 2024). AFP also thanks FCT for funding through Individual Call to Scientific Employment Stimulus (2020.01614.CEECIND/CP1596/CT0007).

Conflicts of Interest: The authors declare no conflict of interest.

References

1. Favas, R.; Morone, J.; Martins, R.; Vasconcelos, V.; Lopes, G. Cyanobacteria and microalgae bioactive compounds in skin-ageing: Potential to restore extracellular matrix filling and overcome hyperpigmentation. *J. Enzym. Inhib. Med. Chem.* **2021**, *36*, 1829–1838. [CrossRef] [PubMed]
2. Augusto, A.; Lemos, M.F.L.; Silva, S.F.J. Exploring Marine-Based Food Production: The Challenges for a Sustainable and Fast Biotechnology-Based Development. *Appl. Sci.* **2024**, *14*, 8255. [CrossRef]
3. Pateiro, M.; Domínguez, R.; Munekata, P.E.S.; Barba, F.J.; Lorenzo, J.M. Lipids and fatty acids. *Innov. Therm. Non-Therm. Process. Bioaccess. Bioavailab. Nutr. Bioact. Compd.* **2019**, 107–137. [CrossRef]
4. Burdge, G.C.; Calder, P.C. Introduction to Fatty Acids and Lipids. *Intraven. Lipid Emuls.* **2014**, *112*, 1–16. [CrossRef]
5. Roy, V.C.; Islam, M.R.; Sadia, S.; Yeasmin, M.; Park, J.-S.; Lee, H.-J.; Chun, B.-S. Trash to Treasure: An Up-to-Date Understanding of the Valorization of Seafood By-Products, Targeting the Major Bioactive Compounds. *Mar. Drugs* **2023**, *21*, 485. [CrossRef]
6. Misawa, N. Comprehensive Natural Products II. In *Chemistry, Molecular Sciences and Chemical Engineering*; Elsevier: Amsterdam, The Netherlands, 2010; Volume 1, pp. 733–753.
7. Chuyen, H.V.; Eun, J.-B. Marine carotenoids: Bioactivities and potential benefits to human health. *Crit. Rev. Food Sci. Nutr.* **2015**, *57*, 2600–2610. [CrossRef]
8. Carpenter, K.; Baigent, M.J. Vitamin. Available online: https://www.britannica.com/science/vitamin (accessed on 23 January 2024).
9. Rucker, R.B.; Morris, J.G. The Vitamins. *Clin. Biochem. Domest. Anim.* **1997**, 703–739. [CrossRef]
10. Thawabteh, A.; Swaileh, Z.; Ammar, M.; Jaghama, W.; Yousef, M.; Karaman, R.; Bufo, S.; Scrano, L. Antifungal and Antibacterial Activities of Isolated Marine Compounds. *Toxins* **2023**, *15*, 93. [CrossRef]
11. Venkatesan, J.; Anil, S.; Kim, S.-K.; Shim, M. Marine Fish Proteins and Peptides for Cosmeceuticals: A Review. *Mar. Drugs* **2017**, *15*, 143. [CrossRef]
12. Hu, D.; Jin, Y.; Hou, X.; Zhu, Y.; Chen, D.; Tai, J.; Chen, Q.; Shi, C.; Ye, J.; Wu, M.; et al. Application of Marine Natural Products against Alzheimer’s Disease: Past, Present and Future. *Mar. Drugs* **2023**, *21*, 43. [CrossRef]
13. Fonseca, S.; Amaral, M.N.; Reis, C.P.; Custódio, L. Marine Natural Products as Innovative Cosmetic Ingredients. *Mar. Drugs* **2023**, *21*, 170. [CrossRef] [PubMed]
14. Almeida, H.; Amaral, M.H.; Lobão, P.; Silva, A.C.; Loboa, J.M.S. Applications of Polymeric and Lipid Nanoparticles in Ophthalmic Pharmaceutical Formulations: Present and Future Considerations. *J. Pharm. Pharm. Sci.* **2014**, *17*, 278–293. [CrossRef]
15. Seo, Y.; Lim, H.; Park, H.; Yu, J.; An, J.; Yoo, H.Y.; Lee, T. Recent Progress of Lipid Nanoparticles-Based Lipophilic Drug Delivery: Focus on Surface Modifications. *Pharmaceutics* **2023**, *15*, 772. [CrossRef]
16. González-Peña, M.A.; Ortega-Regules, A.E.; Anaya de Parrodi, C.; Lozada-Ramírez, J.D. Chemistry, Occurrence, Properties, Applications, and Encapsulation of Carotenoids—A Review. *Plants* **2023**, *12*, 313. [CrossRef]
17. Garcês, A.; Amaral, M.H.; Sousa Lobo, J.M.; Silva, A.C. Formulations based on solid lipid nanoparticles (SLN) and nanostructured lipid carriers (NLC) for cutaneous use: A review. *Eur. J. Pharm. Sci.* **2018**, *112*, 159–167. [CrossRef]
18. Preeti; Sambhakar, S.; Saharan, R.; Narwal, S.; Malik, R.; Gahlot, V.; Mohan, S. Exploring LIPIDs for their potential to improves bioavailability of lipophilic drugs candidates: A review. *Saudi Pharm. J.* **2023**, *31*, 101870. [CrossRef]
19. Tenchov, R.; Bird, R.; Curtze, A.E.; Zhou, Q. Lipid Nanoparticles horizontal line From Liposomes to mRNA Vaccine Delivery, a Landscape of Research Diversity and Advancement. *ACS Nano* **2021**, *15*, 16982–17015. [CrossRef] [PubMed]
20. Mohammed, A.S.A.; Naveed, M.; Jost, N. Polysaccharides; Classification, Chemical Properties, and Future Perspective Applications in Fields of Pharmacology and Biological Medicine (A Review of Current Applications and Upcoming Potentialities). *J. Polym. Environ.* **2021**, *29*, 2359–2371. [CrossRef] [PubMed]
21. Ju, H.; Yu, C.; Liu, W.; Li, H.-H.; Fu, Z.; Wu, Y.-C.; Gong, P.-X.; Li, H.-J. Polysaccharides from marine resources exhibit great potential in the treatment of tumor: A review. *Carbohydr. Polym. Technol. Appl.* **2023**, *5*, 100308. [CrossRef]

22. Tan, K.; Lu, S.-Y.; Tan, K.; Ransangan, J.; Cai, X.; Cheong, K.-L. Bioactivity of polysaccharides derived from bivalves. *Int. J. Biol. Macromol.* **2023**, *250*, 126096. [CrossRef]
23. Liu, Q.; Hu, L.; Wang, C.; Cheng, M.; Liu, M.; Wang, L.; Pan, P.; Chen, J. Renewable marine polysaccharides for microenvironment-responsive wound healing. *Int. J. Biol. Macromol.* **2023**, *225*, 526–543. [CrossRef] [PubMed]
24. Wang, Z.; Xu, Z.; Yang, X.; Li, M.; Yip, R.C.S.; Li, Y.; Chen, H. Current application and modification strategy of marine polysaccharides in tissue regeneration: A review. *Biomater. Adv.* **2023**, *154*, 213580. [CrossRef] [PubMed]
25. Sahraee, S.; Ghanbarzadeh, B.; Mohammadi, M.; Pezeshki, A.; Hoseini, M. Development of heat-stable gelatin-coated nanostructured lipid carriers (NLC): Colloidal and stability properties. *LWT* **2022**, *160*, 113265. [CrossRef]
26. Muller, R.H.; Radtke, M.; Wissing, S.A. Solid lipid nanoparticles (SLN) and nanostructured lipid carriers (NLC) in cosmetic and dermatological preparations. *Adv. Drug Deliv. Rev.* **2002**, *54* (Suppl. 1), S131–S155. [CrossRef]
27. Muller, R.H.; Shegokar, R.; Keck, C.M. 20 years of lipid nanoparticles (SLN and NLC): Present state of development and industrial applications. *Curr. Drug Discov. Technol.* **2011**, *8*, 207–227. [CrossRef]
28. Costa, C.P.; Barreiro, S.; Moreira, J.N.; Silva, R.; Almeida, H.; Sousa Lobo, J.M.; Silva, A.C. In Vitro Studies on Nasal Formulations of Nanostructured Lipid Carriers (NLC) and Solid Lipid Nanoparticles (SLN). *Pharmaceuticals* **2021**, *14*, 711. [CrossRef]
29. Correia, A.C.; Monteiro, A.R.; Silva, R.; Moreira, J.N.; Sousa Lobo, J.M.; Silva, A.C. Lipid nanoparticles strategies to modify pharmacokinetics of central nervous system targeting drugs: Crossing or circumventing the blood–brain barrier (BBB) to manage neurological disorders. *Adv. Drug Deliv. Rev.* **2022**, *189*, 114485. [CrossRef]
30. Patil, T.S.; Deshpande, A.S. Nanostructured lipid carriers-based drug delivery for treating various lung diseases: A State-of-the-Art Review. *Int. J. Pharm.* **2018**, *547*, 209–225. [CrossRef]
31. Torres, J.; Silva, R.; Farias, G.; Sousa Lobo, J.M.; Ferreira, D.C.; Silva, A.C. Enhancing Acute Migraine Treatment: Exploring Solid Lipid Nanoparticles and Nanostructured Lipid Carriers for the Nose-to-Brain Route. *Pharmaceutics* **2024**, *16*, 1297. [CrossRef]
32. Mussi, S.V.; Torchilin, V.P. Recent trends in the use of lipidic nanoparticles as pharmaceutical carriers for cancer therapy and diagnostics. *J. Mater. Chem. B* **2013**, *1*, 5201–5209. [CrossRef]
33. Silva, A.C.; González-Mira, E.; García, M.L.; Egea, M.A.; Fonseca, J.; Silva, R.; Santos, D.; Souto, E.B.; Ferreira, D. Preparation, characterization and biocompatibility studies on risperidone-loaded solid lipid nanoparticles (SLN): High pressure homogenization versus ultrasound. *Colloids Surf. B Biointerfaces* **2011**, *86*, 158–165. [CrossRef] [PubMed]
34. Pardeike, J.; Hommoss, A.; Müller, R.H. Lipid nanoparticles (SLN, NLC) in cosmetic and pharmaceutical dermal products. *Int. J. Pharm.* **2009**, *366*, 170–184. [CrossRef] [PubMed]
35. Lima, S.G.M.; Freire, M.C.L.C.; Oliveira, V.d.S.; Solisio, C.; Converti, A.; de Lima, Á.A.N. Astaxanthin Delivery Systems for Skin Application: A Review. *Mar. Drugs* **2021**, *19*, 511. [CrossRef] [PubMed]
36. Serini, S.; Cassano, R.; Corsetto, P.A.; Rizzo, A.M.; Calviello, G.; Trombino, S. Omega-3 PUFA Loaded in Resveratrol-Based Solid Lipid Nanoparticles: Physicochemical Properties and Antineoplastic Activities in Human Colorectal Cancer Cells In Vitro. *Int. J. Mol. Sci.* **2018**, *19*, 586. [CrossRef]
37. Li, Z.; Yin, Z.; Li, B.; He, J.; Liu, Y.; Zhang, N.; Li, X.; Cai, Q.; Meng, W. Docosahexaenoic Acid-Loaded Nanostructured Lipid Carriers for the Treatment of Peri-Implantitis in Rats. *Int. J. Mol. Sci.* **2023**, *24*, 1872. [CrossRef]
38. Maretti, E.; Leo, E.; Rustichelli, C.; Truzzi, E.; Siligardi, C.; Iannuccelli, V. In vivo β-carotene skin permeation modulated by Nanostructured Lipid Carriers. *Int. J. Pharm.* **2021**, *597*, 120322. [CrossRef]
39. Rohmah, M.; Rahmadi, A.; Raharjo, S. Bioaccessibility and antioxidant activity of β-carotene loaded nanostructured lipid carrier (NLC) from binary mixtures of palm stearin and palm olein. *Heliyon* **2022**, *8*, e08913. [CrossRef]
40. Geng, Q.; Zhao, Y.; Wang, L.; Xu, L.; Chen, X.; Han, J. Development and Evaluation of Astaxanthin as Nanostructure Lipid Carriers in Topical Delivery. *AAPS PharmSciTech* **2020**, *21*, 318. [CrossRef]
41. Vu, N.B.D.; Pham, N.D.; Tran, T.N.M.; Pham, X.H.; Ngo, D.N.; Nguyen, M.H. Possibility of nanostructured lipid carriers encapsulating astaxanthin from Haematococcus pluvialis to alleviate skin injury in radiotherapy. *Int. J. Radiat. Biol.* **2023**, *100*, 209–219. [CrossRef]
42. Mao, X.; Tian, Y.; Sun, R.; Wang, Q.; Huang, J.; Xia, Q. Stability study and in vitro evaluation of astaxanthin nanostructured lipid carriers in food industry. *Integr. Ferroelectr.* **2019**, *200*, 208–216. [CrossRef]
43. Torres, J.; Pereira, J.M.; Marques-Oliveira, R.; Costa, I.; Gil-Martins, E.; Silva, R.; Remião, F.; Peixoto, A.F.; Sousa Lobo, J.M.; Silva, A.C. An In Vitro Evaluation of the Potential Neuroprotective Effects of Intranasal Lipid Nanoparticles Containing Astaxanthin Obtained from Different Sources: Comparative Studies. *Pharmaceutics* **2023**, *15*, 1035. [CrossRef] [PubMed]
44. Shehata, M.; Ismail, A.; Kamel, M. Nose to Brain Delivery of Astaxanthin–Loaded Nanostructured Lipid Carriers in Rat Model of Alzheimer's Disease: Preparation, in vitro and in vivo Evaluation. *Int. J. Nanomed.* **2023**, *18*, 1631–1658. [CrossRef] [PubMed]
45. Okonogi, S.; Riangjanapatee, P. Physicochemical characterization of lycopene-loaded nanostructured lipid carrier formulations for topical administration. *Int. J. Pharm.* **2015**, *478*, 726–735. [CrossRef] [PubMed]
46. Malgarim Cordenonsi, L.; Faccendini, A.; Catanzaro, M.; Bonferoni, M.C.; Rossi, S.; Malavasi, L.; Raffin, R.; Schapoval, E.; Lanni, C.; Sandri, G.; et al. The role of chitosan as coating material for nanostructured lipid carriers for skin delivery of fucoxanthin. *Int. J. Pharm.* **2019**, *567*, 118487. [CrossRef]

47. Lee, Y.-J.; Nam, G.-W. Sunscreen Boosting Effect by Solid Lipid Nanoparticles-Loaded Fucoxanthin Formulation. *Cosmetics* **2020**, *7*, 14. [CrossRef]
48. Jenning, V.; Gysler, A.; Schäfer-Korting, M.; Gohla, S.H. Vitamin A loaded solid lipid nanoparticles for topical use: Occlusive properties and drug targeting to the upper skin. *Eur. J. Pharm. Biopharm. Off. J. Arbeitsgemeinschaft Fur Pharm. Verfahrenstechnik E.V.* **2000**, *49*, 211–218. [CrossRef]
49. Pople, P.V.; Singh, K.K. Development and evaluation of topical formulation containing solid lipid nanoparticles of vitamin A. *AAPS PharmSciTech* **2006**, *7*, E63–E69. [CrossRef]
50. Resende, D.; Costa Lima, S.A.; Reis, S. Nanoencapsulation approaches for oral delivery of vitamin A. *Colloids Surf. B Biointerfaces* **2020**, *193*, 111121. [CrossRef]
51. Sabzichi, M.; Mohammadian, J.; Mohammadi, M.; Jahanfar, F.; Movassagh Pour, A.; Hamishehkar, H.; Ostad-Rahimi, A. Vitamin D-Loaded Nanostructured Lipid Carrier (NLC): A New Strategy for Enhancing Efficacy of Doxorubicin in Breast Cancer Treatment. *Nutr. Cancer* **2017**, *69*, 840–848. [CrossRef]
52. Zai, K.; Hirota, M.; Yamada, T.; Ishihara, N.; Mori, T.; Kishimura, A.; Suzuki, K.; Hase, K.; Katayama, Y. Therapeutic effect of vitamin D3-containing nanostructured lipid carriers on inflammatory bowel disease. *J. Control. Release* **2018**, *286*, 94–102. [CrossRef]
53. Eiras, F.; Amaral, M.H.; Silva, R.; Martins, E.; Lobo, J.M.S.; Silva, A.C. Characterization and biocompatibility evaluation of cutaneous formulations containing lipid nanoparticles. *Int. J. Pharm.* **2017**, *519*, 373–380. [CrossRef] [PubMed]
54. Vaz, S.; Silva, R.; Amaral, M.H.; Martins, E.; Sousa Lobo, J.M.; Silva, A.C. Evaluation of the biocompatibility and skin hydration potential of vitamin E-loaded lipid nanosystems formulations: In vitro and human in vivo studies. *Colloids Surf. B Biointerfaces* **2019**, *179*, 242–249. [CrossRef] [PubMed]
55. Ijaz, M.; Akhtar, N. Fatty acids based α-Tocopherol loaded nanostructured lipid carrier gel: In vitro and in vivo evaluation for moisturizing and anti-aging effects. *J. Cosmet. Dermatol.* **2020**, *19*, 3067–3076. [CrossRef] [PubMed]
56. Gambaro, R.C.; Berti, I.R.; Cacicedo, M.L.; Gehring, S.; Alvarez, V.A.; Castro, G.R.; Seoane, A.; Padula, G.; Islan, G.A. Colloidal delivery of vitamin E into solid lipid nanoparticles as a potential complement for the adverse effects of anemia treatment. *Chem. Phys. Lipids* **2022**, *249*, 105252. [CrossRef] [PubMed]
57. Liu, C.-H.; Wu, C.-T.; Fang, J.-Y. Characterization and formulation optimization of solid lipid nanoparticles in vitamin K1 delivery. *Drug Dev. Ind. Pharm.* **2010**, *36*, 751–761. [CrossRef]
58. Almeida, E.D.P.; Santos Silva, L.A.; de Araujo, G.R.S.; Montalvao, M.M.; Matos, S.S.; da Cunha Gonsalves, J.K.M.; de Souza Nunes, R.; de Meneses, C.T.; Oliveira Araujo, R.G.; Sarmento, V.H.V.; et al. Chitosan-functionalized nanostructured lipid carriers containing chloroaluminum phthalocyanine for photodynamic therapy of skin cancer. *Eur. J. Pharm. Biopharm.* **2022**, *179*, 221–231. [CrossRef]
59. Costa-Fernandez, S.; Matos, J.K.R.; Scheunemann, G.S.; Salata, G.C.; Chorilli, M.; Watanabe, I.S.; de Araujo, G.L.B.; Santos, M.F.; Ishida, K.; Lopes, L.B. Nanostructured lipid carriers containing chitosan or sodium alginate for co-encapsulation of antioxidants and an antimicrobial agent for potential application in wound healing. *Int. J. Biol. Macromol.* **2021**, *183*, 668–680. [CrossRef]
60. Malekmohammadi, M.; Ghanbarzadeh, B.; Hanifian, S.; Samadi Kafil, H.; Gharekhani, M.; Falcone, P.M. The Gelatin-Coated Nanostructured Lipid Carrier (NLC) Containing Salvia officinalis Extract: Optimization by Combined D-Optimal Design and Its Application to Improve the Quality Parameters of Beef Burger. *Foods* **2023**, *12*, 3737. [CrossRef] [PubMed]
61. Osanlou, R.; Emtyazjoo, M.; Banaei, A.; Hesarinejad, M.A.; Ashrafi, F. Preparation of solid lipid nanoparticles and nanostructured lipid carriers containing zeaxanthin and evaluation of physicochemical properties. *Colloids Surf. A Physicochem. Eng. Asp.* **2022**, *641*, 128588. [CrossRef]
62. Oliveira, C.; Coelho, C.; Teixeira, J.A.; Ferreira-Santos, P.; Botelho, C.M. Nanocarriers as Active Ingredients Enhancers in the Cosmetic Industry—The European and North America Regulation Challenges. *Molecules* **2022**, *27*, 1669. [CrossRef]
63. European Commission. Regulation on Cosmetic Products No 1223/2009. Available online: https://eur-lex.europa.eu/legal-content/EN/TXT/?uri=CELEX:32009R1223 (accessed on 15 February 2024).
64. FDA. FDA's Approach to Regulation of Nanotechnology Products. Available online: https://www.fda.gov/science-research/nanotechnology-programs-fda/nanotechnology-fact-sheet (accessed on 15 February 2024).
65. Ferraris, C.; Rimicci, C.; Garelli, S.; Ugazio, E.; Battaglia, L. Nanosystems in Cosmetic Products: A Brief Overview of Functional, Market, Regulatory and Safety Concerns. *Pharmaceutics* **2021**, *13*, 1408. [CrossRef] [PubMed]
66. European Commission. Regulation (EU) 2015/2283. Available online: https://eur-lex.europa.eu/legal-content/EN/TXT/?uri=CELEX:02015R2283-20210327 (accessed on 15 February 2024).
67. European Commission. Regulation (EC) No 178/2002. Available online: https://eur-lex.europa.eu/legal-content/EN/TXT/?uri=CELEX:02002R0178-20220701 (accessed on 15 February 2024).
68. Maurya, V.; Shakya, A.; Aggarwal, M.; Gothandam, K.; Bohn, T.; Pareek, S. Fate of β-Carotene within Loaded Delivery Systems in Food: State of Knowledge. *Antioxidants* **2021**, *10*, 426. [CrossRef] [PubMed]
69. Committee, E.S.; More, S.; Bampidis, V.; Benford, D.; Bragard, C.; Halldorsson, T.; Hernandez-Jerez, A.; Hougaard Bennekou, S.; Koutsoumanis, K.; Lambre, C.; et al. Guidance on risk assessment of nanomaterials to be applied in the food and feed chain: Human and animal health. *EFSA J.* **2021**, *19*, e06768. [CrossRef]

70. Gunawan, M.; Boonkanokwong, V. Current applications of solid lipid nanoparticles and nanostructured lipid carriers as vehicles in oral delivery systems for antioxidant nutraceuticals: A review. *Colloids Surf. B Biointerfaces* **2024**, *233*, 113608. [CrossRef]
71. Silva, A.C.; Santos, D.; Ferreira, D.; Lopes, C.M. Lipid-based nanocarriers as an alternative for oral delivery of poorly water-soluble drugs: Peroral and mucosal routes. *Curr. Med. Chem.* **2012**, *19*, 4495–4510. [CrossRef]

cosmetics

MDPI

Review

The Potential of Coffee and Olive by Products as Ingredient in Cosmetics Formulations and Their Extraction Techniques

Ana Matilde Ferreira [1,2], Rita C. Alves [1], Bernardo Bastos [2,3,4], Maria Beatriz P. P. Oliveira [1], Ana Casas [2] and Hugo Almeida [2,5,6,*]

1 LAQV/REQUIMTE, Department of Chemical Sciences, Faculty of Pharmacy, University of Porto, 4050-313 Porto, Portugal; anaferreira23456@gmail.com (A.M.F.); rcalves@ff.up.pt (R.C.A.); beatoliv@ff.up.pt (M.B.P.P.O.)
2 Mesosystem Investigação & Investimentos by Spinpark, 4805-017 Guimarães, Portugal; bernardo.bastos@universidadeeuropeia.pt (B.B.); ana@mesosystem.com (A.C.)
3 Department of Hospitality and Tourism, Faculty of Social Sciences and Technology (FSCT), European University, Oriente Green Campus, 1886-502 Lisbon, Portugal
4 CETRAD-UE—Center for Transdisciplinary Development Studies, Universidade Europeia, 1500-210 Lisbon, Portugal
5 UCIBIO, Laboratory of Pharmaceutical Technology, Faculty of Pharmacy, University of Porto, 4050-313 Porto, Portugal
6 Associate Laboratory i4HB-Institute for Health and Bioeconomy, Faculty of Pharmacy, University of Porto, 4050-313 Porto, Portugal
* Correspondence: hperas5@hotmail.com

Abstract: This review concentrates on the bioactive potential of two significant agri-food by-products: coffee by-products (coffee pulp and husk, spent coffee grounds, and silverskin) and olive by-products (olive mill wastewater, pomace, stones, and leaves). These residues are produced in substantial quantities, and despite their considerable application potential, they remain predominantly underutilized, thereby contributing to environmental burdens and economic losses. Their richness in bioactive compounds is unequivocal. Specifically, coffee by-products are abundant in caffeine and chlorogenic acids, whereas olive by-products serve as excellent sources of oleuropein, hydroxytyrosol, and tyrosol. Such compounds possess health-promoting properties and are promising active ingredients for cosmetic formulations, owing to their antioxidant, anti-aging, UV protective, antimicrobial, emollient, and moisturizing effects. This review not only compiles the bioactive compounds present in these by-products and explores their potential applications but also examines the extraction methods employed for their recovery. Both conventional techniques (solvent extraction) and green extraction technologies (ultrasound-assisted extraction, microwave-assisted extraction, and supercritical fluid extraction) are discussed. These innovative and environmentally friendly approaches enhance extraction efficiency and are aligned with sustainability objectives. In this context, the importance of incorporating natural ingredients into cosmetic products is emphasized, both to meet regulatory and environmental standards and to satisfy the increasing consumer demand for safer, more effective, and environmentally sustainable formulations.

Keywords: circular economy; sustainability; bioactive compounds; extraction methods; coffee silverskin; coffee pulp; coffee husk; spent coffee grounds; olive pomace; olive mill wastewater; olive leaves; olive stones

1. Introduction

Cosmetic and hygiene products have been an integral part of human culture and self-care practices for many years. However, in recent decades, substantial improvements have been realized owing to advancements in cosmetic formulation and technology, alongside continuous growth in the global market. The cosmetic industry targets specific demographics and now constitutes a significant sector of the global economy, with anticipated further development in the coming years [1,2]. In recent times, consumer concerns have markedly increased regarding skin aging, dermatological imperfections, and conditions such as acne and skin cancer [3]. This heightened awareness has driven research and innovation within the industry, leading to the development of new products that address consumer demands for aesthetically pleasing and healthy skin [4]. Concurrently, consumers are increasingly adopting eco-conscious lifestyles and displaying greater concern for the ingredients contained in cosmetic products, seeking items with natural components and lower environmental impact. Terms such as "natural," "absence of chemical toxic substances," or "product tolerability" significantly enhance a product's appeal. Consequently, it is imperative to consider consumer sensitivity toward environmentally friendly and biodegradable products, while also ensuring product efficacy [3,5]. Consumer preferences, combined with rising interest from organizations and the scientific community regarding environmental sustainability, have spurred innovation in the cosmetic industry, which has historically been associated with considerable ecological impacts. Indeed, the industry is actively pursuing more environmentally sustainable raw materials for product formulation [6]. Within the European regulatory framework, the European Commission (EC) has demonstrated an interest in promoting regulations that involve nations and citizens in the transition towards a sustainable future [7].

This review article offers a comprehensive evaluation of potential solutions for this notably significant issue by incorporating recent data on natural products, specifically coffee and olive by-products, as well as their biological composition, extraction techniques, and applications in cosmetic formulations—an aspect that is seldom examined.

2. Methodology of Search

A comprehensive literature review was undertaken utilizing PubMed and Google Scholar to identify scholarly articles, reviews, and monographs relevant to the subject matter. Various search term combinations were employed, with the key words such as: cosmetic industry, cosmetic products; circular economy; natural ingredients, plant extracts; coffee by-products, coffee silverskin, coffee pulp, parchment, spent coffee grounds, coffee husk, coffee by-products bioactive compounds, polyphenols, caffeine, chlorogenic acids, cosmetic applications, photoprotection, antiaging, antioxidant activity; extraction methods, green methods, conventional methods; olive by-products, olive pomace, olive mill wastewater, olive stones; olive by-products bioactive compounds, hydroxytyrosol, oleuropein, tyrosol. Priority was given to publications from 2020 onward to incorporate recent advancements; however, earlier studies of notable scientific significance were also evaluated. Some relevant articles were identified through references cited in the selected publications.

3. Circular Economy

The European Commission's new agenda for sustainable growth, known as the European Green Deal, encompasses a series of proposals aimed at addressing climate change and environmental degradation—issues of significant global concern. In March 2020, a revised Circular Economy Action Plan (CEAP) was introduced, with the objective of alleviating environmental pressures and enhancing sustainability [8]. The circular economy model is designed to transform products typically destined for waste by applying the principles of reduce, reuse, and recycle (3Rs) to their material components, thereby establishing a closed-loop system and assigning value to discarded materials [9]. This strategy emphasizes the sustainable management of resources, wherein waste is reintegrated into the production cycle, becoming a resource for subsequent manufacturing processes. Such an approach prolongs the lifespan of products, maximizes their value, and minimizes waste [10]. Empirical evidence indicates that this model is efficacious not only in decreasing the volume of final waste but also in reducing the dependence on virgin natural resources, thereby contributing to a more sustainable future [7] (Figure 1). Waste upcycling yields economic advantages, including job creation, cost savings, and mitigation of supply related risks such as price volatility, resource scarcity, and import dependency. Furthermore, it benefits the environment by conserving natural resources for future generations and decreasing landscape and habitat disruption, as the development of more efficient and sustainable products reduces energy and resource consumption [11].

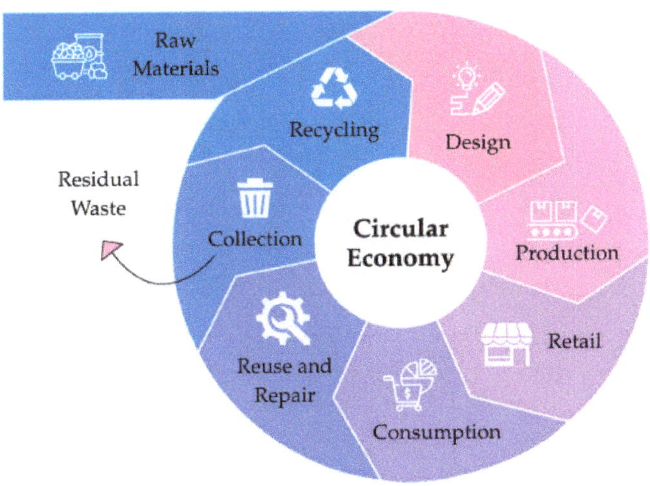

Figure 1. Circular economy model.

The concept of a circular economy, exemplified by the utilization or extraction of by-products from waste materials, has been implemented within the cosmetic industry prior to its widespread adoption [2]. Extracts of natural origin, obtained from waste generated during food and plant processing, have been extensively studied to extract bioactive compounds such as vitamins, enzymes, essential oils, antioxidants, among others [3]. These constituents are highly valuable in the cosmetic sector, enhancing the functional properties of products through the provision of antioxidant, antimicrobial, and antiaging effects, among other benefits. This approach not only adheres to the principles of a circular economy but also promotes the development of greener, value-added cosmetic products [12, 13]. For example, antioxidants that prevent product degradation and diminish oxidative damage to cellular components can be derived from plants, fruits, and grains. Polyphenols, flavonoids, and flavonols are noteworthy active compounds sourced from nature that offer multiple health benefits [14]. In fact, a broad range of natural ingredients can be incorporated into cosmetics, serving as replacements for synthetic substances. Additionally, cosmetics formulated with natural ingredients are generally preferred by consumers and support more sustainable practices [3].

4. Natural Products in Cosmetics

The utilization of natural ingredients in cosmetics originated in ancient civilizations across diverse cultures. Primarily derived from plants, minerals, and animals, these ingredients have been extensively employed. For example, Egyptians (5000 BC–30 BC) utilized oils, creams, and perfumes made from myrrh, lavender, rose, and almond oil. Kohl and henna were utilized for cosmetic purposes and hair coloring [15,16]. In Mesopotamia, now referred to as the Middle East (3000 BC), herbs, oils, and wood were used to produce various types of incense, while pigments served as color cosmetics, notably for eye makeup. The Greeks and Romans (800 BC–500 AD) employed chalk, lead-based face powders, and other ingredients such as soft coral, narcissus bulbs, honey, and orris root for skincare and dermatological applications. Over time, cosmetic formulations evolved, leading to the replacement of plant and mineral-based products with synthetic and chemical counterparts. Currently, however, the cosmetics industry is increasingly returning to natural and sustainable products, illustrating a cyclical pattern [17].

Plant extracts, which are a rich source of biologically active compounds, are increasingly employed due to their safety profile, compatibility with all skin types, minimal side effects, and greater availability, offering advantages over synthetic ingredients [18]. These extracts possess numerous valuable properties, including medicinal benefits, that can aid in the management of certain skin disorders, particularly inflammatory conditions such as acne, psoriasis, or atopic dermatitis. When incorporated into cosmetic formulations, plant extracts can provide antioxidant, antibacterial, astringent, moisturizing, regenerating, cleansing, smoothing, and/or lightening effects [19,20]. They are obtained from various plant components, including fruits, leaves, roots, bark, stems, branches, seeds, flowers, or even the entire plant. Factors such as cultivation practices, plant health, harvesting methods, and post-harvest processing techniques (including fragmentation, drying, and extraction methods) can critically influence the composition and bioactivity of the extracts [21]. Standard approaches to ensure the consistency of bioactive content involve monitoring the raw material using markers (e.g., hydroxytyrosol for olive by-products); this monitoring can be conducted employing NIR or fluorescence spectroscopy combined with chemometrics to identify variations, as documented in [22,23].

Importantly, agri-food by-products generated during food processing, such as peels, seeds, pomace, or husks, offer a sustainable and cost-effective alternative source of valuable plant extracts. The valorization of these by-products not only contributes to waste reduction and environmental sustainability but also enables the recovery of high-value bioactive compounds with high potential to be incorporated in cosmetic products [21].

Aging is characterized as a progressive and continual decline in cellular and organismal functions over the lifespan, representing a complex biological process influenced by both extrinsic and intrinsic factors [24]. Manifestations of skin aging include hyperpigmentation, diminished elasticity and increased laxity, the appearance of fine lines and wrinkles, telangiectasia, uneven skin texture, enlarged pores, periorbital puffiness, and keratosis [24,25]. The exploration of active substances to inhibit or mitigate these signs has been a longstanding area of research. In recent times, plant extracts have gained considerable prominence as sources of such active compounds, given their significant potential to delay the skin aging process. Consequently, plant extracts are increasingly incorporated into anti-aging formulations due to their capacity to protect the skin from oxidative stress, enhance immune responses, and absorb UV radiation, among other beneficial effects [25]. For example, residues from the coffee industry, such as defective coffee beans, silverskin, and spent coffee grounds, are produced in substantial quantities and contain high-value bioactive compounds, notably phenolic constituents such as chlorogenic acids and caffeine. The application of these residues has demonstrated safety, stability, and efficacy in improving skin qualities, alongside favorable consumer acceptance, thereby supporting the valorization of coffee by-product extracts as novel natural ingredients [26].

Consequently, olive by-products (e.g., olive mill wastewater, olive pomace, and leaves) contribute to the generation of significant waste quantities, thereby raising environmental concerns due to their elevated content of phytotoxic phenolics, high organic load, and low pH. Nonetheless, research has demonstrated that extracts derived from these by-products possess valuable properties, and their incorporation into cosmetic formulations offers advantageous benefits. Simultaneously, the utilization of these by-products presents an environmentally sustainable solution that alleviates their phytotoxic effects [27].

5. Coffee By-Products

5.1. Bioactive Compounds of Coffee By-Products

Coffee stands as one of the most widely consumed beverages globally, resulting in substantial quantities of coffee by-products annually. The bioactive compounds present in this plant hold potential applications in skincare products [28]. According to Ribeiro et al. [29], there are 124 species within the genus Coffea (*Coffea* sp.), of which only two are of significant commercial importance: *Coffea canephora* (commonly known as robusta) and *Coffea arabica*. In *C. arabica* beans, up to 17% of the matrix comprises a complex lipid fraction that yields coffee oil, whereas in robusta, this fraction can constitute up to 10%. The concentration of phenolic compounds in green coffee beans varies by species, with Arabica containing lower levels (4–8%) and Robusta exhibiting higher levels (7–14%). Despite robusta possessing higher caffeine content, Arabica coffee is generally regarded as of superior quality due to its sensoritonal properties [30]. The leading producers—Brazil, Vietnam, Colombia, Indonesia, Ethiopia, India, and Honduras—account for approximately 80% of global production [31].

There are two primary methods for processing coffee cherries: dry processing and wet processing, each producing distinct by-products. The dry method, also referred to as the natural process, involves drying the coffee cherry immediately after harvesting, followed by mechanical de-husking to remove the outer layers and extract the green coffee beans [32]. Conversely, the wet processing, also known as the washing method, encompasses a series of steps—including pulping, fermentation, washing, and drying—that systematically remove the fruit layers, necessitating significant water consumption [33]. Husks are the by-product of the dry method, accounting for approximately 45% of the coffee cherry's weight, comprising the fruit's outer layers such as skin, pulp, mucilage, and parchment. From the wet processing, the principal by-products are coffee pulp and parchment [34] (Figure 2). Although exceptions exist, Robusta coffee is generally processed using the dry method, whereas the wet method is predominantly employed for Arabica coffee.

During the roasting process, the beans undergo expansion and the removal of coffee silverskin. Following grinding and brewing, the residual coffee grounds are ultimately collected [35]. In on-farm processing, the coffee beans possess a moisture content ranging from 10% to 12%, thereby ensuring the preservation of quality during transportation [36].

In addition to the aforementioned coffee by-products, defective coffee beans and coffee sieving residues may also contain valuable compounds of high interest. It is noteworthy that various coffee by-products exhibit distinct chemical compositions, which are influenced by the coffee species and the processing methods employed [37]. Nonetheless, the literature indicates that coffee by-products are generally abundant in carbohydrates, proteins, vitamins, nutrients, and a range of bioactive compounds [38]. In a study conducted by Machado et al. [37], the nutritional profiles of different coffee by-products were analyzed. Except for parchment, all samples demonstrated significant protein content, with silverskin exhibiting the highest concentration at 16.3% dry weight (dw), followed by sieving residue at 14.6% dw, and defective coffee beans at 13.3% dw. Coffee silverskin also displayed notably high levels of dietary fiber (56–62%), particularly insoluble fiber predominantly composed of cellulose (18%) and hemicellulose (13%), as well as total minerals (8%) and protein (19%). Other pertinent bioactive compounds identified include chlorogenic acids (1–6%), caffeine (0.8–1.25%), and melanoidins (17–23%), among other antioxidant compounds [39]. Coffee pulp (CP) and husks (CH), aside from their carbohydrate, mineral, and protein content, also contain significant quantities of caffeine, chlorogenic acid, and tannins. The presence of these compounds restricts their application as animal feed or fertilizer due to concerns regarding animal health, seed germination, and plant growth [40–42]. However, these by-products possess valuable applications, such as in biogas and bioethanol pro-duction, mushroom cultivation, and the extraction of bioactive compounds. Utilizing a simple and low-cost enzymatic extraction method, chlorogenic acid was identified as the predominant compound (36.1%) obtained from CP and CH, with caffeic acid closely following at 33% [40].

Figure 2. Coffee by-products. (**A**) Coffee silverskin; (**B**) Parchment layer surrounding the raw coffee bean; (**C**) Coffee husk and (**D**) Coffee pulp.

Spent coffee grounds (SCGs), akin to other mentioned by-products, have also been explored for their potential applications owing to their high concentrations of organic constituents such as carbohydrates (60.3–82.0%), lipids (6.0–38.6%), proteins (11.5–18.0%), and additional compounds including caffeine (0.02–0.4%), chlorogenic acid (1.8–11.5%), caffeic acid, cafestol, and kahweol [30,43]. Consequently, SCGs may be transformed into high-value products, encompassing biofuels, bioactive compounds, and biomaterials. When compared to other lignocellulosic residues, SCGs are distinctly abundant in hemicellulose (30–40 wt%) and lignin (25–30 wt%), while their cellulose content remains relatively modest (8.6–13.3 wt%) [30]. Cellulose and hemicellulose can undergo hydrolysis into mannose and galactose, serving as substrates for microbial fermentation and the synthesis of other valuable chemicals [30].

The extraction methodology employed is of paramount importance to ensure the proper processing of coffee by-products as well as their storage conditions. Inadequate storage conditions may result in microbial contamination, the development of mycotoxins, and the degradation of bioactive compounds. Such issues could restrict their application in cosmetic formulations unless proper processing, safety protocols, and quality control measures are implemented [44]. A study examining spoilage mechanisms during storage revealed that anaerobic storage of spent coffee grounds (SCG) for up to 14 days facilitated significant bacterial proliferation, predominantly Bacilli, lactic acid bacteria, and acetic acid bacteria [44]. Conversely, extensive fungal colonization, including contamination by Aspergillus flavus—a species known for its capacity to synthesize aflatoxins—was observed under aerobic storage conditions. Post-processing, these by-products exhibit minimal microbial contamination, thereby supporting their potential valorization for cosmetic applications. Nonetheless, during storage and processing at recycling facilities, they remain highly susceptible to biochemical degradation and microbial proliferation. Therefore, establishing comprehensive storage and processing guidelines is essential [45]. Another study investigated chemical and microbial contamination in coffee silverskin (CS) extracts, with some samples testing positive for Ochratoxin A [46]. Mycotoxins are expected to undergo substantial degradation during the roasting process [47]. Effective mitigation of Ochratoxin A during coffee transportation and storage necessitates maintaining low temperatures (3–7 °C) and water activity (aw) $\leq$ 0.60 to prevent bean rehydration [48].

5.2. Extraction Methods for Coffe By-Products

The chemical profile of extracts derived from coffee by-products is substantially dependent on the extraction methodology employed, as variations in specific parameters, such as the choice of solvent, result in the recovery of differing quantities and types of bioactive compounds [49]. Indeed, the solvent exerts a predominant influence on the extraction process, given that solvents with varying polarities interact distinctively with the chemical structures of the constituents, thereby leading to variations in the biological activities associated with each by-product and solvent combination [40].

Conventional methods, such as solid–liquid extraction, remain widely employed for the recovery of bioactive compounds from coffee by-products owing to their simplicity and minimal operational costs. Nonetheless, these techniques are considered less environmentally sustainable, as they frequently necessitate substantial quantities of organic solvents [50]. Soxhlet extraction, a representative of solid–liquid extraction predominantly utilized at laboratory scale, facilitates efficient and selective extraction of target compounds; however, it demands large solvent volumes and is time-consuming [29] (see Figure 3). Mechanical pressing, a traditional solvent-free method, does not generate potentially toxic residues but generally yields lower extraction efficiencies and diminished oil purity [51].

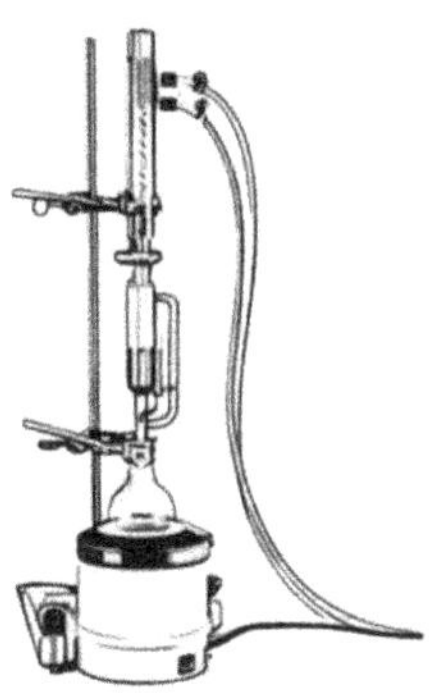

Figure 3. Soxhlet extraction, an example of solid–liquid extraction.

To enhance the recovery yield of compounds, parameters such as temperature, duration, solvent type, and liquid-solid ratio can be optimized according to the specific by-product. Typically, solvents such as alcohols and acetone, with varying water ratios, are employed for the extraction of phenolic compounds [52]. In the study conducted by Gokhan Zengin et al. [40], various solvents—including H_2O, MeOH, MeOH:H_2O (50:50), and EtOH:H_2O (70:30)—were utilized to obtain extracts of CS and SCG. As anticipated, the phenolic compound content in both by-products was affected by the choice of solvent. The SCG extracts had higher levels of bioactive compounds, with the highest antioxidant activity observed using EtOH:H_2O, followed by MeOH:H_2O, MeOH, and H_2O, in that order. The fact that water was the least effective solvent can be explained by the non-polar nature of plant cell walls, which restricts the extraction of certain compounds. In a study conducted by Ramón-Gonçalves et al. [53], ethanol-water mixtures with low ethanol ratios, ranging from 15 to 20% (v), proved to be a simple and cost-effective method for extracting polyphenols from SCGs, such as chlorogenic acid (9.69 mg/g), p-coumaric acid (0.155 to 0.348 mg/g), trans-ferulic acid (0.110 to 0.177 mg/g), rutin (0.086 mg/g), naringin (0.096 to 0.423 mg/g), and resveratrol (0.066 to 0.093 mg/g).

The presence of residual solvents can influence skin safety and the claims of natural products, as methanol, acetone, or hexane may be toxic and can alter the molecular organization of skin lipids and proteins. This alteration could compromise the skin's protective barrier function and induce skin irritation [54]. Regulatory frameworks highlight that solvents can interfere with the classification or commercialization of an extract as natural. Consequently, green solvents such as water, ethanol, and CO_2 are regarded as the most acceptable options in terms of safety, product quality, and consumer perception [55,56].

Conventional methods are increasingly being replaced by greener and more efficient alternatives, which offer shorter processing times, reduced use of solvents, and improved extraction yields [49]. Nonetheless, some of these alternatives may necessitate substantial investment for scaling up into industrial practices, thus presenting a barrier for certain industries. Microwave-assisted extraction (MAE) is less time-consuming and requires smaller volumes of solvent; however, it faces limitations in terms of extraction uniformity and equipment costs. MAE has achieved an extraction yield of 84 ± 2.8 mg/g for chlorogenic acid and approximately 72.5 mg/g of caffeine [57]. Supercritical fluid extraction (SFE), particularly with carbon dioxide, facilitates the extraction of oil with high purity, representing an efficient and environmentally friendly method, notwithstanding the high costs associated with the equipment [29]. SFE has yielded extraction amounts of 6.45 mg/g for caffeine and 0.55 mg/g for chlorogenic acid [56].

Finally, in the study conducted by Ibtissam Bouhzam et al [58], three uncomplicated and straightforward extraction techniques were compared at a laboratory scale for the extraction of chlorogenic acid and caffeine. This investigation demonstrated a comparison among Water Ultrasound-Assisted Extraction (UAE) (Figure 4), Water and Vortex Extraction, and Supramolecular Solvent Extraction. All three procedures lasted 1 min, and ultrasound-assisted water extraction at ambient temperature was identified as the most effective, producing the highest concentrations of chlorogenic acid (1.15 mg/g) and caffeine (0.972 mg/g). Conversely, the utilization of supramolecular solvent resulted in a lower recovery of chlorogenic acid in the supra-phase, likely owing to its stronger affinity for the aqueous lower phase (Table 1). The extraction conditions of the various methods are summarized in Table 2.

Table 1. Comparison between conventional and green extraction methods with pros, cons, and typical yields for coffee by-products extraction.

Category	Method	Pros	Cons	Yield
Conventional	Solid–Liquid Extraction (e.g., Soxhlet)	Simple; Economic; Accuracy/reproducibility [49]	Not environmentally friendly; Time consuming; Poor processing adaptability [49]	(ethanol-water solvent) Chlorogenic acid (9.69 mg/g) p-coumaric acid (0.155 to 0.348 mg/g) rutin (0.086 mg/g) [53]
Green	Microwave-assisted (MAE)	Less time consuming; Efficient and environmentally friendly; [29,57]	Expensive equipment; Limitations concerning extraction uniformity [29,57]	Chlorogenic acid (84 ± 2.8 mg/g) Caffeine (approx. 72.5 mg/g) [57]
	Supercritical fluid Extraction (SFE)			Chlorogenic acid (0.55 mg/g) Caffeine (19.49 mg/g) [57]
	Water Ultrasound-Assisted Extraction (UAE)			Chlorogenic acid (1.15 mg/g) Caffeine (0.972 mg/g) [58]
	Water and Vortex Extraction			Chlorogenic acid (0.827 mg/g) Caffeine (0.766 mg/g) [58]
	Supramolecular Solvent Extraction			Chlorogenic acid (0.116 mg/g) Caffeine (0.885 mg/g) [58]

Table 2. Comparison of extraction conditions for coffee by-products.

Category	Method	Temperature or Power	Time	Solvent Ratio	Ref.
Conventional	Solid–Liquid Extraction (e.g., Soxhlet)	60 °C	30 min	EtOH:H_2O (50:50)	[59]
Green	Microwave-assisted (MAE)	500 W	2 min	H_2O 15 mL/g	[60]
	Supercritical fluid Extraction (SFE)	Room temperature	5 min	EtOH: H_2O 4:1 *v*/*w*	[61]
	Water Ultrasound-Assisted Extraction (UAE)	30–50 °C 100–300 W	5–45 min	EtOH 5:1–30:1	[52]
	Water and Vortex Extraction	50/60 Hz, 195 W	1 min	0.7 g of SCG to 4 mL of ultrapure water.	[58]
	Supramolecular Solvent Extraction	-	1 min	0.7 g of SCG to ethanol (1.2 mL), 1-hexanol (0.96 mL), and water until a total volume of 4 mL	[58]

Extraction of bioactive compounds from plant by-products for cosmetic applications has yielded promising results in laboratory settings. However, studies focusing on process optimization and industrial implementation remain relatively limited. Further research is required to address scalability, process economics, solvent recovery, and integration into existing production lines. While most extraction methods currently produce low quantities of bioactive compounds from coffee by-products, these materials have been demonstrated to be valuable ingredients for cosmetic formulations. Derivatization of these bioactive compounds is feasible, enabling the creation of new ingredients with high commercial value for industrial use. A recent study demonstrated the modification of spent coffee oils through cost-effective processes suitable for scaling-up. This approach also aligns with the increasing consumer preference for eco-friendly and naturally derived cosmetics [62].

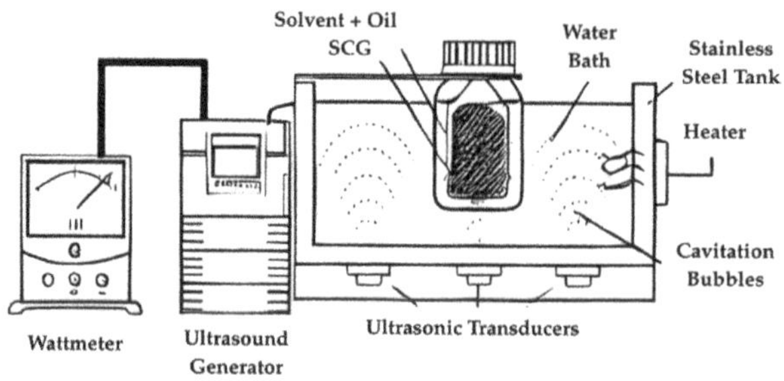

Figure 4. Water Ultrasound-Assisted Extraction (UAE) of SCG oil. Adapted from Miladi et al. [63].

5.3. Applications in Cosmetic Formulations

As previously mentioned, coffee by-products possess potential applications within the cosmetic industry, owing to their content of caffeine and phenolic compounds, such as tannins, flavonoids, and chlorogenic acids, notably 5-caffeoylquinic acid. These constituents are primarily responsible for antioxidant, anti-inflammatory, and antibacterial activities [64].

One of the many external factors that contribute to skin aging is UV exposure. UVA rays penetrate deeply into the epidermis and dermis, causing collagen and elastin fiber degradation through oxidative stress and activation of matrix metalloproteinases [65]. A study conducted by Cho Y-H. et al. [66] investigated the anti-wrinkle effects of chlorogenic acid and other compounds extracted from *Coffea arabica* on UV-B-stimulated mouse fibroblasts, demonstrating the anti-wrinkle potential of chlorogenic acid. Therefore, extracts of coffee by-products can be used to protect against UV light, which damages the skin and accelerates skin aging. Applying these extracts as sunscreen additives would be effective, since the presence of double bonds or aromatic rings in the molecular structure of phenolic compounds provides them with UV absorption properties in the range of 200–400 nm [66]. Polyphenols have protective and healing effects on the skin, and according to the literature, caffeic acid reduces dermatitis and pigmentation resulting from previous UVB radiation exposure [67,68].

Overall, phenolics have been reported as relatively photostable, maintaining their activity under UV exposure and demonstrating no significant phototoxicity in cellular models. Consequently, phenolic compounds are promising candidates for products exposed to sunlight; however, their photostability varies considerably depending on the specific compound and is often enhanced through encapsulation, combination with UV filters, or synergistic antioxidants to ensure long-term safety and efficacy [69,70].

As previously noted, coffee by-products are a source of caffeine and chlorogenic acids. Caffeine ($C_8H_{10}N_4O_2$) possesses notable applications owing to its ability to penetrate the skin barrier, thereby enhancing the condition of skin and hair through their biological effects. Its utilization is prevalent due to its antioxidant, photoprotective, and lipolytic properties [26,71]. Additionally, caffeine may promote hair growth by boosting scalp microcirculation, which enhances the delivery of nutrients and oxygen to hair follicles through skin cell and follicular proliferation, increases metabolic activity, and, owing to its antioxidant properties and capacity to stimulate IGF-1 expression, triggers and sustains the anagen phase of hair growth [72,73]. Due to its antioxidant properties, this compound is capable of neutralizing free radicals, thereby diminishing oxidative stress and decelerating the processes associated with skin aging. Consequently, the inclusion of these extracts in cosmetic formulations can improve skin protection and vitality [26]. Furthermore, caffeine has been shown to enhance microcirculation within blood vessels, which may reduce the visibility of cellulite. A study conducted by Rodrigues et al. [74] involved the isolation of caffeine from CS, which was subsequently incorporated into nanostructured lipid carriers (NLCs) to develop a formulation for cellulite treatment. In vivo assessments using pig ear skin demonstrated that caffeine in CS extracts exhibited improved nanoparticle penetration. Caffeine is extensively utilized in topical skincare formulations and is generally regarded as safe and well-tolerated, with no substantial reports of irritation [75]. Nevertheless, in vitro evidence suggests that excessive caffeine exposure may diminish collagen synthesis in human cultured skin fibroblasts and may also exert inhibitory effects on DNA biosynthesis [76]. Chlorogenic acid ($C_{16}H_{18}O_9$) is the primary phenolic compound found in extracts of coffee by-products, holding significance for topical formulations owing to its antioxidant properties observed in both in vitro and in vivo studies. Additionally, this compound is recognized as a potential therapeutic agent for combating photoaging and exerting anti-inflammatory effects. The therapeutic efficacy of chlorogenic acid was assessed in Wistar rats, demonstrating an acceleration in wound healing processes. Furthermore, this compound is regarded as a photostable molecule effective against UVA and UVB radiation, exhibiting notable photoprotective activity. The antimicrobial properties of chlorogenic acid are particularly valuable for formulation purposes, as it can serve as a preservative, thereby replacing synthetic alternatives [64,77].

Previous studies have detailed potential applications of bioactive compounds derived from coffee in cosmetic products. Grigolon et al. [78] conducted a chemical profiling of a coffee seed (CS) extract, which identified the presence of cafestol and kahweol fatty acid esters, along with acylglycerols, β-sitosterol, and caffeine. The authors conducted in vitro experiments on keratinocytes, which demonstrated heightened expression of genes associated with oxidative stress response and skin barrier function. In a study involving volunteers, after 28 days, the extract was observed to increase skin hydration. Additionally, the extract facilitated skin regeneration following irritation, evidenced by reductions in redness and transepidermal water loss. A research study conducted by Rodrigues et al. [79], involving twenty volunteers who utilized a base cream containing a CS extract, revealed results in skin hydration and firmness comparable to those achieved with an equivalent base formulation containing 1.5% hyaluronic acid. In a study conducted by Rodrigues et al. [80], a formulation containing 2.5% CS extract demonstrated no cytotoxic effects in vitro, thereby indicating its safety for topical application when evaluated on reconstituted human epidermis. As previously noted, the constituents of coffee's lipid fraction provide beneficial attributes for the development of products with photoprotective properties.

The antioxidant activity of coffee by-products, ranging from strong to weak, depends on the specific by-product and the processing method employed [28,81]. Phenolic compounds such as chlorogenic acids and tannins, along with caffeine present in coffee by-products, could serve as active agents against oxidative stress and damage, owing to their redox properties, which enable them to scavenge and neutralize free radicals [82,83]. Consequently, coffee by-products possess numerous potential applications in topical cosmetic formulations [28] (Table 3). Several commercial cosmetic products incorporate extracts of coffee by-products; for instance, the VICHY brand offers an under-eye serum targeting dark circles, with caffeine as a key ingredient, attributable to its antioxidant and anti-inflammatory proper-ties [84].

Table 3. Bioactive compounds from coffee by-products and their potential applications in cosmetic formulations.

Bioactive Compounds	Properties	Cosmetic Formulations	References
Chlorogenic acid	Photoprotective; Antioxidant; Anti-wrinkle; Anti-aging; Anti-inflammatory; Antimicrobial	Anti-aging products; Products with therapeutic effects; Suncreams	[64,66,77]
Caffeine	Antioxidant; Anti-aging; Photoprotective; Lipolytic activity; Microcirculation enhancer	Products for cellulite treatment, hair growth, oxidative damage, photoprotection and skin protection and vitality	[26,71–74,82]
Cafestol; Kahweol; Acylglycerols; β-sitosterol	Antioxidant; Moisturizer; Regenerator; Photoprotective	Skin hydration and firmness creams; Products for skin regeneration;	[78–80]

The presence of multiple compounds can influence each other's activity. Although there is a lack of recent primary studies specifically investigating the combined effects of bioactive compounds from by-products in cosmetic applications, evidence shows complexation between caffeine and chlorogenic acids in aqueous solutions [85]. In a study conducted by Xu et al., [86] the results demonstrated that the combination of chlorogenic acid and caffeine had anti-obesity effects and regulated lipid metabolism in high-fat diet-induced obese mice. Studies have reported that caffeine interacts with many biomolecules, including nucleic acids, polyphenols, and various drug molecules, thereby affecting their physiological properties (e.g., solubility, bioavailability, pharmacology) [87].

6. Olive By-Products

6.1. Bioactive Compounds of Olive By-Products

Olive trees are regarded as among the earliest cultivated plants in human history, with over thirty species documented. Only *Olea europaea* L. is consumable [88], a fruit tree originating from Asia Minor and Syria [89]. Currently, the principal producers of olives and olive oil are Spain, Italy, and Greece. In recent decades, the consumption of olive oil has risen, attributed to its organoleptic qualities and health advantages [90]. Olive oil is notably rich in monounsaturated fatty acids, as well as phenolics, phytosterols, tocopherols, and squalene. In the composition of extra virgin olive oil, triglycerides constitute 98% to 99%, while minor components account for 1% to 2%, with phenolic compounds representing the most significant constituents of this minor fraction. Olive oil predominantly comprises unsaturated fatty acids, constituting up to 85%, chiefly oleic acid, which varies from 70% to 85% [89,91].

The waste and by-products generated from the olive processing industry have experienced an increase. Consequently, environmental issues have emerged due to their elevated content of phytotoxic compounds (e.g., phenolics), despite these compounds also offering health benefits to humans. Therefore, it is imperative to develop strategies for their recovery and reuse in other industries, in alignment with circular economy policies and sustainability objectives [92]. Given that these bioactive compounds possess notable technological and pharmaceutical properties, they can be utilized as active ingredients in cosmetic applications following extraction and/or purification [93]. The olive fruit is structurally composed of three primary parts: the skin, or epicarp, which accounts for approximately 1–3% of the total fruit weight and is rich in chlorophylls, carotenoids, and anthocyanins crucial for the fruit's coloration; the pulp, or mesocarp, which comprises the majority of the fruit, representing 70–80% of its weight and serving as the reservoir of nutritional and bioactive constituents; and the stone, or woody endocarp, containing the seed and the woody shell, which constitutes 18–22% of the olive's weight [88].

Olive oil production is principally a mechanical process, which can be classified into three distinct types: the traditional pressing method, characterized by low efficiency and high labor costs; and two continuous methods, specifically the two-phase and three-phase centrifugal separation techniques, which are the most widely employed. The two-phase centrifugal separation has gained favor in recent years, as the three-phase method necessitates large volumes of water during the centrifugation and decantation stages [94]. The three-phase extraction process yields not only olive oil but also substantial quantities of olive mill wastewater (OMWW) and olive pomace. Conversely, the two-phase process omits the need for additional water, thereby considerably reducing the volume of OMWW produced. This process also results in a semi-solid residue known as olive pomace [95,96]. Overall, the principal by-products derived from olive processing include leaves and pruning biomass, OMWW, olive pomace, and olive stones [93] (see Figure 5). Among these, olive pomace represents the most polluting agricultural by-product in the Mediterranean region [97].

Figure 5. Olive by-products. (**A**) Olive stones; (**B**) Olive pomace; (**C**) Olive leaves; (**D**) Olive mill wastewater.

While the content of bioactive compounds is influenced by factors such as the type of by-product, olive variety or cultivar, harvesting period, processing conditions, and extraction techniques [95], olive by-products are generally abundant in phenolic compounds. These include phenolic alcohols (tyrosol and hydroxytyrosol), phenolic acids (caffeic acid, gallic acid, and verbascoside), flavonoids (luteolin, apigenin, and rutin), secoiridoids (oleuropein and ligstroside), and lignans (pinoresinol) [90,93]. Madureira et al. [98] identified hydroxytyrosol as the predominant phenolic compound in olive pomace (25 mg/g of extract), followed by hydroxytyrosol 1-β-glucoside (9.8 mg/g of extract) and tyrosol (5.9 mg/g of extract). The composition of olive pomace varies, comprising cellulose (30%), pectic polysaccharides (39%), hemicellulosic polymers rich in xylans and glucuronoxylans (14%), xyloglucans (15%), and mannans (2%). Additionally, squalene has been reported in olive pomace [90].

Olive leaves are regarded as a by-product of olive cultivation, arriving in substantial quantities at the olive mill. They contain valuable compounds, including flavones such as luteolin-7-glucoside (0.145–0.165 mg/g), apigenin-7-glucoside (0.197–0.217 mg/g), and luteolin (0.216–0.224 mg/g); flavonols like rutin (0.289–0.651 mg/g); substituted phenols such as tyrosol (3–14 mg/g), hydroxytyrosol (6–24 mg/g), vanillic acid (0.010–0.020 mg/g), and caffeic acid (0.012–0.018 mg/g); and secoiridoids including oleuropein (0.13–21.34 mg/g). Oleuropein, the predominant phenolic compound in olives, is a complex molecule derived from elenolic acid, glucose, and hydroxytyrosol [99,100].

Olive mill wastewater is an organic-rich effluent containing a significant concentration of phytotoxic phenolic compounds. Additionally, it comprises sugars and organic acids, along with mineral nutrients such as nitrogen (N), phosphorus (P), potassium (K), and magnesium (Mg), among others. In the study conducted by Cuffaro et al. [96], the polyphenolic profile of OMWW extract from two different olive cultivars, Leccino (CL) and Frantoio (CF), was reported. Two samples of CL and CF (CL1, CL2, and CF1, CF2) were collected and characterized, revealing that, although present in varying quantities, these samples were particularly rich in various phenolic compounds. CL1 and CF1 exhibited high concentrations of oleacein (314.628 ± 19.535 and 227.273 ± 3.974 μg/mg, respectively), as well as tyrosol and hydroxytyrosol (17.602 ± 0.792, 18.808 ± 0.447 μg/mg, and 55.875 ± 1.511, 71.919 ± 3.943 μg/mg, respectively). Notably, CL1 contained higher levels of oleocanthalic acid compared to CF1 (10.628 ± 0.484 μg/mg and 2.149 ± 0.091 μg/mg, respectively). In addition to these compounds, present in lower quantities yet remaining significant, were caffeic acid (2.791 ± 0.101 μg/mg and 1.453 ± 0.044 μg/mg for CL1 and CL2, respectively), and (2.234 ± 0.112 μg/mg and 3.144 ± 0.142 μg/mg for CF1 and CF2, respectively), as well as verbascoside (1.624 ± 0.035 μg/mg and 0.958 ± 0.018 μg/mg for CL1 and CL2, respectively), and (18.182 ± 0.437 μg/mg and 0.909 ± 0.035 μg/mg for CF1 and CF2, respectively).

The olive stones constitute a lignocellulosic material predominantly composed of hemicellulose (21.9%), cellulose (31.29%), and lignin (26.5%) [100]. Consequently, this by-product holds significant interest owing to its chemical and physical properties as well as its calorific value. Additionally, minor proportions of fat (5.53%), proteins (3.20%), and sugars (0.48%) have been documented [88]. This by-product is among the most abundant sources of secondary metabolites. According to Saad et al. [101], thirty-one metabolites were identified via methanolic extraction, encompassing phenolic alcohols, hydroxycinnamic acid derivatives, secoiridoids, flavonoids, terpenes, and fatty acids.

Therefore, the reutilization of olive waste as an innovative, renewable, and economical source of high-value bioactive compounds holds significant potential owing to its numerous applications as functional ingredients within the food, pharmaceutical, and cosmetics sectors, in accordance with the principles of a circular economy [96] (Figure 6). Given their high phytotoxic properties, their valorization additionally aids in mitigating environmental concerns [96].

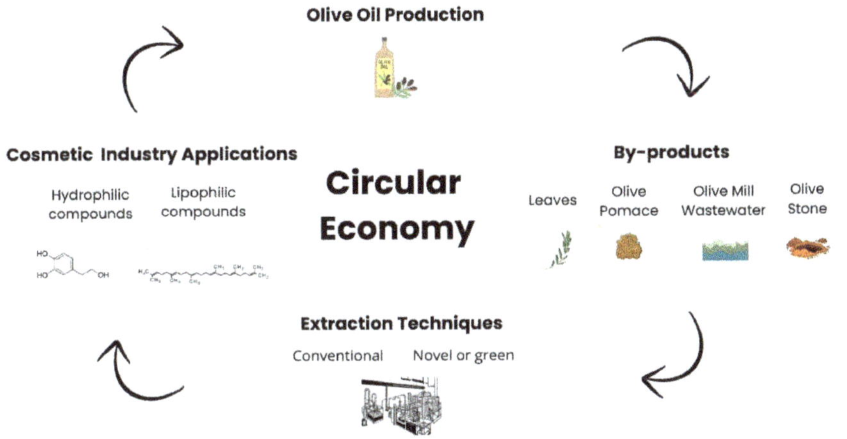

Figure 6. Olive oil by-products following a circular economy.

6.2. Extraction Methods for Olive By-Products

According to the literature, the methods used to extract bioactive compounds from olive by-products are very similar to those applied to coffee by-products. The extraction techniques for olive by-products also include conventional technologies, such as solid–liquid extractions evolving organic solvents, distillation, heat reflux, and filtration processes to separate and obtain the bioactive compounds [102]. When the extracts are to be applied in the food or cosmetic industry, the solvents employed are mainly hydroalcoholic mixtures, since ethanol is recognized as safe for consumers [27]. The solid-liquid extraction method is straightforward, cost-effective, and exhibits high accuracy and reproducibility. However, it is not environmentally sustainable, is time-consuming, and demonstrates limited adaptability in processing. The extraction yield of hydroxytyrosol using a water-ethanol mixture as a solvent ranges from 10 to 15 mg per gram [103,104].

The procedures encompass the recovery of phenolics through the condensation of steps (thermal concentration, filtration, or lyophilization), followed by subsequent extraction stages utilizing methanol, ethanol, or hydro-alcoholic solutions [93].

The employment of innovative or environmentally friendly technologies has been on the rise, including high-hydrostatic pressure, ultrasound-assisted extraction (UAE), microwave-assisted water extraction (MAE), pulsed electric field (PEF), radiofrequency drying, high voltage electrical discharge, supercritical fluid extraction (SFE), and pressurized liquid extraction (PLE) [93,105].

In a study conducted by Madureira et al. [106], ultrasound-assisted extraction (UAE) was utilized to recover hydroxytyrosol and tyrosol, yielding results of 36 ± 2 mg g^{-1} and 14 ± 1 mg g^{-1}, respectively. In another investigation employing microwave-assisted extraction (MAE), it was possible to solubilize 5.87 mg of hydroxytyrosol per gram and 46.70 mg of mannitol per gram from exhausted olive pomace [107]. These methods, in comparison to conventional techniques, are characterized by reduced time requirements and greater environmental sustainability. Presently, UAE is acknowledged as a sustainable technique that facilitates energy-efficient processes and produces notable improvements in extraction yield and economic viability at an industrial scale. Recent innovations in ultrasound technology have been developed to meet current industrial demands, including the availability of continuous-flow devices suitable for large-scale operations. Consequently, the optimization of extraction parameters at the laboratory level remains a crucial step prior to the implementation at an industrial scale [108]. A research study aimed to establish an optimized, reproducible, and industrially scalable UAE protocol for olive oil by-products, specifically olive leaves and olive pomace, with emphasis on maintaining the chemical stability and integrity of key phenolic compounds. Among the evaluated variables, the ethanol concentration in the extraction solvent was identified as the most significant factor exerting the highest impact [109].

The optimization of operational parameters for each extraction method is essential to prevent the degradation of macromolecules and the oxidation of labile compounds [94].

The application of deep eutectic solvents (DES), multicomponent systems comprising hydrogen bond donors and acceptors that interact to form a stable liquid, has recently emerged as a sustainable technique for extracting bioactive compounds from olive by-products [110]. These solvents are easy to prepare, cost-effective, and environmentally benign. They can be synthesized from various substances, permitting the development of a "tailor-made" solvent tailored to specific applications [27,111]. This methodology yielded an extraction of 4.98 mg/g for hydroxytyrosol [112].

Natural Deep Eutectic Solvents (NaDES) constitute a subset of deep eutectic solvents composed of naturally occurring, biomolecular components such as amino acids, sugars, organic acids, and other bio-based molecules [100]. These solvents are engineered to be biocompatible and environmentally sustainable, thereby aligning with the principles of green chemistry [113]. When optimized for specific applications, deep eutectic solvents have demonstrated considerable efficiency; however, they are characterized by high viscosity, which may present a limitation [111,114]. In the study conducted by Cabrera et al. [115], the extraction of phenolic compounds utilizing NADES was investigated, and the findings indicated a hydroxytyrosol content of 1.24 mg/g dry basis.

Enzyme-Assisted Extraction (EAE) is a process that uses hydrolytic enzymes to break down cell walls, enabling the separation of bioactive compounds from plant raw materials. It is environmentally friendly and cost-effective, integrating well with existing industrial processes [116]. Cellulase and viscoenzyme are two enzymes commonly used in EAE. These enzymes can produce new derivatives from the extracts by cleaving bonds in macromolecules, such as cell walls. For example, cellulase hydrolyzes vegetable cell walls by breaking down cellulose into glucose. Meanwhile, viscoenzyme is a multicomponent carbohydrase capable of breaking glycosidic bonds in plant cell walls through hydrolysis, releasing intercellular components. Therefore, EAE has shown superior extraction yields compared to hydroalcoholic solvents, indicating its potential for large-scale use [117]. The downside of this method is that enzymes can be costly, and extraction efficiency may vary depending on the substrate and enzyme used [56]. Using cellulase and viscoenzyme, it is possible to obtain 0.512 mg/g of hydroxytyrosol during extraction [117] (Table 4).

Conventional solvent extractions generally provide the highest yields; however, they are not highly sustainable. Green technologies such as Ultrasound-Assisted Extraction (UAE) and Microwave-Assisted Extraction (MAE) often attain higher yields with satisfactory purity, while utilizing reduced amounts of solvent. The optimal balance for extracting polar bioactive compounds, such as polyphenols, may be achieved through MAE or UAE employing water/ethanol as solvents [118]. The extraction conditions of the various methods are summarized in Table 5.

Table 4. Comparison of conventional and green extraction methods, including their pros, cons, and typical yields for extracting olive by-products.

Category	Method	Pros	Cons	Yield
Conventional	Solid–Liquid Extraction (e.g., Soxhlet)	Simple; Economic; Accuracy/reproducibility [104]	Not environmentally friendly; Time consuming; Poor processing adaptability [104]	(ethanol-water solvent) Hydroxytyrosol (10–15 mg/g) [103]
Green	Microwave-assisted (MAE)	Less time consuming; Efficient and environmentally friendly; [107]	Expensive equipment; Limitations concerning extraction uniformity; [107]	Hydroxytyrosol (5.87 mg/g) Manitol (46.70 mg/g) [107]
	Deep eutectic solvents (DES)	Solvents easy to prepare; Efficient; Inexpensive, and environmentally friendly; [27,111]	Challenges in compound and solvent recovery; [119]	Hydroxytyrosol (4.98 mg/g) [112]
	Water Ultrasound-Assisted Extraction (UAE)	Less time consuming; Efficient and environmentally friendly; [108]	Expensive equipment; Limitations concerning extraction uniformity [108]	Hydroxytyrosol (36 ± 2 mg/g) Tyrosol (14 ± 1 mg/g) [106]
	Natural Deep Eutectic Solvents (NaDES)	Biocompatible and environmentally friendly; Efficient; [113]	High viscosity; [114]	Hydroxytyrosol (1.24 mg/g) [115]
	Enzyme-Assisted Extraction (EAE)	Environmentally friendly and cost-effective; [116]	Enzymes can be expensive; Extraction efficiency may vary with substrate and enzyme; [57]	(Using cellulase and viscoenzyme) Hydroxytyrosol (0.512 mg/g) [117]

Table 5. Comparison of extraction conditions for olive by-products.

Category	Method	Temperature or Power	Time	Solvent Ratio	Ref.
Conventional	Solid–Liquid Extraction (e.g., Soxhlet)	25 °C	180 min	EtOH: H_2O 5:1 v/w	[120]
Green	Microwave-assisted (MAE)	250–350 W	2–3 min	Solvent free Sample Amount 5–10 g.	[121]
	Deep eutectic solvents (DES)	55 °C	5 min	ChCl:AA (1:2) ethanol (80:20 w/w)	[122]
	Water Ultrasound-Assisted Extraction (UAE)	500 W 30 °C	10 min	EtOH: H_2O 50% v/v	[123]
	Natural Deep Eutectic Solvents (NaDES)	60 °C	10 min	Choline chloride and caffeic acid (1:2)	[124]
	Enzyme-Assisted Extraction (EAE)	50 °C	120 min	Cellulase or viscoenzyme 1.0%	[125]

6.3. Applications in Cosmetic Formulations

Olive trees can supply compounds for the cosmetic industry, including both hydrophilic and lipophilic agents. Hydrophilic compounds primarily consist of polyphenols, whereas lipophilic compounds encompass fatty acids, liposoluble vitamins, and squalene [27,125]. In the hydrophilic fraction, polyphenols are characterized by their multifunctionality, exhibiting antioxidant properties along with anti-aging, photoprotective, and antimicrobial activities [126,127]. The lipophilic compounds also possess various valuable properties, including antioxidant, emollient, and moisturizing effects. They facilitate healing by supporting the repair of the skin's lipid barrier function; can be employed as lubricants to achieve soft, elastic, and well-lubricated skin; and have a remarkable ability to adhere to the skin surface, thereby providing prolonged activity [27,128,129].

The application of olive by-products has been investigated for skin treatments employing various topical delivery systems. For instance, creams are recognized for their efficacy in hydration and emollience. The utilization of natural ingredients in creams has witnessed an increasing trend [130]. In a study conducted by Nunes et al. [131], oil-in-water creams incorporating three distinct extracts derived from olive oil by-products were produced without the use of organic solvents and were compared to a control cream devoid of extracts. The authors concluded that these extracts might serve as UV-protection enhancers by promoting the absorption of specific synthetic UV filters. The results further suggest their potential use as preservative enhancers. Moreover, in an in vitro assessment conducted on human keratinocytes and corneal epithelium cells, these extracts did not induce skin or eye irritation. In vivo efficacy assessments also demonstrated that these formulations exhibit excellent compatibility and acceptability with human skin. Another investigation conducted by Wanitphakdeedecha et al. [130] evaluated the efficacy of a cream containing olive leaf extract in facial rejuvenation. A pilot study was carried out involving 36 participants with photoaged skin, who applied the olive leaf extract-containing cream. Observations revealed changes in the biophysical properties of the skin, including melanin and erythema indices, transepidermal water loss, skin hydration, skin pH, sebum levels, texture, and wrinkles. Notably, significant improvements in the reduction in wrinkles were identified after merely one month of treatment, while enhancements in skin barrier function, hydration, and texture were observed after two months. In conclusion, the cream containing olive leaf extract demonstrated benefits in skin rejuvenation in human subjects.

As previously stated, hydroxytyrosol ($C_8H_{10}O_3$), a phenolic compound and the predominant product derived from the hydrolysis of oleuropein ($C_{25}H_{32}O_{13}$), is abundantly present in olive by-products and constitutes one of the most potent naturally occurring antioxidants [131]. Oleuropein, obtained from olive leaves, exhibits antibacterial efficacy comparable to that of synthetic preservatives, as reported by Fuad Al-Rimawi et al. [132]. In this study, a microbiological analysis was conducted to evaluate two fungal species and three bacterial species; the results indicated that the product containing natural preservatives performed similarly to the formulation utilizing synthetic preservatives.

Hydroxytyrosol, which develops throughout olive maturation, oil preservation, and table olive processing, is the major contributor to the sensory diversity of olive products [133]. Hydroxytyrosol can be used in formulations that suppress melanin production, thereby reducing skin pigmentation, and acting as a whitening, lightening, and depigmenting compound. This compound has not only antioxidant, but also anticancer, anti-inflammatory, and antimicrobial activities [95,134–136]. Smeriglio et al. [137] examined the safety and efficacy of a cream containing hydroxytyrosol as an active ingredient, verifying its ability to enhance the epidermal barrier, mitigate inflammatory responses, and promote skin repair.

Nunes et al. [138] examined the nutritional, chemical, and antioxidant profiles of functional ingredients derived from various olive pomace samples, as well as their antimicrobial activity against *S. aureus*, *E. coli*, and *C. albicans*. Among the evaluated ingredients, the sample containing the highest concentration of hydroxytyrosol (220 mg/100 g) exhibited the most potent antibacterial activity, as evidenced by the lowest minimal inhibitory concentration (MIC) values against the strains tested. The antibacterial mechanisms of hydroxytyrosol are linked to its capacity to chelate transition metals, thereby diminishing the reactivity of iron and copper through the formation of inert metal–ligand complexes, which subsequently reduce their bioavailability for bacterial growth. Additionally, it may exert its effects by decreasing intracellular ATP concentrations, disrupting membrane potential, and/or reducing bacterial protein content [138,139].

Tyrosol ($C_8H_{10}O_2$), as a hydrophilic compound, presents certain limitations in its incorporation into oil-based cosmetic products. Nevertheless, it is feasible, under green chemistry conditions, to synthesize lipophilic esters of tyrosol, thereby enhancing its suitability for emulsions, creams, and other oil-based cosmetic formulations. These lipophilic derivatives exhibit superior antioxidant and surfactant properties and also possess anti-inflammatory effects [140]. Although tyrosol demonstrates valuable properties for inclusion in cosmetic formulations, unlike hydroxytyrosol, the literature lacks controlled clinical or targeted in vitro investigations regarding its dermal tolerability, particularly for sensitive or atopic skin. Therefore, the topical application of tyrosol should be supported by formal dermal safety assessments, such as reconstructed human epidermis assays and trials involving volunteers with sensitive or atopic skin, prior to broader cosmetic use [137,141].

Regarding lipophilic compounds, oleic acid ($C_{18}H_{34}O_2$) stands as the most abundant fatty acid in olive oil by-products and can be employed in cosmetic formulations as an enhancer of transdermal penetration for specific active compounds [128,142]. Vitamin E, specifically α-tocopherol, is the predominant form found in olive oil and serves as a potent non-enzymatic antioxidant. It is extensively incorporated into topical formulations due to its capacity to neutralize free radicals, prevent lipid peroxidation of fatty acids, and safeguard cellular membranes [143,144]. Vitamin E ($C_{29}H_{50}O_2$) also possesses the ability to modulate pathways involved in inflammation, apoptosis, and cell differentiation. Carotenoids, natural pigments with photoprotective properties, are capable of preventing UVA-induced dermal damage, reducing oxidative stress, preserving enzyme activity, and inhibiting fibroblast apoptosis, while also exerting anti-inflammatory effects [145]. Squalene ($C_{30}H_{50}$), a polyunsaturated triterpenic hydrocarbon characterized by six carbon double bonds [146], is found in olive pomace and offers notable advantages for skin. In addition to its antioxidant properties, squalene serves as a sink for highly lipophilic xenobiotics, aiding in their elimination from the organism. This compound has potential applications as an ingredient in cosmetic and dermo-protective creams due to its emollient properties [90,147] (Table 6). Several commercial cosmetic products incorporate extracts derived from olive by-products. For instance, OLIVEDA, a distinguished skincare brand, offers various product lines, including cleansing balms, hair serums, facial masks, and face creams containing olive by-product ingredients—most notably, olive leaf cell extracts rich in hydroxytyrosol—owing to their antioxidant and anti-aging properties [148].

Cosmetic formulations must undergo multiple physical-chemical and microbiological evaluations prior to market release. Safety assessment reports have been prepared for the utilization of olive-derived ingredients in various cosmetic formulations, including leave-on products [149]. Such documentation underscores the importance of comprehensive safety assessments, encompassing dermatological compatibility testing, to safeguard consumer welfare and ensure product efficacy. Various studies have conducted dermatological compatibility tests to identify potential adverse skin reactions, with most yielding favorable outcomes that support the safe integration of these innovative ingredients into cosmetic products [131]. These evaluations enable companies to achieve clear regulatory compliance and substantiate the stability and efficacy of their final products [56,150].

Table 6. Olive by-products' bioactive compounds and their potential applications in cosmetic formulations.

Bioactive Compounds	Properties	Cosmetic Formulations	References
Hydroxytyrosol	Whitening, lightning and depigmenting Antioxidant Anti-aging Photoprotective Anti-inflammatory Antimicrobial	Formulations for reducing skin pigmentation; Creams to enhance the epidermal barrier, inflammatory responses and for skin repair;	[95,126,127,131,134,136]
Tyrosol	Antioxidant Anti-aging Photoprotective Anti-inflammatory Surfactant	Emulsions, creams and oil-based formulations.	[126,140]
Oleic acid Vitamin E Squalene	Antioxidant Emolient Moisturizer Anti-inflammatory	Formulations with healing and lubricant activity; Formulations to prevent oxidative stress; Dermo-protective creams	[27,125,128,129,142,147]
Carotenoids	Antioxidant Photoprotective Anti-inflammatory	Formulations to prevent UVA induced dermal damage and reduce oxidative stress.	[145]

Phenolic compounds derived from coffee and olive by-products, such as hydroxytyrosol and chlorogenic acids, are prone to degradation under unfavorable conditions such as elevated temperatures, exposure to light, and alkaline pH levels. Such degradation can impair their antioxidant and antimicrobial properties. To mitigate this issue, cosmetic formulations commonly utilize stabilization strategies, including pH adjustment, incorporation of antioxidants or chelating agents, and encapsulation techniques such as liposomes, nanoemulsions, and biopolymeric carriers, which have demonstrated significant improvements in the shelf-life and bioactivity of these compounds [151,152].

Coffee and olive by-products possess valuable properties individually and can be utilized concurrently to achieve a synergistic effect. A recent study investigated the development of a hair care formulation employing lipid nanocarriers loaded with enriched coffee silverskin extract and olive pomace. The findings indicated that the synergistic combination of these two by-products elicited multiple beneficial effects, including the stimulation of hair growth, anti-dandruff activity, and improved nourishment of the hair [153].

7. Conclusions and Future Perspectives

Throughout this work, it became clear that the cosmetic industry can significantly benefit from valuing natural ingredients, especially those derived from agro-industrial by-products like coffee and olive residues. What is often seen as waste contains a wide range of compounds with real potential for skincare applications. Several studies confirm the role of by-product extracts as active ingredients since they possess antioxidant, antimicrobial, moisturizing, antiaging, and UV-protective properties, demonstrating their effectiveness both in vitro and in vivo by improving skin hydration, elasticity, barrier function, and appearance. Using these by-products in cosmetic products not only introduces new active ingredients but also encourages more environmentally friendly production models, reducing environmental impact and addressing the demand for products that are effective, safe, and eco-friendly. This approach aligns with the principles of a circular economy and sustainability. Consumers increasingly care about what they apply to their skin, and ingredients from olive and coffee by-products are becoming a key point of differentiation. The extraction methods for these by-products are becoming more efficient and environmentally friendly, however the consistency of olive and coffee by-products as ingredients remains a challenge. Despite that, some industries have already commercialized some products using hydroxytyrol and caffeine as ingredients, with promising results and consumers approval.

Future research should prioritize optimizing extraction methodologies, establishing standardized extracts with uniform bioactive profiles, and increasing the number of clinical and dermatological investigations. Standardizing bioactive content is crucial to maintain consistent quality across batches, considering the seasonal and geographical variability of raw materials. Further research on formulation compatibility is also imperative to evaluate stability and interactions with other cosmetic constituents. Moreover, addressing industrial scalability and the cost-efficiency of eco-friendly extraction methods such as UAE, MAE, DES, and EAE remains a significant challenge requiring further exploration. Quantitative data demonstrating carbon savings and estimates of waste reuse within the cosmetic industry would further substantiate sustainability claims.

Author Contributions: Conceptualization, A.M.F. and H.A.; methodology, R.C.A.; software, A.M.F.; validation, B.B., M.B.P.P.O.; formal analysis, R.C.A.; investigation, A.M.F.; resources, B.B.; data curation, H.A.; writing—original draft preparation, A.M.F.; writing—review and editing, H.A., R.C.A., M.B.P.P.O.; visualization, A.C.; supervision, H.A., R.C.A., M.B.P.P.O.; project administration, H.A., R.C.A., M.B.P.P.O. All authors have read and agreed to the published version of the manuscript.

Funding: This work was supported by PT national funds (FCT/MECI, Fundação para a Ciência e Tecnologia/Ministério da Educação, Ciência e Inovação) through the project UID/50006—Laboratório Associado para a Química Verde—Tecnologias e Processos Limpos.

Institutional Review Board Statement: Not applicable.

Informed Consent Statement: Not applicable.

Data Availability Statement: All articles and data referenced in this review are publicly available online.

Conflicts of Interest: Author Ana Matilde Ferreira, Bernardo Bastos, Ana Casas, and Hugo Almeida were employed by the company Mesosystem Investigação & Investimentos. The authors declare that the research was conducted in the absence of any commercial or financial relationships that could be construed as a potential conflict of interest. The remaining authors declare that the research was conducted in the absence of any commercial or financial relationships that could be construed as a potential conflict of interest.

Abbreviations

The following abbreviations are used in this manuscript:

C. arabica	*Coffea arabica*
C. robusta	*Coffea canephora*
CH	Coffee Husk
CP	Coffee Pulp
CS	Coffee Silver skin
CPSR	Cosmetic Product Safety Report
DES	Deep Eutectic Solvents
IGF-1	Insulin-like growth factor 1

EAE	Enzyme-Assisted Extraction
EC	European Commission
EVOO	Extra Virgin Olive Oil
GMP	Good Manufacturing Practices
MAE	Microwave-Assisted Extraction
MF	Microfiltration
NF	Nanofiltration
OMWW	Olive Mill Wastewater
O/W	Oil-in-Water
RO	Reverse Osmosis
SCGs	Spent Coffee Grounds
SFE	Supercritical Fluid Extraction
UAE	Ultrasound-Assisted Extraction
UF	Ultrafiltration
W/O	Water-in-Oil

References

1. Baki, G.; Alexander, K.S. *Introduction to Cosmetic Formulation and Technology*, 2nd ed.; John Wiley & Sons: Hoboken, NJ, USA, 2023.
2. Mondello, A.; Salomone, R.; Mondello, G. Exploring circular economy in the cosmetic industry: Insights from a literature review. *Environ. Impact Assess. Rev.* **2024**, *105*, 107443. [CrossRef]
3. Ferreira, S.M.; Falé, Z.; Santos, L. Sustainability in Skin Care: Incorporation of Avocado Peel Extracts in Topical Formulations. *Molecules* **2022**, *27*, 1782. [CrossRef]
4. Nhani, G.B.B.; Di Filippo, L.D.; de Paula, G.A.; Mantovanelli, V.R.; da Fonseca, P.P.; Tashiro, F.M.; Monteiro, D.C.; Fonseca-Santos, B.; Duarte, J.L.; Chorilli, M. High-Tech Sustainable Beauty: Exploring Nanotechnology for the Development of Cosmetics Using Plant and Animal By-Products. *Cosmetics* **2024**, *11*, 112. [CrossRef]
5. Fortunati, S.; Martiniello, L.; Morea, D. The strategic role of the corporate social responsibility and circular economy in the cosmetic industry. *Sustainability* **2020**, *12*, 5120. [CrossRef]
6. Martins, A.M.; Marto, J.M. A sustainable life cycle for cosmetics: From design and development to post-use phase. *Sustain. Chem. Pharm.* **2023**, *35*, 101178. [CrossRef]
7. Khajuria, A.; Atienza, V.A.; Chavanich, S.; Henning, W.; Islam, I.; Kral, U.; Liu, M.; Liu, X.; Murthy, I.K.; Oyedotun, T.D.T.; et al. Accelerating circular economy solutions to achieve the 2030 agenda for sustainable development goals. *Circ. Econ.* **2022**, *1*, 100001. [CrossRef]
8. Circular Economy Action Plan. Available online: https://environment.ec.europa.eu/strategy/circular-economy-action-plan_en (accessed on 28 April 2025).
9. Khajuria, A. Integrated Approach between DPSIR, Planetary Boundaries and Sustainable Development Goals towards 3Rs and Resource Efficiency. *World Environ.* **2020**, *10*, 52–56. [CrossRef]
10. Kirchherr, J.; Reike, D.; Hekkert, M. Conceptualizing the circular economy: An analysis of 114 definitions. *Resour. Conserv. Recycl.* **2017**, *127*, 221–232. [CrossRef]
11. Circular Economy: Definition, Importance and Benefits. Available online: https://www.europarl.europa.eu/topics/en/article/20151201STO05603/circular-economy-definition-importance-and-benefits#:~:text=What%20is%20the%20circular%20economy,products%20as%20long%20as%20possible (accessed on 8 February 2025).
12. Tlais, A.Z.A.; Fiorino, G.M.; Polo, A.; Filannino, P.; Di Cagno, R. High-value compounds in fruit, vegetable and cereal byproducts: An overview of potential sustainable reuse and exploitation. *Molecules* **2020**, *25*, 2987. [CrossRef]
13. Faria-Silva, C.; Ascenso, A.; Costa, A.M.; Marto, J.; Carvalheiro, M.; Ribeiro, H.M.; Simões, S. Feeding the skin: A new trend in food and cosmetics convergence. *Trends Food Sci. Technol.* **2020**, *95*, 21–32. [CrossRef]
14. Hoang, H.T.; Moon, J.Y.; Lee, Y.C. Natural antioxidants from plant extracts in skincare cosmetics: Recent applications, challenges and perspectives. *Cosmetics* **2021**, *8*, 108. [CrossRef]
15. Scott, D.A. A review of ancient Egyptian pigments and cosmetics. *Stud. Conserv.* **2016**, *61*, 185–202. [CrossRef]
16. González-Minero, F.J.; Bravo-Díaz, L. The use of plants in skin-care products, cosmetics and fragrances: Past and present. *Cosmetics* **2019**, *5*, 50. [CrossRef]
17. McMullen, R.L.; Dell'Acqua, G. History of Natural Ingredients in Cosmetics. *Cosmetics* **2023**, *10*, 71. [CrossRef]
18. Mansoor, K.; Aburjai, T.; Al-Mamoori, F.; Schmidt, M. Plants with cosmetic uses. *Phytother. Res.* **2023**, *37*, 5755–5768. [CrossRef]
19. Menaa, F. Skin Anti-Aging Benefits of Phytotherapeutics-based Emulsions. *Pharm. Anal. Acta* **2014**, *5*, e168. [CrossRef]

20. Jadoon, S.; Karim, S.; Asad, M.H.H.; Bin Akram, M.R.; Kalsoom Khan, A.; Malik, A.; Chen, C.; Murtaza, G. Anti-aging potential of phytoextract loaded-pharmaceutical creams for human skin cell longetivity. *Oxid. Med. Cell. Longev.* **2015**, *2015*, 709628. [CrossRef] [PubMed]
21. Michalak, M. Plant Extracts as Skin Care and Therapeutic Agents. *Int. J. Mol. Sci.* **2023**, *24*, 15444. [CrossRef] [PubMed]
22. Ugolini, T.; Cecchi, L.; Sani, G.; Digiglio, I.; Adinolfi, B.; Ciaccheri, L.; Zanoni, B.; Melani, F.; Mulinacci, N. Seasonal and Cultivar-Dependent Phenolic Dynamics in Tuscan Olive Leaves: A Two-Year Study by HPLC-DAD-MS for Food By-Product Valorization. *Separations* **2025**, *12*, 192. [CrossRef]
23. Dias, E.A.; Miho, H.; Escobar, L.; Moral, J.; Diez, C.M.; Capote, P.F. Influence of genetic and interannual factors on bioactive compounds of olive pomace determined through a germplasm survey. *Food Chem.* **2022**, *378*, 1322107. [CrossRef]
24. Yang, Y.; Wu, Y.; Xiang, L.; Picardo, M. Deciphering the role of skin aging in pigmentary disorders. *Free Radic. Biol. Med.* **2025**, *227*, 638–655. [CrossRef]
25. Saewan, N.; Jimtaisong, A.; Panyachariwat, N.; Chaiwut, P. In Vitro and In Vivo Anti-Aging Effect of Coffee Berry Nanoliposomes. *Molecules* **2023**, *28*, 6830. [CrossRef]
26. Rodrigues, R.; Oliveira, M.B.P.P.; Alves, R.C. Chlorogenic Acids and Caffeine from Coffee By-Products: A Review on Skincare Applications. *Cosmetics* **2023**, *10*, 12. [CrossRef]
27. Dauber, C.; Parente, E.; Zucca, M.P.; Gámbaro, A.; Vieitez, I. Olea europea and By-Products: Extraction Methods and Cosmetic Applications. *Cosmetics* **2023**, *10*, 112. [CrossRef]
28. Hasballah, K.; Lestari, W.; Listiawan, M.Y.; Sofia, S. Coffee by-products as the source of antioxidants: A systematic review. *F1000Research* **2022**, *11*, 220. [CrossRef]
29. Ribeiro, R.C.; Mota, M.F.S.; Silva, R.M.V.; Silva, D.C.; Novaes, F.J.M.; da Veiga, V.F.; Bizzo, H.R.; Teixeira, R.S.S.; Rezende, C.M. Coffee Oil Extraction Methods: A Review. *Foods* **2024**, *13*, 2601. [CrossRef]
30. Lee, Y.G.; Cho, E.J.; Maskey, S.; Nguyen, D.T.; Bae, H.J. Value-Added Products from Coffee Waste: A Review. *Molecules* **2023**, *28*, 3562. [CrossRef]
31. Oussou, K.F.; Guclu, G.G.; Kelebek, H.; Selli, S. Valorization of cocoa, tea and coffee processing by-products-wastes. In *Advances in Food and Nutrition Research*; Çapanoğlu Güven, E., Navarro-Hortal, M.D., Forbes-Hernández, T.Y., Battino, M., Eds.; Academic Press: Cambridge, MA, USA, 2023; Volume 107, pp. 91–130.
32. De Bruyn, F.; Zhang, S.J.; Pothakos, V.; Torres, J.; Lambot, C.; Moroni, A.V.; Callanan, M.; Sybesma, W.; Weckx, S.; De Vuyst, L. Exploring the impacts of postharvest processing on the microbiota and metabolite profiles during green coffee bean production. *Appl. Environ. Microbiol.* **2017**, *83*, e02398-16. [CrossRef]
33. Figueroa Campos, G.A.; Sagu, S.T.; Celis, P.S.; Rawel, H.M. Comparison of batch and continuous wet-processing of coffee: Changes in the main compounds in beans, by-products and wastewater. *Foods* **2020**, *9*, 1135. [CrossRef]
34. Oliveira, G.; Passos, C.P.; Ferreira, P.; Coimbra, M.A.; Gonçalves, I. Coffee by-products and their suitability for developing active food packaging materials. *Foods* **2021**, *10*, 683. [CrossRef]
35. Galali, Y.; Omar, Z.A.; Sajadi, S.M. Biologically active components in by-products of food processing. *Food Sci. Nutr.* **2020**, *8*, 3004–3022. [CrossRef] [PubMed]
36. de Melo Pereira, G.V.; de Carvalho Neto, D.P.; Magalhães Júnior, A.I.; do Prado, F.G. Chemical composition and health properties of coffee and coffee by-products. In *Advances in Food and Nutrition Research*; Toldrá, F., Ed.; Academic Press: Cambridge, MA, USA, 2020; Volume 91, pp. 65–96.
37. Machado, M.; Espírito Santo, L.; Machado, S.; Lobo, J.C.; Costa, A.S.G.; Oliveira, M.B.P.P.; Ferreira, H.; Alves, R.C. Bioactive Potential and Chemical Composition of Coffee By-Products: From Pulp to Silverskin. *Foods* **2023**, *12*, 2354. [CrossRef]
38. Martínez-Inda, B.; Jiménez-Moreno, N.; Esparza, I.; Ancín-Azpilicueta, C. Coffee and Cocoa By-Products as Valuable Sources of Bioactive Compounds: The Influence of Ethanol on Extraction. *Antioxidants* **2025**, *14*, 42. [CrossRef] [PubMed]
39. Bessada, S.M.F.; Alves, R.C.; Oliveira, M.B.P.P. Coffee silverskin: A review on potential cosmetic applications. *Cosmetics* **2018**, *5*, 5. [CrossRef]
40. Zengin, G.; Sinan, K.I.; Mahomoodally, M.F.; Angeloni, S.; Mustafa, A.M.; Vittori, S.; Maggi, F.; Caprioli, G. Chemical composition, antioxidant and enzyme inhibitory properties of different extracts obtained from spent coffee ground and coffee silverskin. *Foods* **2020**, *9*, 713. [CrossRef] [PubMed]
41. Janissen, B.; Huynh, T. Chemical composition and value-adding applications of coffee industry by-products: A review. *Resour. Conserv. Recycl.* **2018**, *128*, 110–117. [CrossRef]
42. Costa, A.S.G.; Peixoto, J.A.B.; Machado, S.; Espírito Santo, L.; Soares, T.F.; Andrade, N.; Azevedo, R.; Almeida, A.; Costa, H.S.; Oliveira, M.B.P.P.; et al. Coffee Pulp from Azores: A Novel Phytochemical-Rich Food with Potential Anti-Diabetic Properties. *Foods* **2025**, *14*, 306. [CrossRef]

43. Maiyah, N.; Kerdpiboon, S.; Supapvanich, S.; Kerr, W.L.; Sriprom, P.; Chotigavin, N.; Klaypradit, W.; Puttongsiri, T. Recovering bioactive compounds and antioxidant capacity of medium roasted spent coffee grounds through varied hydrothermal brewing cycles. *J. Agric. Food Res.* **2025**, *20*, 101789. [CrossRef]
44. Lu, L.; Tibpromma, S.; Karunarathna, S.C.; Jayawardena, R.S.; Lumyong, S.; Xu, J.; Hyde, K.D. Comprehensive Review of Fungi on Coffee. *Pathogens* **2022**, *11*, 411. [CrossRef]
45. Hermansen, C.; Chong, Q.K.; Ho, S.; Natali, F.; Weingarten, M.; Peterson, E.C. Microbiome Evolution of Brewer's Spent Grain and Spent Coffee Ground Solid Sidestreams Under Industrial Storage Conditions. *Appl. Sci.* **2024**, *14*, 9759. [CrossRef]
46. Nolasco, A.; Squillante, J.; Esposito, F.; Velotto, S.; Romano, R.; Aponte, M.; Giarra, A.; Toscanesi, M.; Montella, E.; Cirillo, T. Coffee Silverskin: Chemical and Biological Risk Assessment and Health Profile for Its Potential Use in Functional Foods. *Foods* **2022**, *11*, 2834. [CrossRef]
47. Bessaire, T.; Perrin, I.; Tarres, A.; Bebius, A.; Reding, F.; Theurillat, V. Mycotoxins in green coffee: Occurrence and risk assessment. *Food Control* **2019**, *96*, 59–67. [CrossRef]
48. López-Rodríguez, C.; Verheecke-Vaessen, C.; Strub, C.; Fontana, A.; Schorr-Galindo, S.; Medina, A. Reduction in Ochratoxin A Occurrence in Coffee: From Good Practices to Biocontrol Agents. *J. Fungi* **2024**, *10*, 590. [CrossRef]
49. Felten Bondam, A.; da Silveira, D.D.; Pozzada dos Santos, J.; Hoffmann, J.F. Phenolic compounds from coffee by-products: Extraction and application in the food and pharmaceutical industries. *Trends Food Sci. Technol.* **2022**, *123*, 172–186. [CrossRef]
50. Pyrzynska, K. Useful Extracts from Coffee By-Products: A Brief Review. *Separations* **2024**, *11*, 334. [CrossRef]
51. Mwaurah, P.W.; Kumar, S.; Kumar, N.; Attkan, A.K.; Panghal, A.; Singh, V.K.; Garg, M.K. Novel oil extraction technologies: Process conditions, quality parameters, and optimization. *Compr. Rev. Food Sci. Food Saf.* **2020**, *19*, 3–20. [CrossRef]
52. Kumar, K.; Srivastav, S.; Sharanagat, V.S. Ultrasound assisted extraction (UAE) of bioactive compounds from fruit and vegetable processing by-products: A review. *Ultrason. Sonochem.* **2020**, *70*, 105325. [CrossRef]
53. Ramón-Gonçalves, M.; Gómez-Mejía, E.; Rosales-Conrado, N.; León-González, M.E.; Madrid, Y. Extraction, identification and quantification of polyphenols from spent coffee grounds by chromatographic methods and chemometric analyses. *Waste Manag.* **2019**, *96*, 15–24. [CrossRef]
54. Pham, Q.D.; Topgaard, D.; Sparr, E. Tracking solvents in the skin through atomically resolved measurements of molecular mobility in intact stratum corneum. *Proc. Natl. Acad. Sci. USA* **2017**, *114*, E112–E121. [CrossRef]
55. *ISO 16128-2:2017*; Cosmetics—Guidelines on Technical Definitions and Criteria for Natural and Organic Cosmetic Ingredients Part 2: Criteria for Ingredients and Products. ISO Copyright Office: Geneva, Switzerland, 2017.
56. Regulation (EC) No 1223/2009 of the European Parliament and of the Council of 30 November 2009 on Cosmetic Products. Available online: https://eur-lex.europa.eu/legal-content/EN/TXT/?uri=CELEX:32009R1223 (accessed on 19 June 2025).
57. Ramadan, M.; Fadel, A.; Bahi, A.; Alajlani, S.; Al Neyadi, M.S.S.; Alshamsi, E.S.H.; Naser, S.; Shahbaz, H.M.; Samuel, R.; Ranneh, Y. Optimizing Chlorogenic Acid Extraction From Spent Coffee Grounds: A Comparative Review of Conventional and Non-Conventional Techniques. *Food Sci. Nutr.* **2025**, *13*, e70315. [CrossRef]
58. Bouhzam, I.; Cantero, R.; Balcells, M.; Margallo, M.; Aldaco, R.; Bala, A.; Fullana-i-Palmer, P.; Puig, R. Environmental and Yield Comparison of Quick Extraction Methods for Caffeine and Chlorogenic Acid from Spent Coffee Grounds. *Foods* **2023**, *12*, 779. [CrossRef]
59. Brzezińska, R.; Wirkowska-Wojdyła, M.; Piasecka, I.; Górska, A. Application of Response Surface Methodology to Optimize the Extraction Process of Bioactive Compounds Obtained from Coffee Silverskin. *Appl. Sci.* **2023**, *13*, 5388. [CrossRef]
60. Savić, I.M.; Savić, G.I.M.; Gajić, D.G. Optimization of the Microwave-Assisted Extraction of Caffeine from Roasted Coffee Beans. *Foods* **2024**, *13*, 2333. [CrossRef] [PubMed]
61. Valenzuela, L.S.; Gómez, A.B.; Rubio, S. Supramolecular solvent extraction of bioactives from coffee cherry pulp. *J. Food Eng.* **2020**, *278*, 109933. [CrossRef]
62. Kanlayavattanakul, M.; Lourith, N.; Chaikul, P. Valorization of spent coffee grounds as the specialty material for dullness and aging of skin treatments. *Chem. Biol. Technol. Agric.* **2021**, *8*, 55. [CrossRef]
63. Miladi, M.; Martins, A.A.; Mata, T.M.; Vegara, M.; Pérez, I.M.; Remmani, R.; Ruiz-Canales, A.; Núñez-Gómez, D. Optimization of Ultrasound-Assisted Extraction of Spent Coffee Grounds Oil Using Response Surface Methodology. *Processes* **2021**, *9*, 2085. [CrossRef]
64. dos Santos, É.M.; de Macedo, L.M.; Tundisi, L.L.; Ataide, J.A.; Camargo, G.A.; Alves, R.C.; Oliveira, M.B.P.P.; Mazzola, G. Coffee by-products in topical formulations: A review. *Trends Food Sci. Technol.* **2021**, *111*, 280–291. [CrossRef]
65. Dias, R.; Oliveira, H.; Fernandes, I.; Simal-Gandara, J.; Perez-Gregorio, R. Recent advances in extracting phenolic compounds from food and their use in disease prevention and as cosmetics. *Crit. Rev. Food Sci. Nutr.* **2020**, *61*, 1130–1151. [CrossRef]
66. Cho, Y.-H.; Bahuguna, A.; Kim, H.-H. Potential effect of compounds isolated from Coffea arabica against UV-B induced skin damage by protecting fibroblast cells. *J. Photochem. Photobiol. B* **2017**, *174*, 323–332. [CrossRef] [PubMed]

67. Costa, E.F.; Magalhães, W.V.; Di Stasi, L.C. Recent Advances in Herbal-Derived Products with Skin Anti-Aging Properties and Cosmetic Applications. *Molecules* **2022**, *27*, 7518. [CrossRef] [PubMed]
68. Hahn, H.J.; Kim, K.B.; Bae, S.; Choi, B.G.; An, S.; Ahn, K.J.; Kim, S.Y. Pretreatment of ferulic acid protects human dermal fibroblasts against ultraviolet a irradiation. *Ann. Dermatol.* **2016**, *28*, 740–748. [CrossRef]
69. Stevanato, R.; Bertelle, M.; Fabris, S. Photoprotective characteristics of natural antioxidant polyphenols. *Regul. Toxicol. Pharmacol.* **2014**, *69*, 71–77. [CrossRef]
70. Ng, S.Y.; Eh Suk, V.R.; Gew, L.T. Plant polyphenols as green sunscreen ingredients: A systematic review. *J. Cosmet. Dermatol.* **2022**, *21*, 5409–5444. [CrossRef] [PubMed]
71. Toschi, T.G.; Cardenia, V.; Bonaga, G.; Mandrioli, M.; Rodriguez, E.M.T. Coffee silverskin: Characterization, possible uses, and safety aspects. *J. Agric. Food Chem.* **2014**, *62*, 10836–10844. [CrossRef]
72. Szendzielorz, E.; Spiewak, R. Caffeine as an Active Molecule in Cosmetic Products for Hair Loss: Its Mechanisms of Action in the Context of Hair Physiology and Pathology. *Molecules* **2025**, *30*, 167. [CrossRef] [PubMed]
73. Fischer, T.W.; Herczeg, L.E.; Funk, W.; Zillikens, D.; Bíró, T.; Paus, R. Differential effects of caffeine on hair shaft elongation, matrix and outer root sheath keratinocyte proliferation, and transforming growth factor-β2/insulin-like growth factor-1-mediated regulation of the hair cycle in male and female human hair follicles in vitro. *Br. J. Dermatol.* **2014**, *171*, 1031–1043. [CrossRef] [PubMed]
74. Rodrigues, F.; Alves, A.C.; Nunes, C.; Sarmento, B.; Amaral, M.; Reis, S.; Oliveira, M.B.P.P. Permeation of topically applied caffeine from a food by—Product in cosmetic formulations: Is nanoscale in vitro approach an option? *Int. J. Pharm.* **2016**, *513*, 496–503. [CrossRef] [PubMed]
75. Cherian, P.A.; Bergfeld, W.F.; Belsito, D.V.; Hill, R.A.; Klaassen, C.D.; Liebler, D.C. Safety Assessment of Methylxanthines as Used in Cosmetics. *Int. J. Toxicol.* **2024**, *43*, 42–77. [CrossRef]
76. Donejko, M.; Przylipiak, A.; Rysiak, E.; Głuszuk, K.; Surażyński, A. Influence of caffeine and hyaluronic acid on collagen biosynthesis in human skin fibroblasts. *Drug Des. Dev. Ther.* **2014**, *8*, 1923–1928. [CrossRef]
77. Rivelli, D.P.; Filho, C.A.H.; Almeida, R.L.; Ropke, C.D.; Sawada, T.C.H.; Barros, S.B.M. Chlorogenic acid UVA-UVB photostability. *Photochem. Photobiol.* **2010**, *86*, 1005–1007. [CrossRef]
78. Grigolon, G.; Nowak, K.; Poigny, S.; Hubert, J.; Kotland, A.; Waldschütz, L.; Wandrey, F. From Coffee Waste to Active Ingredient for Cosmetic Applications. *Int. J. Mol. Sci.* **2023**, *24*, 8516. [CrossRef]
79. Rodrigues, F.; Matias, R.; Ferreira, M.; Amaral, M.H.; Oliveira, M.B.P.P. In vitro and in vivo comparative study of cosmetic ingredients Coffee silverskin and hyaluronic acid. *Exp. Dermatol.* **2016**, *25*, 572–574. [CrossRef] [PubMed]
80. Rodrigues, F.; Sarmento, B.; Amaral, M.H.; Oliveira, M.B. Cytotoxicity evaluation of a new ingredient with cosmetic proposals incorporated into a hand cream. *Toxicol. Lett.* **2015**, *238*, S354. [CrossRef]
81. Al-Yousef, H.M.; Amina, M. Essential oil of Coffee arabica L. husks: A brilliant source of antimicrobial and antioxidant agents. *Biomed. Res.* **2018**, *29*, 174–180. [CrossRef]
82. Andrade, C.; Perestrelo, R.; Câmara, J.S. Bioactive Compounds and Antioxidant Activity from Spent Coffee Grounds as a Powerful Approach for Its Valorization. *Molecules* **2022**, *27*, 7504. [CrossRef]
83. Silva, V.; Igrejas, G.; Falco, V.; Santos, T.P.; Torres, C.; Oliveira, A.M.P.; Pereira, J.E.; Amaral, J.S.; Poeta, P. Chemical composition, antioxidant and antimicrobial activity of phenolic compounds extracted from wine industry by-products. *Food Control* **2018**, *92*, 516–522. [CrossRef]
84. VITCHY Laboratoires. Available online: https://www.vichy.pt (accessed on 26 August 2025).
85. Erskine, E.; Gültekin, S.B.; Vahapoglu, B.; Capanoglu, E. Coffee Phenolics and Their Interaction with Other Food Phenolics: Antagonistic and Synergistic Effects. *ACS Omega* **2022**, *7*, 1595–1601. [CrossRef] [PubMed]
86. Xu, M.; Yang, L.; Zhu, Y.; Liao, M.; Chu, L.; Li, X.; Lin, L.; Zheng, G. Collaborative effects of chlorogenic acid and caffeine on lipid metabolism via the AMPKα-LXRα/SREBP-1c pathway in high-fat diet-induced obese mice. *Food Funct.* **2019**, *10*, 7489–7497. [CrossRef]
87. Nemzer, B.; Edwards, J.; Kalita, D. Matrix-Specific Effects on Caffeine and Chlorogenic Acid Complexation in a Novel Extract of Whole Coffea arabica Coffee Cherry by NMR Spectroscopy. *Molecules* **2022**, *27*, 7803. [CrossRef]
88. Valvez, S.; Maceiras, A.; Santos, P.; Reis, P.N.B. Olive stones as filler for polymer-based composites: A review. *Materials* **2021**, *14*, 845. [CrossRef]
89. Romani, A.; Ieri, F.; Urciuoli, S.; Noce, A.; Marrone, G.; Nediani, C.; Bernini, R. Health effects of phenolic compounds found in extra-virgin olive oil, by-products, and leaf of *Olea europaea* L. *Nutrients* **2019**, *11*, 1776. [CrossRef]
90. Rodrigues, F.; Pimentel, F.B.; Oliveira, M.B.P.P. Olive by-products: Challenge application in cosmetic industry. *Ind. Crops Prod.* **2015**, *70*, 116–124. [CrossRef]

91. Rodrigues, R.; Alves, R.C.; Oliveira, M.B.P.P. Exploring Olive Pomace for Skincare Applications: A Review. *Cosmetics* **2023**, *10*, 35. [CrossRef]
92. Rodrigues, R.; Lobo, J.C.; Ferreira, D.M.; Senderowicz, E.; Nunes, M.A.; Amaral, M.H.; Alves, R.C.; Oliveira, M.B.P.P. Chemical and Rheological Characterization of a Facial Mask Containing an Olive Pomace Fraction. *Cosmetics* **2023**, *10*, 64. [CrossRef]
93. Otero, P.; Oliveira, P.G.; Carpena, M.; Martinez, M.B. Applications of by-products from the olive oil processing: Revalorization strategies based on target molecules and green extraction technologies. *Trends Food Sci. Technol.* **2021**, *116*, 1084–1104. [CrossRef]
94. Domingues, E.; Fernandes, E.; Gomes, J.; Castro-Silva, S.; Martins, R.C. Olive oil extraction industry wastewater treatment by coagulation and Fenton's process. *J. Water Process Eng.* **2021**, *39*, 101818. [CrossRef]
95. Madureira, J.M.A.; Margaça, F.; Santos-Buelga, C. Applications of bioactive compounds extracted from olive industry wastes: A review. *Compr. Rev. Food Sci. Food Saf.* **2021**, *21*, 453–476. [CrossRef]
96. Cuffaro, D.; Bertolini, A.; Bertini, S.; Ricci, C.; Cascone, M.G.; Danti, S.; Saba, A.; Macchia, M.; Digiacomo, M. Olive Mill Wastewater as Source of Polyphenols with Nutraceutical Properties. *Nutrients* **2023**, *15*, 3746. [CrossRef]
97. Sosa, A.; Armienta, M.A.; Aguayo, A.; Cruz, O. Evaluation of the influence of main groundwater ions on arsenic removal by limestones through column experiments. *Sci. Total Environ.* **2020**, *727*, 138459. [CrossRef]
98. Madureira, J.; Dias, M.I.; Pinela, J.; Calhelha, R.C.; Barros, L.; Santos-Buelga, C.; Margaça, F.M.A.; Ferreira, I.C.F.R.; Cabo Verde, S. The use of gamma radiation for extractability improvement of bioactive compounds in olive oil wastes. *Sci. Total Environ.* **2020**, *727*, 138706. [CrossRef]
99. Abbattista, R.; Ventura, G.; Calvano, C.D.; Cataldi, T.R.I.; Losito, I. Bioactive compounds in waste by-products from olive oil production: Applications and structural characterization by mass spectrometry techniques. *Foods* **2021**, *10*, 1236. [CrossRef]
100. Mir-Cerdà, A.; Ramos-Calero, V.; Beltrán, J.L.; Cortina, J.L.; Saurina, J.; Sentellas, S.; Granados, M. Towards a sustainable recovery of polyphenols from agrifood waste: Performance of polymeric sorbents with natural deep eutectic solvent extracts. *LWT* **2024**, *207*, 116632. [CrossRef]
101. Ben Saad, A.; Tiss, M.; Keskes, H.; Chaari, A.; Sakavitsi, M.E.; Hamden, K.; Halabalaki, M.; Allouche, N. Antihyperlipidemic, Antihyperglycemic, and Liver Function Protection of Olea europaea var. Meski Stone and Seed Extracts: LC-ESI-HRMS-Based Composition Analysis. *J. Diabetes Res.* **2021**, *2021*, 6659415. [CrossRef]
102. Clodoveo, M.L.; Crupi, P.; Annunziato, A.; Corbo, F. Innovative Extraction Technologies for Development of Functional Ingredients Based on Polyphenols from Olive Leaves. *Foods* **2022**, *11*, 103. [CrossRef] [PubMed]
103. Papageorgiou, C.S.; Lyri, P.; Xintaropoulou, I.; Diamantopoulos, I.; Zagklis, D.P.; Paraskeva, C.A. High-Yield Production of a Rich-in-Hydroxytyrosol Extract from Olive (*Olea europaea*) Leaves. *Antioxidants* **2022**, *11*, 1042. [CrossRef]
104. Markhali, F.S.; Teixeira, J.A.; Rocha, C.M.R. Olive tree leaves—A source of valuable active compounds. *Processes* **2020**, *8*, 1177. [CrossRef]
105. Panzella, L.; Moccia, F.; Nasti, R.; Marzorati, S.; Verotta, L.; Napolitano, A. Bioactive Phenolic Compounds from Agri-Food Wastes: An Update on Green and Sustainable Extraction Methodologies. *Front. Nutr.* **2020**, *7*, 60. [CrossRef]
106. Madureira, J.; Albuquerque, B.; Dias, M.I.; Pinela, J.; Calhelha, R.C.; Santos-Buelga, C.; Margaça, F.M.A.; Ferreira, I.C.F.R.; Cabo Verde, S.; Barros, L. Ultrasound-assisted extraction of hydroxytyrosol and tyrosol from olive pomace treated by gamma radiation: Process optimization and bioactivity assessment. *Food Funct.* **2023**, *14*, 3038–3050. [CrossRef]
107. Gómez, C.I.; Del Mar, C.M.; Romero, I.; Castro, E. Optimization of Microwave-Assisted Water Extraction to Obtain High Value-Added Compounds from Exhausted Olive Pomace in a Biorefinery Context. *Foods* **2022**, *11*, 2002. [CrossRef]
108. Gómez, C.I.; Del Mar, C.M.; Carvalheiro, F.; Duarte, L.C.; Roseiro, L.B.; Romero, I.; Castro, E. Recovery of Bioactive Compounds from Industrial Exhausted Olive Pomace through Ultrasound-Assisted Extraction. *Biology* **2021**, *10*, 514. [CrossRef] [PubMed]
109. Expósito, A.X.; Munguía, U.Á.; Duque, S.C.; Borrás, L.I.; Quirante, P.R.; Lozano, S.J. Optimized Ultrasound-Assisted Extraction for Enhanced Recovery of Valuable Phenolic Compounds from Olive By-Products. *Antioxidants* **2025**, *14*, 938. [CrossRef] [PubMed]
110. Contreras, A.L.A.; Laaroussi, H.; Ousaaid, D.; Bakour, M.; Lyoussi, B.; Ferreira, S.P. Sustainable Valorization of Olive Oil By-Products: Green Extraction of Phytochemicals, Encapsulation Strategies, and Food Applications. *J. Food Sci.* **2025**, *90*, e70412. [CrossRef]
111. de Almeida Pontes, P.V.; Ayumi Shiwaku, I.; Maximo, G.J.; Caldas Batista, E.A. Choline chloride-based deep eutectic solvents as potential solvent for extraction of phenolic compounds from olive leaves: Extraction optimization and solvent characterization. *Food Chem.* **2021**, *352*, 129346. [CrossRef]
112. Hu, M.; Han, B.; Xie, L.; Lu, B.; Bai, D.; Shi, N.; Liao, Y.; Wang, Y.; Liu, L.; Wu, S.; et al. Ultrasonic assisted natural deep eutectic solvents as a green and efficient approach for extraction of hydroxytyrosol from olive leaves. *Ind. Chem. Mater.* **2024**, *2*, 309–320. [CrossRef]
113. Oyoun, F.; Toncheva, A.; Henríquez, L.C.; Grougnet, R.; Laoutid, F.; Mignet, N.; Alhareth, K.; Corvis, Y. Deep Eutectic Solvents: An Eco-friendly Design for Drug Engineering. *ChemSusChem* **2023**, *16*, 938. [CrossRef]

114. Hikmawanti, N.P.E.; Ramadon, D.; Jantan, I.; Mun'im, A. Natural deep eutectic solvents (Nades): Phytochemical extraction performance enhancer for pharmaceutical and nutraceutical product development. *Plants* **2021**, *10*, 2091. [CrossRef] [PubMed]
115. Cabrera, L.; Xavier, L.; Zecchi, B. Extraction of phenolic compounds with antioxidant activity from olive pomace using natural deep eutectic solvents: Modelling and optimization by response surface methodology. *Discov. Food* **2024**, *4*, 29. [CrossRef]
116. Łubek-Nguyen, A.; Ziemichód, W.; Olech, M. Application of Enzyme-Assisted Extraction for the Recovery of Natural Bioactive Compounds for Nutraceutical and Pharmaceutical Applications. *Appl. Sci.* **2022**, *12*, 3232. [CrossRef]
117. Martins, V.F.R.; Ribeiro, T.B.; Lopes, A.I.; Pintado, M.E.; Morais, R.M.S.C.M.; Morais, A.M.M.B. Comparison among Different Green Extraction Methods of Polyphenolic Compounds from Exhausted Olive Oil Pomace and the Bioactivity of the Extracts. *Molecules* **2024**, *29*, 1935. [CrossRef]
118. Dobrinčić, A.; Repajić, M.; Garofulić, I.E.; Tuden, L.; Dragović, U.V.; Levaj, B. Comparison of different extraction methods for the recovery of olive leaves polyphenols. *Processes* **2020**, *8*, 1008. [CrossRef]
119. Ivanović, M.; Razboršek, M.I.; Kolar, M. Innovative extraction techniques using deep eutectic solvents and analytical methods for the isolation and characterization of natural bioactive compounds from plant material. *Plants* **2020**, *9*, 1428. [CrossRef]
120. Lafka, T.I.; Lazou, A.E.; Sinanoglou, V.J.; Lazos, E.S. Phenolic extracts from wild olive leaves and their potential as edible oils antioxidants. *Foods* **2013**, *2*, 18–31. [CrossRef]
121. Sahin, S.; Samli, R.; Birteks, Z.; Tan, A.S.; Barba, F.J.; Chemat, F.; Cravotto, G.; Lorenzo, J.M. Solvent-free microwave-assisted extraction of polyphenols from olive tree leaves: Antioxidant and antimicrobial properties. *Molecules* **2017**, *22*, 1056. [CrossRef]
122. Boli, E.; Prinos, N.; Louli, V.; Pappa, G.; Stamatis, H.; Magoulas, K.; Voutsas, E. Recovery of Bioactive Extracts from Olive Leaves Using Conventional and Microwave-Assisted Extraction with Classical and Deep Eutectic Solvents. *Separations* **2022**, *9*, 255. [CrossRef]
123. Niknam, S.M.; Kashaninejad, M.; Escudero, I.; Sanz, M.T.; Beltrán, S.; Benito, J.M. Valorization of olive mill solid residue through ultrasound-assisted extraction and phenolics recovery by adsorption process. *J. Clean. Prod.* **2021**, *316*, 128340. [CrossRef]
124. Chanioti, S.; Katsouli, M.; Tzia, C. Novel processes for the extraction of phenolic compounds from olive pomace and their protection by encapsulation. *Molecules* **2021**, *26*, 1781. [CrossRef] [PubMed]
125. Difonzo, G.; Aresta, A.; Cotugno, P.; Ragni, R.; Squeo, G.; Summo, C.; Massari, F.; Pasqualone, A.; Faccia, M.; Zambonin, C.; et al. Supercritical CO2 Extraction of Phytocompounds from Olive Pomace Subjected to Different Drying Methods. *Molecules* **2021**, *26*, 598. [CrossRef]
126. Lecci, R.M.; D'antuono, I.; Cardinali, A.; Garbetta, A.; Linsalata, V.; Logrieco, A.F.; Leone, A. Antioxidant and pro-oxidant capacities as mechanisms of photoprotection of olive polyphenols on uva-damaged human keratinocytes. *Molecules* **2021**, *26*, 2153. [CrossRef]
127. Pourzand, C.; Albieri-Borges, A.; Raczek, N.N. Shedding a New Light on Skin Aging, Iron-and Redox-Homeostasis and Emerging Natural Antioxidants. *Antioxidants* **2022**, *11*, 471. [CrossRef]
128. Viola, P.; Viola, M. Virgin olive oil as a fundamental nutritional component and skin protector. *Clin. Dermatol.* **2009**, *27*, 159–165. [CrossRef]
129. Nunes, M.A.; Costa, A.S.G.; Bessada, S.; Santos, J.; Puga, H.; Alves, R.C.; Freitas, V.; Oliveira, M.B.P.P. Olive pomace as a valuable source of bioactive compounds: A study regarding its lipid- and water-soluble components. *Sci. Total Environ.* **2018**, *644*, 229–236. [CrossRef]
130. Wanitphakdeedecha, R.; Ng, J.N.C.; Junsuwan, N.; Phaitoonwattanakij, S.; Phothong, W.; Eimpunth, S.; Manuskiatti, W. Efficacy of olive leaf extract–containing cream for facial rejuvenation: A pilot study. *J. Cosmet. Dermatol.* **2020**, *19*, 1662–1666. [CrossRef]
131. Nunes, A.; Gonçalves, L.; Marto, J.; Martins, A.M.; Silva, A.N.; Pinto, P.; Martins, M.; Fraga, C.; Ribeiro, H.M. Investigations of olive oil industry by-products extracts with potential skin benefits in topical formulations. *Pharmaceutics* **2021**, *13*, 465. [CrossRef]
132. Al-Rimawi, F.; Sbeih, M.; Amayreh, M.; Rahhal, B.; Mudalal, S. Evaluation of the effectiveness of natural extract as a substituent for synthetic preservatives and antioxidants in pharmaceutical preparations. *Saudi Pharm. J.* **2024**, *32*, 102014. [CrossRef]
133. Ramírez, E.; Brenes, M.; Garcia, P. Oleuropein Hydrolysis in Natural Green Olives: Importance of the Endogenous Enzymes. *Food Chem.* **2016**, *206*, 204–209. [CrossRef] [PubMed]
134. Silva, A.F.R.; Resende, D.; Monteiro, M.; Coimbra, M.A.; Silva, A.M.S.; Cardoso, S.M. Application of hydroxytyrosol in the functional foods field: From ingredient to dietary supplements. *Antioxidants* **2020**, *9*, 1246. [CrossRef]
135. Albini, A.; Corradino, P.; Morelli, D.; Albini, F.; Noonan, D. Cosmeceutical and Dermatological Potential of Olive Mill Wastewater: A Sustainable and Eco-Friendly Source of Natural Ingredients. *Cosmetics* **2025**, *12*, 142. [CrossRef]
136. Jeon, S.; Choi, M. Anti-inflammatory and anti-aging effects of hydroxytyrosol on human dermal fibroblasts (HDFs). *Biomed. Dermatol.* **2018**, *2*, 21. [CrossRef]

137. Smeriglio, A.; Denaro, M.; Mastracci, L.; Grillo, F.; Cornara, L.; Shirooie, S.; Nabavi, S.M.; Trombetta, D. Safety and efficacy of hydroxytyrosol-based formulation on skin inflammation: In vitro evaluation on reconstructed human epidermis model. *DARU J. Pharm. Sci.* **2019**, *27*, 283–293. [CrossRef]
138. Nunes, M.A.; Palmeira, J.D.; Melo, D.; Machado, S.; Lobo, J.C.; Costa, A.S.G.; Alves, R.C.; Ferreira, H.; Oliveira, M.B.P.P. Chemical composition and antimicrobial activity of a new olive pomace functional ingredient. *Pharmaceuticals* **2021**, *14*, 913. [CrossRef]
139. Guo, L.; Gong, S.; Wang, Y.; Sun, Q.; Duo, K.; Fei, P. Antibacterial Activity of Olive Oil Polyphenol Extract Against Salmonella Typhimurium and Staphylococcus aureus: Possible Mechanisms. *Foodborne Pathog. Dis.* **2020**, *17*, 396–403. [CrossRef]
140. Bernini, R.; Carastro, I.; Santoni, F.; Clemente, M. Synthesis of lipophilic esters of tyrosol, homovanillyl alcohol and hydroxytyrosol. *Antioxidants* **2019**, *8*, 174. [CrossRef]
141. González, A.A.; Ramos, T.J.; Illescas, M.R.; Costela, R.V.J.; Ruiz, C.; Melguizo, R.L.; García-Martínez, O. The Benefits of Olive Oil for Skin Health: Study on the Effect of Hydroxytyrosol, Tyrosol, and Oleocanthal on Human Fibroblasts. *Nutrients* **2023**, *15*, 2077. [CrossRef]
142. Kováčik, A.; Kopečná, M.; Hrdinová, I.; Opálka, L.; Boncheva, B.M.; Vávrová, K. Time-Dependent Differences in the Effects of Oleic Acid and Oleyl Alcohol on the Human Skin Barrier. *Mol. Pharm.* **2023**, *20*, 6237–6245. [CrossRef]
143. Michalak, M. Plant-Derived Antioxidants: Significance in Skin Health and the Ageing Process. *Int. J. Mol. Sci.* **2022**, *23*, 585. [CrossRef] [PubMed]
144. Oliveira, A.L.S.; Gondim, S.; Gómez-García, R.; Ribeiro, T.; Pintado, M. Olive leaf phenolic extract from two Portuguese cultivars—Bioactivities for potential food and cosmetic application. *J. Environ. Chem. Eng.* **2021**, *9*, 106175. [CrossRef]
145. Aggoun, M.; Arhab, R.; Cornu, A.; Portelli, J.; Barkat, M.; Graulet, B. Olive mill wastewater microconstituents composition according to olive variety and extraction process. *Food Chem.* **2016**, *209*, 72–80. [CrossRef] [PubMed]
146. Hernández, M.L.; Muñoz-Ocaña, C.; Posada, P.; Sicardo, M.D.; Hornero-Méndez, D.; Gómez-Coca, R.B.; Belaj, A.; Moreda, W.; Martínes-Rivas, J.M. Functional Characterization of Four Olive Squalene Synthases with Respect to the Squalene Content of the Virgin Olive Oil. *J. Agric. Food Chem.* **2023**, *71*, 15701–15712. [CrossRef]
147. Oliveira, A.L.S.; Valente, D.; Moreira, H.R.; Pintado, M.; Costa, P. Effect of squalane-based emulsion on polyphenols skin penetration: Ex vivo skin study. *Colloids Surf. B Biointerfaces* **2022**, *218*, 112779. [CrossRef]
148. Olive Tree People. Available online: https://olivetreepeople.com/en/collections/oliveda (accessed on 26 August 2025).
149. Safety Assessment of *Olea europaea* (Olive)-Derived Ingredients as Used in Cosmetics. 2023. Available online: https://www.cir-safety.org/sites/default/files/Olive.pdf (accessed on 26 August 2025).
150. SCCS. Notes of Guidance for the Testing of Cosmetic Ingredients and Their Safety Evaluation, 12th Revision, 22 December 2023. Available online: https://health.ec.europa.eu/publications/sccs-notes-guidance-testing-cosmetic-ingredients-and-their-safety-evaluation-12th-revision_en (accessed on 29 April 2025).
151. De Leo, V.; Maurelli, A.M.; Giotta, L.; Daniello, V.; Di Gioia, S.; Conese, M.; Ingrosso, C.; Ciriaco, F.; Catucci, L. Polymer Encapsulated Liposomes for Oral Co-Delivery of Curcumin and Hydroxytyrosol. *Int. J. Mol. Sci.* **2023**, *24*, 790. [CrossRef]
152. Katsouli, M.; Thanou, I.V.; Raftopoulou, E.; Ntzimani, A.; Taoukis, P.; Giannakourou, M.C. Bioaccessibility and Stability Studies on Encapsulated Phenolics and Carotenoids from Olive and Tomato Pomace: Development of a Functional Fruit Beverage. *Appl. Sci.* **2024**, *14*, 10495. [CrossRef]
153. Nanonourish—Hair Care Ingredient for Nourishing Scalp and Hair Structure Based on Bioactives from Food Byproducts. Available online: https://upin.up.pt/pt-pt/tecnologias/nanonourish-hair-care-ingredient-nourishing-scalp-and-hair-structure-based-bioactives (accessed on 30 August 2025).

Journal of
Functional Biomaterials

MDPI

Review

Antibiotic Action, Drug Delivery, Biodegradability, and Wound Regeneration Characteristics of Surgical Sutures and Cutting-Edge Surgical Suture Manufacturing Technologies

Hye-Ree Han

Department of Beauty Art Care, Dongguk University, Seoul 04620, Republic of Korea; luckyherry@hanmail.net;
Tel.: +82-10-9130-8087

Abstract: (1) Background: With the emergence of various super bacteria, interest in antibacterial properties, drug delivery, and wound regeneration is increasing in the field of surgical materials. There are many studies on surgical sutures, but not many recent ones that have studied structurally subdivided functions. Accordingly, various studies on surgical sutures were classified based on the main functions that are considered important, and studies were conducted by categorizing the latest production technology into 3D printing and electrospinning. (2) Methods: Data from the literature (n = 1077) were collected from databases such as PubMed, Harvard.edu, MDPI, Google Scholar, Web of Science, ACS, Nature, and IOP Publishing. The selected 103 papers were divided into two main groups: cutting-edge characteristics of surgical sutures and the latest technologies for manufacturing surgical sutures. (3) Results: Cutting-edge characteristics of surgical sutures were divided into four major categories: antibacterial, drug delivery, biodegradability, and wound regeneration, and examined in depth. In addition, the final technologies for manufacturing surgical sutures were divided into electrospinning and 3D printing. (4) Conclusions: The results of this study can contribute to the development of multifunctional surgical sutures that promote wound regeneration through antibacterial properties, drug elution, and biodegradability.

Keywords: surgical sutures; biodegradable; antibacterial; drug release; electrospinning; 3D printing

1. Introduction

With the emergence of new viruses, human society is facing a serious threat from the increased resistance of microorganisms to existing antibiotics. According to the World Health Organization (WHO), scientists around the world have argued that general infections and minor injuries can have serious consequences for human health if antibiotic resistance is not addressed [1].

The emergence of new super bacteria has further enhanced the importance of imparting antibacterial properties to surgical sutures. In addition, for patients at high risk of infection or with weak immune systems, surgical suture materials that can minimize infection should be chosen. Surgical sutures are the leading cause of in-hospital infections through surgical wound contamination, and their rates are increasing worldwide. An efficient way to prevent this is to modify the sutures' surface to prevent bacteria from adhering [2].

Figure 1 shows a schematic representation of the main biomaterials data and the timeline representing each discovery.

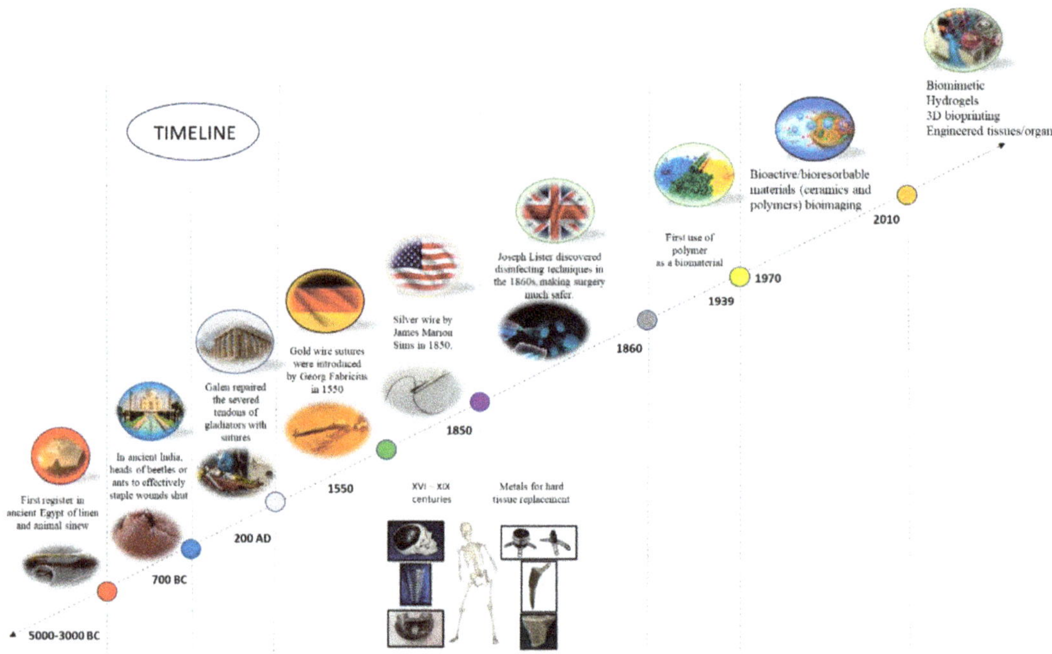

Figure 1. Schematic representation of the timeline indicating some of the main biomaterials data and the respective discoveries. Reprinted from ref. [3].

Antimicrobial sutures reduce the risk of infection at the surgical site through a coating of antimicrobial substances on general sutures.

On the other hand, studies on surgical suture materials have explored poly[(aminoethyl methacrylate)-co-(butyl methacrylate)] (PAMBM), catgut, and polyglycolic acid [PGA] sutures, and expanded polytetrafluoroethylene, silk sutures, collagen, chitosan, polyuronide, and polylactic acid [4,5]. In addition, many studies have been conducted in connection with antibiotic coating, Bio-inspired helical hollow bacterial cellulose fiber, oral surgery, bioelectric surgical suture, natural cellulose fiber, and site infection [6–13].

These materials exhibit human suitability, and it is important not to induce side effects such as human inflammatory reactions. External substances left in the human body are one of the most representative triggers of infection at the surgical site. In addition, for plastic surgery and cosmetic surgery, it is important to ensure natural results by reducing skin irritation and minimizing scars. Accordingly, there is an increasing demand for biodegradable sutures that naturally decompose in the body. For these reasons, biodegradability has also gained research attention.

A new trend in surgical materials is focused on natural resource materials with high biocompatibility, non-toxic properties, and specific properties aimed at the target site. In the past, polymer systems were used for simple sutures, but now their role is even more important because they need to heal surgical lesions or wounds, activate cell proliferation to refill damaged surfaces, protect damaged tissues and wounds from bacterial contamination, suppress infections, stop bleeding, and reabsorb moisture from wounds [14].

In addition, the importance of the drug delivery system for continuous and chronic wound treatment is increasing. Drug delivery is being studied with a focus on reducing wound areas, promoting cell migration, and maintaining continuous drug release, and research is being actively conducted to positively guide human recovery.

Although most sutures can provide sufficient mechanical support, their effects on clinical treatment are not satisfactory because they do not actively participate in wound healing. Therefore, it is very important to utilize biologically active sutures to promote wound healing. It has been reported that, after a wound occurs, a lateral wound electric field (EF) is formed from the surrounding intact tissue toward the center of the wound. Effective transmission of electrical signals generated by the wound EF can accelerate wound healing by regulating cellular behavior, including adhesion, proliferation, migration, and differentiation [15].

The purpose of this study is to examine, in depth, the main properties of surgical sutures that have recently been in the spotlight, including antibacterial properties, drug delivery systems, biodegradability, and wound regeneration. The application of electrospinning technology and 3D printing technology to the manufacture of surgical sutures is then considered. Next, based on the contents of this study, we tried to analyze the methods applied to multifunctional nanosurgical sutures, multifunctional mass production of 3D printing systems, and hybrid material manufacturing based on their use.

There are a lot of data on surgical sutures, but not many studies have structurally subdivided and explored the functions that have gained importance recently. Therefore, various studies on surgical sutures were classified based on the main functions (antibacterial, biodegradable, drug delivery, and wound regeneration), and the cases, strengths, and weaknesses of each were considered in depth. In addition, the latest technologies used for the production of surgical sutures were divided into 3D printing and electrospinning according to the manufacturing method, and a literature review was conducted.

2. Materials and Methods

In this study, an in-depth systematic review was conducted by classifying the main highlighted characteristics of surgical suture materials (Figure 2). Research data (n = 1077) were collected from databases such as PubMed, Harvard.edu, MDPI, Google Scholar, Web of Science, ACS, Nature, and IOP Publishing. Most of the documents used were papers published in the last 4 years (2021–2025). Keywords for the literature survey included 'surgical suture drug delivery, surgical suture antibacterial coating, surgical suture material, surgical suture plant extract, surgical suture 3d printing, surgical suture electrospinning, biodegradability of surgical sutures, wound regeneration'. The databases containing the included studies were evaluated for further eligible publications. A total of 103 selected papers were divided into two major groups: cutting-edge characters of surgical sutures and the latest technologies for manufacturing surgical sutures.

Cutting-edge characteristics of surgical sutures were largely divided into four categories—antibacterial, drug delivery, biodegradability, and wound regeneration—and reviewed in depth.

In addition, the latest technologies for manufacturing surgical sutures were divided into electrospinning and 3D printing. In conclusion, this study aimed to provide a systematic review of the cornerstone applications of cutting-edge surgical suture material design.

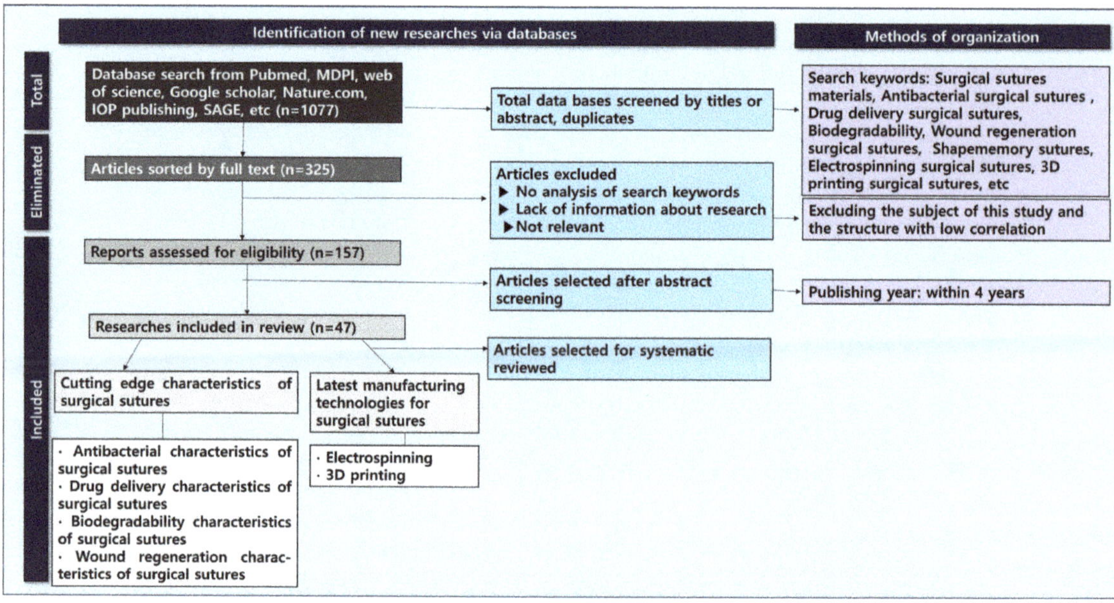

Figure 2. Flowchart of surgical sutures related to antibacterial, drug delivery, biodegradable, and wound regeneration characteristics and their production technologies.

3. Results and Discussion

3.1. Cutting-Edge Characteristics of Surgical Sutures

The main characteristics that have recently gained attention in surgical suture research include 'Antibacterial, Drug delivery, Biodegradability, Wound regeneration'. In this section, these were divided into four topics and explored in depth.

3.1.1. Antibacterial Characteristics of Surgical Sutures

Surgical sutures are used in almost all operations, but due to their lack of physiological signal responsiveness and their susceptibility to surgical site infection (SSI), their clinical treatment effectiveness is less satisfactory [15].

Figures 3 and 4 show images of antimicrobial surgical sutures.

Therefore, studies on surgical sutures and antibacterial agents have been conducted in various fields, exploring chitosan, nanosilver particles, sodium carboxymethyl cellulose/chitosan/chlorhexidine, chitosan–gelatin/tannic acid/polypyrrole composite coating, utilization of supercritical carbon dioxide, curcumin@ ZIF-8, nylon monofilament, etc. [1,2,15–43]. Table 1 presents the articles on the antibacterial characteristics of surgical sutures, with their bibliographic sources.

James et al. argued that surgical sutures are the main cause of in-hospital infections through surgical wound contamination, with increasing rates worldwide. Among the methods for preventing infection is modifying the surface of the joint to prevent bacteria from adhering. In this study, antibacterial properties were obtained by coating a biopolymer polycaprolactone material containing nanosilver on the surface of the sutures. Biocompatible polyethylene glycol was selected as a solvent for dispersing nanosilver particles to improve the mechanical properties of the sutures [2].

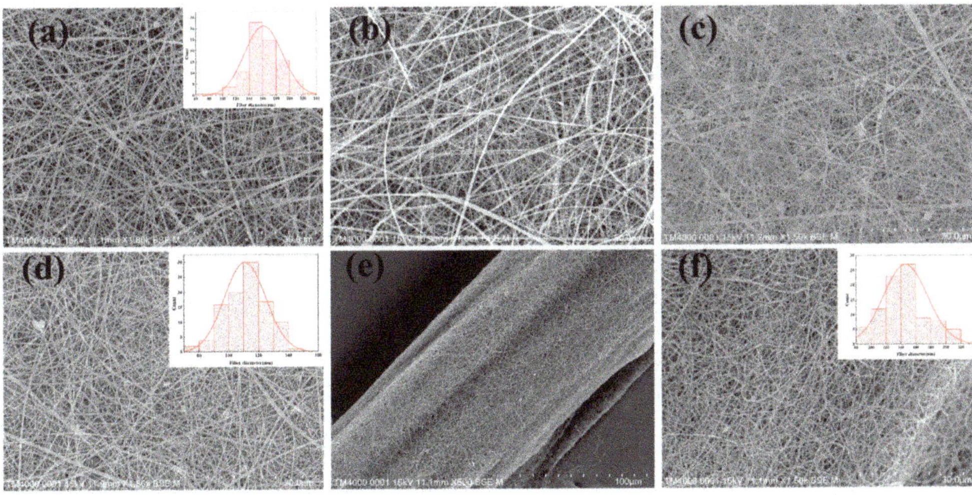

Figure 3. (**a**) SEM images of HPC/PVP electrospun nanofiber membrane; (**b**) SEM images of HPC/PVP-0.06% electrospun nanofiber membrane; (**c**) SEM images of HPC/PVP-0.07% electrospun nanofiber membrane; (**d**) SEM images of HPC/PVP-0.08% electrospun nanofiber membrane; (**e**,**f**) HPC/PVP/Zn electrospun nanofiber membrane and HPC/PVP/Zn electrospun wrapped yarns. Reprinted from ref. [16].

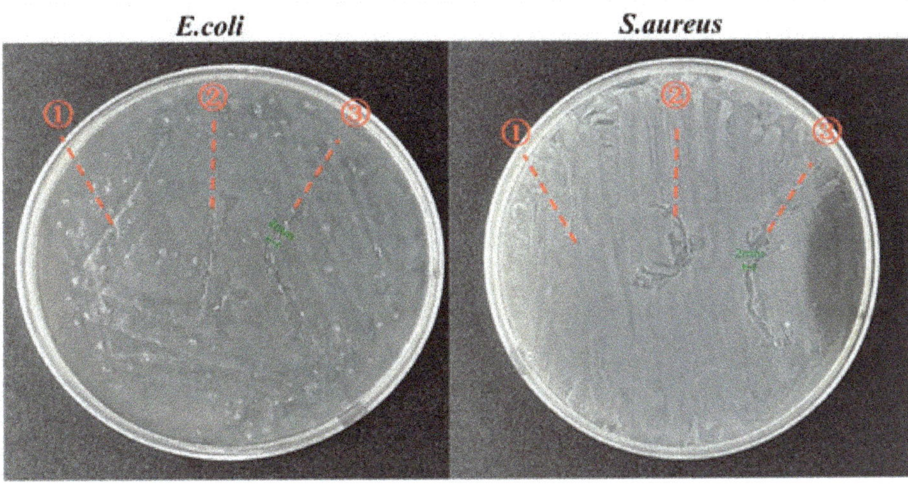

Figure 4. Zone of inhibitions for the in vitro antimicrobial activities of ① pure, ② HPC/PVP-coated, and ③ HPC/PVP-0.08% surgical sutures. Reprinted from ref. [16].

Zhang et al. fabricated a step-by-step controlled antimicrobial suture (SZC) by coating the surface with Cur@ZIF-8(curcumin@zeolitic imidazolate framework-8). Taking advantage of the pH-responsive behavior of ZIF-8, Zn^{2+}, and Cur(curcumin), it can be quickly released from acidic environments, increasing the efficiency of killing bacteria and preventing surgical site infections in a timely manner. The release of Zn^{2+} and Cur is reduced in a normal pH environment, protecting the cell/tissue from damage caused by excessive release [17].

Table 1. Articles on the classified antibacterial characteristics of surgical sutures, with bibliographic sources.

Authors	Title	Main Contents of Antibacterial Characteristics	Effect of the Research	Bibliographic Source
James, B et al.	Development of environmentally safe biodegradable, antibacterial surgical sutures using nanosilver particles.	Antibacterial surgical sutures, coating a biopolymer, nano silver particles, environmentally safe, biodegradable	Antimicrobial properties were obtained by coating a biopolymer polycaprolactone material with nanosilver on the suture surface. Biocompatible polyethylene glycol was selected as the nanosilver particle dispersion solvent to improve the mechanical properties of the suture.	Journal of Polymers and the Environment, 2021. 29(7): p. 2282–2288 [2].
Zhang, Q et al.	Stage-controlled antibacterial surgical sutures based on curcumin@ ZIF-8 functional coating for improved wound healing.	Antibacterial surgical sutures, *E. coli* and *S. aureus*, curcumin@ ZIF-8, SZC, wound healing, organic coatings	SZC exhibited excellent antimicrobial properties against *E. coli* and *S. aureus*. SZC has excellent mechanical properties and handling performance.	Progress in Organic Coatings, 2023. 184: p. 107829 [17].
Zhang, X et al.	Sustainable Antibacterial Surgical Suture Based on Recycled Silk Resource by an Internal Combination of Inorganic Nanomaterials.	Antibacterial surgical sutures, recycled silk resources, inorganic nanomaterials, wound healing.	The surgical suture with 1.25 wt% TiO_2 had a 2.40 N knot strength (diameter 143 μm) and achieved a sustainable antibacterial effect of 93.58%. Surprisingly, this suture significantly reduced the inflammatory response and promoted wound healing.	ACS applied materials & interfaces, 2023. 15(25): p. 29971–29981 [18].
Wang, X et al.	Sustainable antibacterial and anti-inflammatory silk suture with surface modification of combined-therapy drugs for surgical site infection.	Antibacterial, anti-inflammatory, Sprague Dawley (SD) mice, silk suture, surgical site infection.	In vivo evaluation using Sprague Dawley (SD) mice showed that the coating reduced the expression of the inflammatory cytokines interleukin-10 (IL-10) and tumor necrosis factor-α (TNF-α), thus shortening the duration of inflammation and promoting angiogenesis. The results showed that these novel sutures exhibited stable structure, good biocompatibility, sustainable antimicrobial, and anti-inflammatory functions, and had surgical applicability.	ACS Applied Materials & Interfaces, 2022. 14(9): p. 11177–11191 [19].
Pajnik, J et al.	Utilization of supercritical carbon dioxide for development of antibacterial surgical sutures.	Antibacterial, *E. coli, Staphylococcus aureus*, supercritical carbon dioxide, antibacterial surgical sutures.	The SSI process allowed the manufacture of surgical sutures effective against *E. coli* and *Staphylococcus aureus*. Mild process conditions (35 °C, 10 MPa, 1–6 h) allowed up to 5.9% of thymol loading.	The Journal of Supercritical Fluids, 2022. 181: p. 105490 [1].

Table 1. *Cont.*

Authors	Title	Main Contents of Antibacterial Characteristics	Effect of the Research	Bibliographic Source
Altun, E et al.	Pressure-Spun Fibrous Surgical Sutures for Localized Antibacterial Delivery: Development, Characterization, and In Vitro Evaluation.	Pressure-spun fibrous surgical sutures, antibacterial delivery, vitro evaluation.	After 24 h of in vitro antimicrobial testing, sutures with 285 ± 12 μg/mg Tri loading exhibited antimicrobial activity against all tested bacterial strains.	ACS applied materials & interfaces, 2023. 15(39): p. 45561–45573 [20].
Zhang, Q et al.	Electroactive and antibacterial surgical sutures based on chitosan-gelatin/tannic acid/polypyrrole composite coating.	Antibacterial surgical sutures, *E. coli* and *Staphylococcus aureus*, CS-GE/TA/PPy, electroactive, chitosan–gelatin/tannic acid/polypyrrole	Functional sutures showed excellent antibacterial properties against *E. coli* and *Staphylococcus aureus*. Functional sutures are conductive even when knotted, ensuring stable electrical signal transmission.	Composites Part B: Engineering, 2021. 223: p. 109140 [15].

Zhang et al. also argued that the current antimicrobial treatment method of silk sutures has the disadvantages of short antimicrobial effect, slow release, strong toxicity, and vulnerability to drug resistance, and these can only be addressed with surface modification. Recycled regenerated silk fibrin (RRSF) was extracted from waste silk resources to create an RRSF solution. Internal combination with organic titanium dioxide (TiO_2) nanoparticles was carried out to produce an antimicrobial RRSF-based surgical suture. Surgical sutures with 1.25 wt% TiO_2 obtained a knot strength of 2.40 N (diameter 143 μm) and achieved a sustainable antimicrobial effect of 93.58% [18].

Wang et al. prepared silk sutures with a braided core–shell structure from rubberized silk filaments for continuous double-drug delivery to prevent surgical site infections (SSIs), coating them with a silk fibrin (SF) layer loaded with berberine (BB) and artemisinin (ART). In an in vivo evaluation using Sprague Dawley (SD) mice, the coating was shown to reduce the expression of inflammatory cytokines interleukin-10 (IL-10) and tumor necrosis factor-α (TNF-α), thereby shortening the duration of inflammation and promoting angiogenesis [19].

Pajnik et al. were able to manufacture effective surgical sutures against *E. coli* and *Staphylococcus aureus* through the SSI process. All test samples showed antibacterial activity against *E. coli* and *S. aureus*, and this can be applied in the prevention of infection at the surgical site [1].

Altun et al. used a new manufacturing process based on pressure rotation for the first time to produce fiber surgical sutures by physically mixing antimicrobial triclosan (Tri) agents with poly(lactic-co-glycolic acid) (PLGA) and polyethylene oxide (PEO) polymers. All samples produced (5–40 TPP) showed increased inhibitory activity against *E. faecalis* and *E. coli* strains, depending on the Tri concentration loaded on fiber sutures. As expected, 40 TPP samples loaded with higher amounts of Tri exhibited higher antimicrobial activity compared with Vicryl Plus in all strains tested [20].

Zhang et al. made functional sutures by decorating the surface with a CS-GE/TA/PPy composite coating, which showed excellent antibacterial properties against *E. coli* and *Staphylococcus aureus*. Electroactive and antibacterial sutures greatly improved tissue regeneration [15].

As mentioned above, in order to impart antimicrobial properties to the surgical suture, the synthetic suture surface can be modified, or a nanosilver particle coating, coated with Cur@ZIF-8(curcumin@zeolitic imidazolate framework-8), recycled generated silk fibrin and

TiO_2, and mixed with PLGA and PEO, can be used to produce an antimicrobial surgical suture. It has been reported that these methods show excellent antimicrobial activity against bacteria such as *Staphylococcus aureus* or *E. coli* and can positively affect tissue regeneration.

3.1.2. Drug Delivery Characteristics of Surgical Sutures

Deng et al. argued that sutures are the most popular surgical implants in the global surgical equipment market [44].

Studies on surgical sutures and drug delivery include research on multifunctional electronic sutures, anti-inflammatory silk sutures, chitosan/keratin/PCL/PEG drug elution sutures, scalable silk–fibrin-based berberine loading systems, hybrid hydrogel actuator, biopolymer–nanotube nerve guidance conduit drug delivery for peripheral nerve regeneration, etc. [14,19,24,44–58].

Table 2 presents the classified articles on drug delivery characteristics of surgical sutures, together with their bibliographic sources.

Deng et al. argued that natural polymers such as collagen, silk, nylon, and cotton, as well as synthetic polymers such as polycaprolactone, poly(lactic acid-co-glycolic acid), poly(p-dioxanone), etc., contribute to a solid foundation for drug elution suture engineering. In addition, it is possible to effectively address wound healing requirements by controlling drug elution profiles through manufacturing processes and the use of polymer materials [44].

Table 2. Articles on classified drug delivery characteristics of surgical sutures with their bibliographic sources.

Authors	Title	Main Contents of Drug Delivery Characteristics	Effect of the Research	Bibliographic Source
Deng, X et al.	Engineering and polymeric composition of drug-eluting suture: a review.	Drug-eluting, polymeric composition, wound healing	Natural polymers such as lagen, silk, nylon, and cotton, and synthetic polymers such as polycaprolactone, poly(lactic-co-glycolic acid), and poly(p-deoxanone) provide a solid foundation for drug elution suture engineering. Manufacturing processes and polymer materials can effectively address wound healing requirements by controlling drug elution profiles.	Journal of Biomedical Materials Research Part A, 2021. 109(10): p. 2065–2081 [44].
Lee, Y et al.	A multifunctional electronic suture for continuous strain monitoring and on-demand drug release.	Drug release, multifunctional electronic suture, continuous strain monitoring	A thermoreactive shell layer consisting of flexible poly(vinyl alcohol) (PVA) grafted onto poly(*N*-isopropylacrylamide) (PNIPAm) facilitates on-demand drug release through Joule heating. In vitro scratch analysis demonstrated the efficacy of DRESS in stimulation response treatment by reducing the wound area by 66% upon heat activation after 48 h, facilitating cell migration.	Nanoscale, 2021. 13(43): p. 18112–18124 [45].

Table 2. *Cont.*

Authors	Title	Main Contents of Drug Delivery Characteristics	Effect of the Research	Bibliographic Source
Deng, X et al.	Fabrication and characterisation of melt-extruded chitosan/keratin/PCL/PEG drug-eluting sutures designed for wound healing.	Drug-eluting, melt-extruded, chitosan/keratin/PCL/PEG, wound healing	A literature study comparing drug release in pure PCL and PCL/PEG blends found that the pure PCL matrix released only 4% of the drug over a 7-day period. In contrast, in the case of a polymer blend with a hydrophilic component of 30%, more than 80% of the drug was released within 72 h. In the current study, up to 20% of PEG was used, and the duration of rapid release of the drug was related to the time point at which the PEG began to decompose.	Materials Science and Engineering: C, 2021. 120: p. 111696 [46].
Mendez, K et al.	Mechanoresponsive Drug Release from a Flexible, Tissue-Adherent, Hybrid Hydrogel Actuator.	Drug release, hybrid hydrogel actuator, flexible, tissue-adherent	In this study, a new type of hybrid hydrogel actuator (HHA) was introduced to facilitate enhanced drug delivery. Multi-material soft actuators can induce controllable mechanoreactive release of charged drugs in the alginate/acrylamide hydrogel layer through temporal control.	Advanced Materials, 2024. 36(43): p. 2303301 [47].
Parikh, K.S et al.	Ultra-thin, high strength, antibiotic-eluting sutures for prevention of ophthalmic infection.	Antibiotic-eluting sutures, prevention of ophthalmic infection, ultra-thin, high strength.	Nanofiber-based sutures maintained strength even when loaded with a wide range of drugs, administered antibiotics for 30 days in the eyes of mice, and prevented eye infections in a bacterial keratitis mouse model. The nanofiber-based polycaprolactone sutures did not decrease in strength when loaded with 8% levofloxacin, whereas the monofilament sutures decreased in strength by more than 50%.	Bioengineering & Translational Medicine, 2021. 6(2): p. e10204 [48].
Manoukian, O.S et al.	Biopolymer-nanotube nerve guidance conduit drug delivery for peripheral nerve regeneration: In vivo structural and functional assessment.	Drug delivery, biopolymer-nanotube, in vivo structural peripheral nerve regeneration.	Continuous 4-aminopyridine release amplifies neurotrophic factor release by Schwann cells, promoting axon regeneration. The TGA derivative curve shows a drug loading of 7.69% by weight within the range of 5–10% established by successful drug loading of HNT, which has not been modified in the literature. NGC was able to provide a continuous release of 4-AP for 8 weeks.	Bioactive Materials, 2021. 6(9): p. 2881–2893 [49].

Table 2. *Cont.*

Authors	Title	Main Contents of Drug Delivery Characteristics	Effect of the Research	Bibliographic Source
Bibire, T et al.	Biopolymers for surgical applications.	Drug delivery, biopolymer, excellent wet adhesion performance, antimicrobial activity, cell viability	Various combinations of PEG/PCL/chitosan/keratin mixtures achieved rapid and sustained drug release rates. PCL/PEG/chitosan–keratin complex with 30% of diclophenac potassium by weight showed high cell viability and wound healing rates during in vitro cytotoxicity testing.	Coatings, 2022. 12(2): p. 211 [14].

Lee et al. argued that monitoring wound integrity and simultaneously promoting tissue regeneration remain challenging in the field of surgical sutures. To address this issue, they developed a drug-releasing electron sutures system (DRESS) to monitor suture integrity in real time and trigger drug release to improve tissue regeneration. The DRESS was fabricated using a single fiber of the core–shell structure consisting of a flexible conductive fiber core and a thermoreactive polymer shell containing the drugs. A layer of thermoreactive shell composed of flexible poly(vinyl alcohol) (PVA) grafted onto poly(*N*-isopropylacrylamide) (PNIPAm) facilitates drug release on demand through Joule heating. The results of an in vitro scratch assay showed a 66% reduction in wound area upon heat activation after 48 h, which demonstrated the efficacy of DRESS in stimulating response treatment by promoting cell migration [45].

Deng et al. fabricated diclophenac potassium load sutures based on PEG/PCL/chitosan/keratin mixtures using hot melt extrusion techniques. Optimal formulations of tensile strength were obtained when PCL/PEG/chitosan/keratin was combined at a ratio of 80/19/1 *w*/*w*, and rapid and continuous drug release rates were achieved with PEG/PCL/chitosan/keratin mixtures in various combinations. The complex of diclophenac potassium 30 wt% and PCL/PEG/chitosan/keratin exhibited high cell viability and wound healing rates in in vitro cytotoxicity tests [46].

Mendez et al. studied soft robotic drug delivery. Micro-CT imaging of the hybrid hydrogel actuator (HHA) prototype demonstrates robust and flexible adhesion to tissues during dynamic operation. The device enables coordinated mechanically reactive drug delivery directly to the target site and presents an innovative approach that integrates precisely controlled drug delivery with mechanical stimulation for enhanced topical treatment interventions [47].

Parikh et al. developed nanostructured multi-filament sutures capable of loading high-level low-molecular drugs while maintaining the high breaking strength required by the U.S.P. standard. The nanofiber-based sutures were shown to maintain strength even with a wide range of medications, deliver antibiotics to the eyes of mice for 30 days, and prevent ocular infections in a mouse model of bacterial keratitis. The multi-filament nanofiber sutures demonstrated biocompatibility and prevention of ophthalmic infections after multiple inoculations with *Staphylococcus aureus* for 1 week [48].

Manoukian et al. explained that chitosan-based nerve guidance conditions (NGCs) confirmed the continued release of 4-aminopyridine (4AP), a small-molecular drug that promotes neuroconductivity and neurotrophic factor release through aligned microchannel porosity and halosite nanotubes [49].

Bibire et al. recently explained that further modification of biomolecules using advanced technology can improve functions such as drug release capacity, wet adhesion performance, antibacterial activity, cell viability, and mechanical properties [14].

As mentioned above, drug elution profile control through polymer materials, weakly released electronic sutures systems, PEG/PCL/chitosan/keratin mixtures, micro-CT imaging of hybrid hydrogel actuator (HHA) prototypes, biopolymer modification, etc., can be used to produce drug delivery surges. These methods tended to have positive effects on human recovery, such as drug elution, wound site reduction, cell migration promotion, cell survival and wound healing rates, and continuous drug release.

3.1.3. Biodegradability Characteristics of Surgical Sutures

Long-term use of non-degradable fiber implants can release some debris due to wear, which can be detrimental to surrounding cells and tissues, and can cause cell malformation, cell death, and even carcinogenesis [59].

Recent advances in biomedical engineering have highlighted the important role of biodegradable materials in solving challenges related to tissue regeneration therapy. The current spectrum of biodegradable materials includes ceramics, polymers, metals, and composites, each providing distinct benefits for the replacement or repair of damaged human tissues. Despite their utility, these biomaterials are not without limitations, and problems such as suboptimal tissue integration, potential cytotoxicity, and mechanical discrepancy (stress shielding) have emerged as serious concerns [60].

Studies on surgical sutures and biodegradability include research on oral surges, knot strength, orthopedic application, smart materials, self healing, silver nanowire, albumin composites, mechanical properties, absorbable sutures, PLA-based biopolymers, poly(lactic acid)/poly(ethylene glycol) blends, etc. [2,5,60–73].

Table 3 presents classified articles on the biodegradability characteristics of surgical sutures with bibliographic sources.

Table 3. Classified articles on the biodegradability characteristics of surgical sutures with their bibliographic sources.

Authors	Title	Main Contents of Biodegradability Characteristics	Effect of the Research	Bibliographic Source
Antoniac, I et al.	In vitro study on biodegradation of absorbable suture Materials used for surgical applications.	Biodegradation, in vitro study, absorbable suture materials, surgical applications	Depending on the pH value, it is observed that at pH = 7.4, the sample decomposes faster than under acidic conditions. In addition, important aspects were highlighted for samples P2 (lactic acid-co-glycolic acid) and P5 (poly(glycolide-co-ε-caprolactone).	Mater. Plast, 2021. 58: p. 130–139 [61].
Nouri, A et al.	Biodegradable metallic suture anchors: A review.	Biodegradable, metallic suture anchors, MG, implant	Since AM technology was only recently introduced in orthopedics, the issues related to its effectiveness in bone anchor manufacturing are still unresolved. Only a small number of natural polymers, including collagen, keratin, and chitosan, have been found as suitable biodegradable implant materials.	Smart Materials in Manufacturing, 2023. 1: p. 100005 [62].

Table 3. *Cont.*

Authors	Title	Main Contents of Biodegradability Characteristics	Effect of the Research	Bibliographic Source
Naser, M.A et al.	Biodegradable suture development-based albumin composites for tissue engineering applications.	Biodegradable suture, tissue engineering, albumin composites	Preliminary tensile tests described the mechanical profile of filament sutures (FSs), with tensile strengths ranging from 1.3 to 9.616 MPa and fracture elongation rates ranging from 11.5 to 146.64%. These results reveal the mechanical diversity of sutures, suggesting their applicability to a wide range of medical interventions.	Scientific Reports, 2024. 14(1): p. 7912 [60].
Szabelski, J et al.	Short-Term Hydrolytic Degradation of Mechanical Properties of Absorbable Surgical Sutures: A Comparative Study.	Hydrolytic degradation, absorbable surgical sutures, protein composites	SafilQuick+ and MonosynQuick sutures lost their strength within 9 to 12 days, as evidenced by statistically significant changes in tensile strength. Surgical sutures play an important role in wound sutures and facilitate tissue healing processes in various medical fields.	Journal of Functional Biomaterials, 2024. 15(9): p. 273 [63].
Wu, H et al.	Surface coating prolongs the degradation and maintains the mechanical strength of surgical suture in vivo.	Degradation, surgical sutures in vivo, surface coating, mechanical strength	Absorbable sutures undergo decomposition and absorption during tissue repair, whereas nonabsorbable sutures maintain tensile strength and resist absorption in combination with healing. In anterior cruciate ligament repair, which requires a long time to heal, nonabsorbable sutures outperformed absorbent sutures after 15 weeks of observation.	Colloids and Surfaces B: Biointerfaces, 2022. 209: p. 112214 [64].
Alhulaybi, Z.A et al.	Fabrication and Characterization of Poly(lactic acid)-Based Biopolymer for Surgical Sutures.	Biodegradable, PLA, surgical Sutures, biopolymer	According to decomposition studies, the sutures manufactured in this study were demonstrated to degrade in physiological saline. After 15 days, the sutures lost 50% of their weight, and their pH decreased from 6.49 to 4.42.	ChemEngineering, 2023. 7(5): p. 98 [66].

Absorbable sutures can be easily or completely absorbed into the body, and it is important to avoid postoperative wound infections and reduce patient pain. Absorbable sutures have been demonstrated to degrade by losing 50% of the tensile strength in tissues within 60 days. Among them, natural absorbent sutures are degraded by proteolysis, and synthetic absorbent sutures are degraded by hydrolysis. Absorbable suture materials are divided into natural absorbent suture materials and synthetic absorbent suture materials. Currently, commonly used absorbent suture materials include sheep intestines, poly-p-dioxanone (PDO), polyurethane (PU), polylactic acid (PLA), cellulose, polyglycolic acid (PGA), polycaprolactone (PCL), and poly(lactic-co-glycolic acid) (PLGA) [74].

Li et al. explained that compared to conventional non-degradable suture materials, absorbent materials exhibit unique properties that can be degraded in vivo without the need to be removed later, protecting patients from secondary trauma [65].

Antoniac et al. explained that under physiological conditions (pH = 7.4), irradiated surgical sutures with glycolide components degrade more rapidly, whereas surgical sutures made with polydioxanone do not. In addition, increasing pH leads to a higher degree of degradation and lower tensile strength, making it more likely that a wound will occur [61].

Nouri et al. found that biodegradable metals have high fixed strength and a low modulus of elasticity close to the bone, which facilitates bone union and allows the design of thinner and less bulky implants. Meanwhile, Mg has properties such as excellent biocompatibility, mechanical strength, and low modulus of elasticity similar to human bones. There are several competing approaches to slow down the decomposition/corrosion rate of Mg-based biomaterials, including coating, alloying, and surface treatment. Furthermore, Mg with abnormally high purity (>99.99%) has been demonstrated to degrade more slowly than other alloy systems. Using Mg and its alloys reduces MRI interference [62].

Naser et al. initiated the development of protein-based composites to alleviate shortcomings such as cytotoxicity, demonstrating improved biodegradability and biocompatibility; specifically, this study focused on the elaboration and characterization of innovative sutures fabricated with human serum albumin via extrusion methodology. It also shows that the use of proteins from human serum albumin (PHSA) can be utilized in the development of novel biodegradable sutures that can be used in many medical applications [60].

Szabelski et al. developed biopolymer composites by extensively fabricating filament sutures through a hot melt extrusion process based on human serum albumin protein. They studied the potential enhancements provided by the mixing of protein-based composites and biodegradable organic compounds, such as improved biodegradability and biocompatibility [63].

Wu et al. coated polylactic acid (PLLA) and polycaprolactone (PCL) on polydioxanone (PDO) sutures through a single-step method and showed that it increased the decomposition time and mechanical strength of the sutures in vitro and in vivo [64].

Alhulaybi et al. focused on producing absorbent surgical sutures from a biocompatible polymer material called polylactic acid (PLA) with PLA–chitosan composite sutures using an extrusion method for negative changes, followed by stretching methods. The sutures manufactured in this study were demonstrated to degrade in physiological saline. After 15 days, the sutures lost 50% of their weight, and their pH decreased from 6.49 to 4.42 [66].

As mentioned above, there are many studies on biodegradable metals, cellulose, PCL, PLA, etc., in absorbent materials, pH, protein-based composites, human serum albumin, and Mg that examine the biodiversity of surgical sutures. In addition, biodegradable materials made from the materials examined above tended to be more advantageous for protecting the human body or reducing pain than non-degradable materials.

3.1.4. Wound Regeneration Characteristics of Surgical Sutures

Factors influencing wound healing include infection, necrosis, nutrition, medication, moisture, and physiological changes in individuals (age, immune system, stress, obesity, diabetes, etc.). The wound-healing process begins when a wound is formed after mechanical or chemical damage and includes a series of chemical signaling and cellular mechanisms, leading to regeneration and/or repair. Interventions to accelerate wound healing are essential because interruption of the healing process can lead to complications [28,75–78].

Research on wound regeneration and tissue regeneration using sutures includes studies on bioactive glass/graphene oxide-coated surgical sutures, reactive oxygen species-scavenging sutures, hybrid suture coating, and electrospun drug-eluting nanofibers [28,75–78].

Table 4 presents the classified articles on the wound regeneration characteristics of surgical sutures, together with their bibliographic sources.

Table 4. Classified articles on the wound regeneration characteristics of surgical sutures and their bibliographic sources.

Authors	Title	Main Contents of Wound Regeneration Characteristics	Effect of the Research	Bibliographic Source
Öksüz, K.E., et al.	Novel Bioactive Glass/Graphene Oxide-Coated Surgical Sutures for Soft Tissue Regeneration	Soft tissue regeneration, bioactive glass/graphene oxide, surgical sutures	The highest cell activity was observed in the S + BGNs (95.5 ± 1.23%) and S + BGNs/GO (86.73 ± 2.83%) groups. However, there was a noticeable difference in cell activity in the S group (80.87 ± 1.93%), and consequently, it proved beneficial for cell growth because of the presence of a mixed polycrystalline hydroxyl–carbonate–apatite (HCA) layer on the suture surface.	ACS omega, 2023. 8(24): p. 21628–21641 [78].
Zhu, J et al.	Reactive Oxygen Species Scavenging Sutures for Enhanced Wound Sealing and Repair	Reactive oxygen species, scavenging sutures, wound sealing and repair	The obtained sutures coated with GANP can effectively promote wound closure by maintaining tension around the wound and reducing the ROS level. Specifically, the GANP coated on the sutures can effectively eliminate ROS, upregulate anti-inflammatory molecules, and polarize macrophages around the wound site by M2 phenotype due to the efficient antioxidant activity of GA, a type of low molecular weight difference polyphenol.	Small Structures, 2021. 2(7): p. 2100002 [77].
Chen, Y.-G et al.	Hybrid suture coating for dual-staged control over antibacterial actions to match well wound healing progression	Dual-staged control, wound healing progression, hybrid suture coating, antibacterial	Acidic conditions induce the explosive release of antimicrobial TA, mostly from adsorbed TA monomers. In the later stages, TA release is fully tuned to pH conditions, which depend on the degree of healing of the wound, enabling continuous antimicrobial prevention in a biologically controllable manner.	Materials Horizons, 2022. 9(11): p. 2824–2834 [28].
Han, H et al.	A multifunctional surgical suture with electroactivity assisted by oligochitosan/gelatin-tannic acid for promoting skin wound healing and controlling scar proliferation	Electroactivity assisted, oligochitosan/gelatin–tannic acid, skin wound healing, controlling scar proliferation	Through in vivo experiments, S@LC@CGTP was able to alleviate inflammation and promote scar-free wound healing following the suppression of invasive inflammatory cells, downregulation of TGF-β1 and type I collagen expression, and improved collagen arrangement. Cumulatively, we have shown that S@LC@CGTP suture materials have great potential in promoting optimal, almost scar-free healing of surgical incisions.	Carbohydrate Polymers, 2023. 320: p. 121236 [76].

Table 4. *Cont.*

Authors	Title	Main Contents of Wound Regeneration Characteristics	Effect of the Research	Bibliographic Source
Akombaetwa, N et al.	Applications of Electrospun Drug-Eluting Nanofibers in Wound Healing: Current and Future Perspectives	Wound healing, drug-eluting, electrospun nanofibers	Running sutures were inserted into the non-abdominal muscles of male Wistar mice at a concentration of 0.1 or 1 μg. VEGF/poly(L-lactide)(PLA) blend coatings showed increased biological activity and cell viability.	Polymers, 2022. 14(14): p. 2931 [75].

Kerim Emre Öksüz et al. developed a novel bioactive glass/graphene oxide-coated surgical suture for soft tissue regeneration. In this experimental work, they demonstrated that graphene oxide (GO)-doped melt-derived bioactive bioglass nanopowders (BGNs) were synthesized through a sol-gel process. Subsequently, new GO-doped and undoped BGNs were used to coat the surgical sutures with poly[glycolide-co-lactide] surgical sutures to confer bioactive, biocompatible, and accelerated wound healing properties on the sutures. As a result, it was shown that the formation of BGN and GO on the sutures' surface is significantly improved, thereby improving fibroblast attachment, migration, and proliferation and accelerating wound healing by promoting the secretion of angiogenesis growth factor [78].

Jiafei Zhu et al. developed a new ROS erasing suture by coating surgical sutures with gallic acid (GA)-based nanoparticles (GAMPs) for wound healing. Wound healing is a complex biological process involving hemostasis, cell proliferation, and tissue remodeling that is closely related to the level of oxidative stress in the damaged skin. The ROS cleaning sutures developed in this study enhance wound sutures, reduce inflammatory responses, and reduce scar formation, and can serve as promising sutures for wound healing under various conditions [77].

Ying-Ge Chen et al. developed mild-conditioned aqueous solutions without the use of organic solvents and preserved the basic properties of suturing materials based on the pH-dependent reversible self-polymerization of tannic acid (TA) and the strong adhesion of poly (tannic acid) (PTA). In the early postoperative stages, the wound site is susceptible to aseptic and/or bacterial inflammation. The acid conditions produced induce the explosive release of antimicrobial TA, mainly from adsorbed TA monomers. In the early postoperative stages, the wound site is susceptible to aseptic and/or bacterial inflammation. The acid conditions produced induce the explosive release of antimicrobial TA, mainly from adsorbed TA monomers [28].

Hui Han et al. studied a multifunctional surgical suture with electroactivity aided by oligochitosan/gelatin–tannic acid to promote skin wound healing and control scar proliferation. The multifunctional suture, designated S@LC@CGTP, had desirable in vitro persistent drug release properties that enabled Cur to be released for 8 days due to the action of PLGA. It has been shown that S@LC@CGTP suture materials have great potential to promote optimal, almost scar-free healing of surgical incision sites [76].

Nakamwi Akombaetwa et al. explained that in addition to the mechanical support provided by sutures and traditional wound dressing, therapeutic agents play an important role in accelerating wound healing. Drugs known to improve wound healing rate and extent include antibacterial, anti-inflammatory, and proliferation promoters. The development of these agents as eluted nanofibers has the potential to produce wound dressing and sutures that provide mechanical support with the benefit of locally delivering the therapeutic agent to the wound site [75].

As mentioned above, bioactive glass/graphene oxide coating, gallic acid (GA)-based nanoparticle (GAMP) coating, poly (tannic acid) utilization, and oligochitosan/gelatin-tannic acid can be used to produce surgical sutures. These methods tend to promote wound regeneration by improving fibroblast attachment, movement, and proliferation, promoting the secretion of vascular growth factors and enhancing antibacterial properties.

3.2. Latest Manufacturing Technologies for Surgical Sutures

3.2.1. Electrospinning

Electrospinning technology can produce nano-thick fibers through high voltage and Van der Waals attraction. This nanotechnology is being actively studied in medical, fiber, and various engineering fields for reasons such as high surface area, possible core–sheath production, etc. A schematic of the electrospinning process is shown in Figure 5.

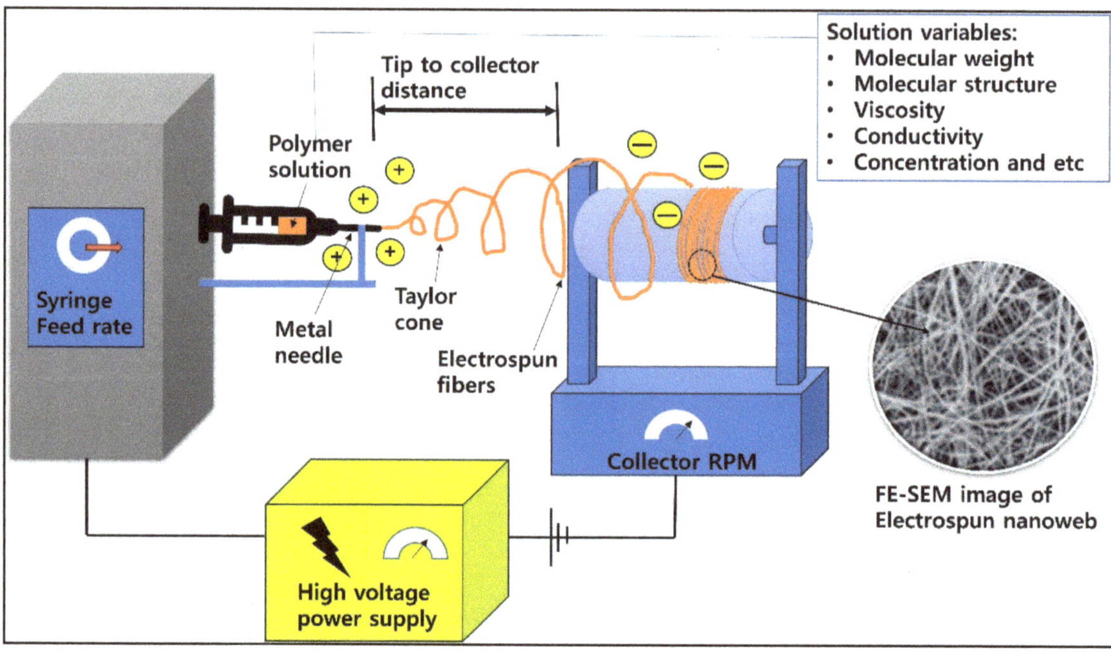

Figure 5. A schematic diagram of the electrospinning process.

Articles on electrospinning surgery sealing include wound healing, inspection prevention, bioactive nano yarn, PLA, supercritical CO_2 impregnation of ibuprofen and naproxen, poly-ε-caprolactone/propolis electrospun yarns, surface biofunctional bFGF-loaded electrospun suture, etc. [59,74,75,79–99].

Table 5 presents the classified articles on electrospinning technologies of surgical sutures, together with their bibliographic sources.

Xu et al. explained that, unlike single-fluid electrospinning techniques, double-fluid electrospinning spheres adopt double-layer composite superposition structures. There is a certain gap between the inner and outer layers so that the polymer solution flows smoothly. Finally, nanofibers of core–sheath structure are obtained. Using this manufacturing method, the performance can be altered by changing the position of the bioactive materials in the medical sutures' nanofibers. For example, adding bioactive materials to the core solution slowly releases nanofibers' control drugs and prolongs the sutures' treatment time for the wound. When biologically active materials are added to the cover solution, the drug comes into direct contact with the wound, resulting in a wound treatment effect within a short period of time [74].

Table 5. Classified articles on the electrospinning technologies of surgical sutures with their bibliographic sources.

Authors	Title	Main Content	Bibliographic Source
Xu, L et al.	Electrospun medical sutures for wound healing: A review.	Electrospinning, wound healing, medical sutures	Polymers, 2022. 14(9): p. 1637 [74].
Teno, J et al.	Development of Ciprofloxacin-Loaded Electrospun Yarns of Application Interest as Antimicrobial Surgical Suture Materials.	Electrospun yarns, ciprofloxacin, antimicrobial, surgical suture materials	Pharmaceutics, 2024. 16(2): p. 220 [80].
Madheswaran, D et al.	Composite yarns with antibacterial nanofibrous sheaths produced by collectorless alternating-current electrospinning for suture applications.	Electrospinning, composite yarns, antibacterial nanofibrous sheaths, suture applications.	Journal of Applied Polymer Science, 2022. 139(13): p. 51851 [81].
Rivera, P et al.	Development of PLA suture materials by extrusion, electrospinning and supercritical CO_2 impregnation of ibuprofen and naproxen.	Electrospinning, PLA suture materials, supercritical CO_2 impregnation, ibuprofen, naproxen	The Journal of Supercritical Fluids, 2023. 194: p. 105854 [82].
Wu, S et al.	State-of-the-art review of advanced electrospun nanofiber yarn-based textiles for biomedical applications.	Electrospun nanofiber yarn, biomedical applications, double-layered coverspun NYs, multilayered coverspun	Applied Materials Today, 2022. 27: p. 101473 [59].
Mohamadinooripoor, R et al.	Poly-ε-Caprolactone/Propolis Electrospun Yarns as Suture.	Electrospun yarns, Poly-ε-Caprolactone, propolis, sutures	Fibers and Polymers, 2023. 24(8): p. 2641–2651 [83].
Li, Y et al.	Surface biofunctional bFGF-loaded electrospun suture accelerates incisional wound healing.	Electrospun suture, incisional wound healing, surface biofunctional bFGF	Materials & Design, 2023. 225: p. 111451 [84].

Teno et al. fabricated yarn based on flying entangled microfibers (1.95 ± 0.22 μm) in the field during the electrospinning process using a specially designed yarn collector. Electrospinning yarn sutures (300–500 μm in diameter) were made from poly(3-hydroxybutyrate-co-3-hydroxyvalerate) with different contents of 3HV units and contained ciprofloxacin hydrochloride (CPX) as an antibacterial activity pharmaceutical ingredient (API). All yarn sutures exhibited antibacterial properties during a 5-day time release against both Gram-positive and Gram-negative pathogenic bacteria [80].

Madheswaran et al. used electrospinning methods to fabricate chlorohexidine (CHX)-containing antimicrobial nanofiber covers around polyamide core threads. Cell conformity tests using 3T3-SA mouse fibroblasts and antimicrobial evaluations using *E. coli* and *Staphylococcus aureus* have suggested that CHX-containing PA6-based CNYs are biocompatible and have antimicrobial properties, suggesting that these threads can be used as functional and mechanically performing sutures [81].

Rivera et al. developed a poly(lactic acid) (PLA)-based suture using extrusion and electrospinning techniques followed by supercritical CO_2 impregnation treatment. Polylactic acid (PLA) electrospinning fibers and films were impregnated with scCO2 with two nonsteroidal anti-inflammatory agents, ibuprofen and naphroxene, to obtain biodegradable suture materials with topical drug delivery capacity. A larger amount of the impregnated drug was obtained from the electrospinning PLA film, and 27% and 20% incorporation rates of naphroxene and ibuprofen, respectively, were obtained [82].

Wu et al. explained that the specific surface area increased significantly as the flow diameter reduced and that the ECM nanofiber yarns' (NYs') mimetic properties made electrospun nanofibers an ideal material for biomedical applications. More complex NY structures, including hollow NYs, double-layered coverspun NYs, and multilayerized coverspun Nys, were successfully produced. Numerous materials, including natural and synthetic polymers, as well as various functional additives, such as conductive materials, nanomaterials, and biologically active ingredients, have been successfully utilized to construct electrospun NYs with predetermined properties and functions [59].

Mohammedinooripoor et al. prepared yarn by applying an electrospinning composite mat consisting of poly-ε-caprolactone (PCL)/propolis ethanolic extrusion (PEE). In vitro cell culture studies on the biocompatibility of MG-63 and human skin fibroblasts in sutures and the antifungal and antifungal effects of sutures were also evaluated. Consequently, PEE-PCL sutures were shown to have elastic, mechanical, antifungal, and antifungal properties, as well as rapid wound-healing properties [83].

Li et al. fabricated a new degradable suture that controls the release of basic fibroblast growth factor (bFGF) for incisional wound healing. The suture (bFGF-DA@PCL) was fabricated by fixing bFGF to the surface of the electrospun polycaprolactone (PCL) suture with the help of dopamine. In vitro data demonstrated that the suture can effectively extend bFGF release by up to 10 days and improve cell adhesion and proliferation. Further in vivo results showed that the suture can increase wound healing and induce complete wound closure within 13 days. In addition, the suture was found to accelerate the mechanical recovery of the skin by restoring maximum intensity in healthy skin by 87.1% [84].

As mentioned above, it has been reported that electrospinning technology uses various substances to change the performance by changing the position of physiologically active substances or to show efficiency in imparting functions such as containing antibacterial active substances and excellent biodegradability. Electrospinning technology is expected to exert an excellent influence on the development of human-friendly surgical suture technology due to its thinness and high specific surface area.

3.2.2. Three-Dimensional Printing

Three-dimensional printing technology has the ability to design products with computer programs without producing wastewater. The schematic diagram for 3D printing is shown in Figure 6.

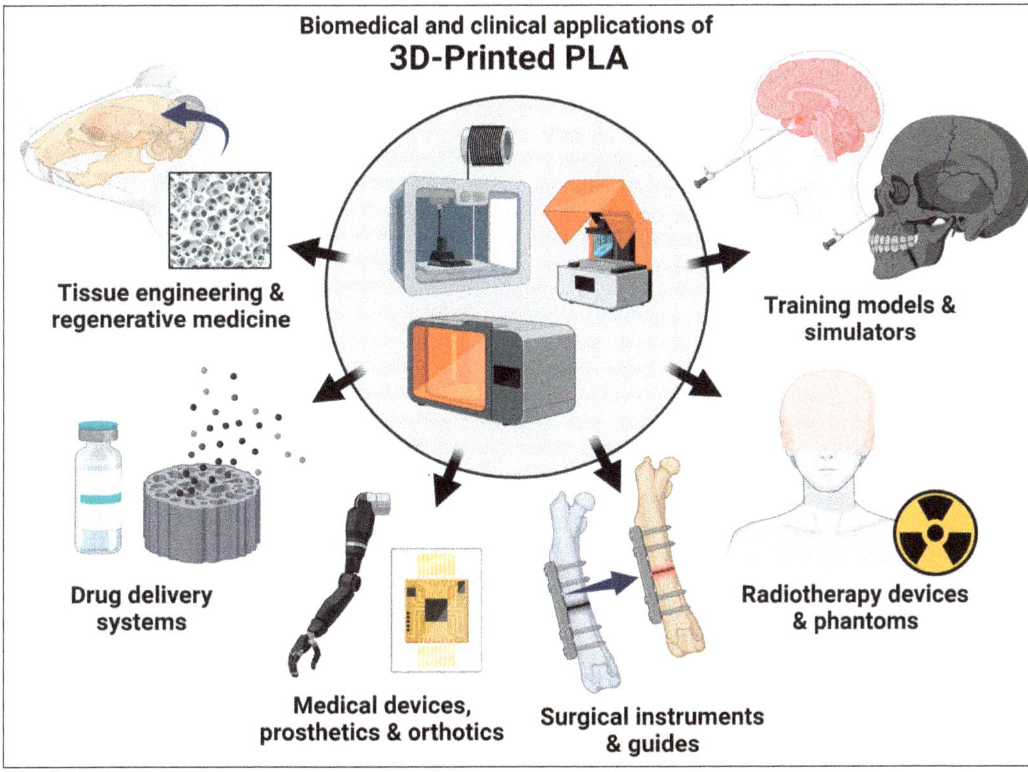

Figure 6. The schematic diagram related to 3D printing. Reprinted from ref. [100].

Studies on 3D printing have explored biomimicry, bending behavior, suture fiber enhancement of gelatin scaffold, osteoporosis, artificial blood vessels, multi-material polymer composites, synthetic biomolecules, and elastic hydrogel conditions [101–113].

Table 6 presents the classified articles on 3D printing technologies for surgical sutures, together with their bibliographic sources.

Wickramasinghe et al. fabricated specimens through fused deposition modeling (FDM) using polylactic acid (PLA). They studied 3D-printed structures inspired by exoskeleton sutures of diabolic ironclad beetles. The sutures had the shape of an ellipse that meshes with each other, and the size was controlled by varying the lengths of the small and large radii of the ellipse, while the ratio between the radii was maintained at 1:1.8 [101].

Wickramasinghe et al. designed and 3D-printed specimens of diabolic ironclad beetle using polylactic acid (PLA) and studied the flexural behavior of the suture structure. The S3 design can be helpful when high energy absorption is required, and the S1 design can be very effective when higher load support is required. The results of this study argued that the bio-inspired suture structure can be further optimized to improve the performance [102].

Nash et al. fabricated a single-wrap shear specimen of a bio-inspired suture interface through a layer-by-layer polymer injection lamination fabrication technique (Stratasys connex) for multi-material printing. As a result, they showed that by increasing the waveform of the suture layer, the effective shear stiffness of the junction can be significantly increased. For certain designs in this paper, wavy sutures can reach an effective shear stiffness 10–20 times greater than that of flat sutures [103].

Table 6. Classified articles on 3D printing technologies for surgical sutures and their bibliographic sources.

Authors	Title	Main Content	Bibliographic Source
Wickramasinghe, S et al.	Analysing fracture properties of bio-inspired 3D printed suture structures.	3D printing, bio-inspired structure fracture properties	Thin-Walled Structures, 2022. 176: p. 109317 [101].
Wickramasinghe, S., T et al.	Flexural behavior of 3D printed bio-inspired interlocking suture structures.	3D printing, bio-inspired, interlocking suture structures, flexural behavior	Mater. Sci. Addit. Manuf, 2022. 1(9) [102].
Nash, R.J et al.	Experimental and numerical analysis of 3D printed suture joints under shearing load.	3D printing, suture joints, shearing load	Engineering Fracture Mechanics, 2021. 253: p. 107912 [103].
Choi, D.J et al.	Suture fiber reinforcement of a 3D printed gelatin scaffold for its potential application in soft tissue engineering.	3D printing, soft tissue engineering, suture fiber, gelatin scaffold	International journal of molecular sciences, 2021. 22(21): p. 11600 [104].
Zhou, L et al.	A 3D printing mold method for rapid fabrication of artificial blood vessels.	3D printing, artificial blood vessels, polyacrylamide/*N,N′*-methylenebisacrylamide/ sodium alginate (PMSA) hydrogel tube	Colloids and Surfaces A: Physicochemical and Engineering Aspects, 2023. 662: p. 130952 [105].
Xu, X., et al. et al.	Multifunctional high-simulation 3D-printed hydrogel model manufacturing engineering for surgical training.	3D printing, hydrogel model, manufacturing engineering for surgical training	International Journal of Bioprinting, 2023. 9(5) [106].
Altuntas, U., D. Coker et al.	Creating tougher interfaces via suture morphology in 3D-printed multi-material polymer composites by fused filament fabrication.	3D printing, multi-material polymer composites, suture morphology, fused filament fabrication.	Additive Manufacturing, 2023. 61: p. 103359 [107].

Choi et al. improved the printability, mechanical strength, and dimensional stability of 3D-printed scaffolds by adding biodegradable suture fibers to gelatin biomaterial ink. The suture fibers increased the mechanical strength of the 3D-printed scaffold by up to sixfold, and the rate of deterioration could be controlled by the suture fiber content. In an in vitro cell study, DNA analysis revealed that the human skin fibroblast proliferation rate of 3D-printed skeletons containing 0.5% suture fibers was 10% higher than that of 3D-printed skeletons without suture fibers after 14 days of incubation [104].

Zhou et al. argued that damaged blood vessels should be replaced when cardiovascular disease is severe. They also developed a new and simple 3D mold fabrication technology that can be used to prepare 3D molds of various sizes. They prepared a double cross-linking (ionic cross-linking between Ca^{2+} and sodium alginate) and free radio polymerization of acrylamide (using *N,N′*-methylenebisacrylamide as a cross-linking agent), resulting in polyacrylamide/*N,N′*-methylenebisacrylamide/sodium alginate (PMSA) hydrogel tube using this method. The manufactured PMSA hydrogel tube has excellent mechanical properties and biocompatibility and can be adjusted to within 6 mm in diameter, so it can be used as a small-diameter vascular graft [105].

Xu et al. adjusted the composition of the hydrogel matrix and the immersion time of the ionic solution to accurately match the different biological soft tissues to complete a dual-network (DN) hydrogel that mimics tissue with custom stiffness. Combined with advanced three-dimensional (3D) printing fabrication technology, they perfected a variety of performance-adjustable biological hydrogel long-term models with structural complexity and fidelity, including kidney, liver, pancreas, and vascular tissue [106].

Altuntas et al. explained that the suture interface formed between the two different polymer phases of PLA (hard) and TPU (soft) was designed and fabricated by a molten filament fabrication technique. As a result, it was argued that lamination fabrication could provide a guideline for producing a multi-material polymer composite material with a stronger and more robust interface [107].

As mentioned above, many studies on mechanical strength and dimensional stability have been reported in 3D printing using lamination manufacturing technology and biomaterials. Three-dimensional printing technology is expected to have a positive effect on the development of human-friendly surgical suture technology due to the convenience of structural design.

4. Conclusions

Antibiotic properties, drug delivery, biodegradability, and wound regeneration, which are the main characteristics of surgical sutures, were examined in depth in this study. In addition, the final technologies for manufacturing surgical sutures were divided into electrospinning and 3D printing.

- Current status

Research on surgical sutures and antibacterial properties proposes methods such as modifying the yarn surface or coating with nanosilver particles, coating with Cur@ZIF-8, utilizing recycled silk–fibrin and TiO_2, and coating the chitosan–gelatin/tannic acid/polypyrrole composite. These methods have been reported to produce excellent antibacterial activity against super bacteria (staphylococcus aureus, *E. coli*, etc.).

Studies on surgical sutures and drug delivery explore the use of anti-inflammatory silk sutures, chitosan/keratin/PCL/PEG drug elution sutures, scalable silk–fibrin-based berberine loading system, and hybrid hydrogel actuators. Some of these methods have been reported to positively affect human recovery, such as drug elution, wound site reduction, cell migration promotion, high cell survival and wound healing rates, and continuous drug release.

Studies on the biodegradability of surgical sutures explore orthopedic applications, knot strength, smart materials, self healing, silver nanowire, albumin composites, mechanical properties, absorbable sutures, PLA-based biopolymers, poly(lactic acid)/poly (ethylene glycol) blends, etc. In addition, it is reported that biodegradable materials are more advantageous in protecting the human body and reducing pain than non-degradable materials.

Reactive oxygen species-scavenging sutures, hybrid suturing coating, and electrospun drug-eluting nanofibers are explored in the studies on wound regeneration and tissue regeneration using sutures. Using these methods, fibroblast attachment, migration, and proliferation are improved, the secretion of vascular growth factors is promoted, and wound healing is reported.

Electrospinning technology has been reported in many cases to change performance by changing the position of physiologically active substances using various substances or to show efficiency in imparting functions such as antibacterial-active substances and excellent biodegradability. Electrospinning technology is expected to exert an excellent influence on the development of human-friendly surgical suture technology due to its thinness and high specific surface area.

Research on 3D printing includes studies on biomimicry, bending behavior, artificial blood vessels, multi-material polymer composites, and elastic hydrogel conditions. As mentioned above, many studies on mechanical strength and dimensional stability using lamination manufacturing technology and biomaterials have been reported for 3D printing. Three-dimensional printing technology is expected to have a positive effect on the development of human-friendly surgical suture technology due to the convenience of structural design.

- The main obstacles to be addressed

There are many studies on antibiotic action, drug delivery, biodegradability, and wound regeneration of surgical sutures; however, side effects include toxicity.

In Korea, it is a violation of the Medical Device Management Act for a doctor to modify or modulate a general medical thread for the purpose of removing facial wrinkles. After the procedure, the skin wrinkle removal procedure usually involves swelling and pain, and since the skin wrinkle removal thread is inserted directly into the body, it is necessary to ensure its stability. If the skin wrinkle removal thread is arbitrarily manufactured using a general surgical thread, the protrusion spacing may not be uniform and may be partially broken, and the production process may be contaminated or deteriorated; thus, the possibility of side effects cannot be excluded. Therefore, the Korean court would convict the prosecution in this case, considering that the act of arbitrarily making protrusions in the general surgical thread and using it for skin wrinkle removal constitutes tampering with and modulation of medical devices. Due to these precedents, attention should be paid to the modification, modulation, and misuse of surgical sutures.

- Future perspectives

With the emergence of new viruses, research on antimicrobial properties, biodegradability for human suitability, and drug delivery systems for rapid healing are expected to be further promoted, along with mass production technology. In addition, the results of this study are considered to have a positive effect on the development of a multifunctional sur-chemical suture that promotes wound regeneration through antibacterial properties, drug elution, and biodegradability.

In this study, the main characteristics required for surgical sutures in the current society where new bacteria appear were considered from various angles in terms of antibacterial, biodegradable, drug delivery, and wound regeneration properties. In addition, electrospinning techniques that can produce nanofibers and 3D printing technologies that are output in a stacked form according to computer design were examined. In the present era of requiring multifunctional surgical sutures, the implications and practical significance of this study are expected to provide the foundation for the manufacture of hybrid surgical sutures. In addition, future research is expected to provide useful information for the surgical sutures industry if the functional materials mentioned in this study are fused to apply appropriate surgical sutures according to the wound site and type, age, race, and gender.

Funding: This research received no external funding.

Data Availability Statement: Not applicable.

Conflicts of Interest: The author declares no conflicts of interest.

References

1. Pajnik, J.; Milovanovic, S.; Stojanovic, D.; Dimitrijevic-Brankovic, S.; Jankovic-Častvan, I.; Uskokovic, P. Utilization of supercritical carbon dioxide for development of antibacterial surgical sutures. *J. Supercrit. Fluids* **2022**, *181*, 105490. [CrossRef]
2. James, B.; Ramakrishnan, R.; Aprem, A.S. Development of environmentally safe biodegradable, antibacterial surgical sutures using nanosilver particles. *J. Polym. Environ.* **2021**, *29*, 2282–2288. [CrossRef]
3. Ornaghi, H.L.; Monticeli, F.M.; Agnol, L.D. A Review on Polymers for Biomedical Applications on Hard and Soft Tissues and Prosthetic Limbs. *Polymers* **2023**, *15*, 4034. [CrossRef]
4. Li, Y.; Kumar, K.N.; Dabkowski, J.M.; Corrigan, M.; Scott, R.W.; Nüsslein, K.; Tew, G.N. New Bactericidal Surgical Suture Coating. *Langmuir* **2012**, *28*, 12134–12139. [CrossRef]
5. Faris, A.; Khalid, L.; Hashim, M.; Yaghi, S.; Magde, T.; Bouresly, W.; Hamdoon, Z.; Uthman, A.T.; Marei, H.; Al-Rawi, N. Characteristics of suture materials used in oral surgery: Systematic review. *Int. Dent. J.* **2022**, *72*, 278–287. [CrossRef] [PubMed]
6. Wang, J.; Liu, Z.; Qiu, H.; Wang, C.; Dong, X.; Du, J.; Li, X.; Yang, X.; Fang, H.; Ding, Y. A robust bio-based polyurethane employed as surgical suture with help to promote skin wound healing. *Biomater. Adv.* **2025**, *166*, 214048. [CrossRef] [PubMed]
7. Chung, D.; Bakmiwewa, S.; Suthananthan, A.; Idrees, M. Suture material in pancreticojejunal anastomosis: A systematic review. *ANZ J. Surg.* **2025**, *95*, 321–327. [PubMed]
8. Paladini, F.; Panico, A.; Masi, A.; Russo, F.; Sannino, A.; Pollini, M. Silver-Treated Sutures for the Prevention of Biofilm-Associated Surgical Site Infections. *Antibiotics* **2025**, *14*, 49. [CrossRef]
9. Rineksa, G.; Whulanza, Y.; Gozan, M. Mechanical characteristics of thermoplastic sago starch-based biopolymer composite reinforced with microcrystalline cellulose (MCC) as a potential surgical suture material. In *Proceedings of the AIP Conference Proceedings*; AIP Publishing LLC: Melville, NY, USA, 2025.
10. Das, S.; Ghosh, D.; Khatib, M.N.; Balaraman, A.K.; Roopashree, R.; Kaur, M.; Srivastava, M.; Barwal, A.; Prasad, G.S.; Rajput, P. Bioelectric surgical sutures: Advancing wound healing through mechano-electrical stimulation. *Int. J. Surg. Open* **2025**, *63*, 28–34.
11. Makrygiannis, I.H.; Nikolaidis, A.K.; Tilaveridis, I.; Kouvelas, A.D.; Lykakis, I.N.; Venetis, G. Coated sutures for use in oral surgery: A comprehensive review. *Clin. Oral Investig.* **2025**, *29*, 109. [CrossRef]
12. Zhou, Y.; Liu, X.; Yang, M.; Song, G.; Wang, Y.; Sun, H.; Yuan, T.; Rao, J.; Lü, B.; Yao, C. Bio-inspired helical-hollow bacterial cellulose fiber for suture materials. *Chem. Eng. J.* **2025**, *505*, 159670. [CrossRef]
13. Abdulagatov, I.M.; Khanaliev, V.Y.; Ragimov, R.M.; Maksumova, A.M.; Khamidov, M.A.; Abdullaeva, N.M.; Mollaeva, N.R. Atomic-layer-deposition application for antibacterial coating of biomedical materials: Surgical sutures. *Biomed. Mater.* **2025**, *20*, 025012. [CrossRef] [PubMed]
14. Bibire, T.; Yilmaz, O.; Ghiciuc, C.M.; Bibire, N.; Dănilă, R. Biopolymers for surgical applications. *Coatings* **2022**, *12*, 211. [CrossRef]
15. Zhang, Q.; Qiao, Y.; Zhu, J.; Li, Y.; Li, C.; Lin, J.; Li, X.; Han, H.; Mao, J.; Wang, F. Electroactive and antibacterial surgical sutures based on chitosan-gelatin/tannic acid/polypyrrole composite coating. *Compos. Part B Eng.* **2021**, *223*, 109140. [CrossRef]
16. Lou, C.-W.; Hung, C.-Y.; Wei, M.; Li, T.; Shiu, B.-C.; Lin, J.-H. Antibacterial Surgical Sutures Developed Using Electrostatic Yarn Wrapping Technology. *J. Funct. Biomater.* **2023**, *14*, 248. [CrossRef] [PubMed]
17. Zhang, Q.; Zou, Y.; Tang, L.; Liu, X.; Hu, M.; Han, H.; Li, Y.; Wang, F.; Wang, L.; Mao, J. Stage-controlled antibacterial surgical sutures based on curcumin@ ZIF-8 functional coating for improved wound healing. *Prog. Org. Coat.* **2023**, *184*, 107829. [CrossRef]
18. Zhang, X.; Yang, Z.; Yang, X.; Zhang, F.; Pan, Z. Sustainable Antibacterial Surgical Suture Based on Recycled Silk Resource by an Internal Combination of Inorganic Nanomaterials. *ACS Appl. Mater. Interfaces* **2023**, *15*, 29971–29981. [CrossRef]
19. Wang, X.; Liu, P.; Wu, Q.; Zheng, Z.; Xie, M.; Chen, G.; Yu, J.; Wang, X.; Li, G.; Kaplan, D. Sustainable antibacterial and anti-inflammatory silk suture with surface modification of combined-therapy drugs for surgical site infection. *ACS Appl. Mater. Interfaces* **2022**, *14*, 11177–11191. [CrossRef]
20. Altun, E.; Bayram, C.; Gultekinoglu, M.; Matharu, R.; Delbusso, A.; Homer-Vanniasinkam, S.; Edirisinghe, M. Pressure-Spun Fibrous Surgical Sutures for Localized Antibacterial Delivery: Development, Characterization, and In Vitro Evaluation. *ACS Appl. Mater. Interfaces* **2023**, *15*, 45561–45573. [CrossRef] [PubMed]
21. Syukri, D.M.; Nwabor, O.F.; Singh, S.; Voravuthikunchai, S.P. Antibacterial functionalization of nylon monofilament surgical sutures through in situ deposition of biogenic silver nanoparticles. *Surf. Coat. Technol.* **2021**, *413*, 127090. [CrossRef]

22. Rakhmatullayeva, D.; Ospanova, A.; Bekissanova, Z.; Jumagaziyeva, A.; Savdenbekova, B.; Seidulayeva, A.; Sailau, A. Development and characterization of antibacterial coatings on surgical sutures based on sodium carboxymethyl cellulose/chitosan/chlorhexidine. *Int. J. Biol. Macromol.* **2023**, *236*, 124024.
23. La Rosa, G.R.M.; Scapellato, S.; Cicciù, M.; Pedullà, E. Antimicrobial Activity of Antibacterial Sutures in Oral Surgery: A Scoping Review. *Int. Dent. J.* **2024**, *74*, 688–695. [PubMed]
24. Wu, Q.; He, C.; Wang, X.; Zhang, S.; Zhang, L.; Xie, R.; Li, Y.; Wang, X.; Han, Z.; Zheng, Z. Sustainable antibacterial surgical suture using a facile scalable silk-fibroin-based berberine loading system. *ACS Biomater. Sci. Eng.* **2021**, *7*, 2845–2857.
25. Li, H.; Wang, Z.; Robledo-Lara, J.A.; He, J.; Huang, Y.; Cheng, F. Antimicrobial surgical sutures: Fabrication and application of infection prevention and wound healing. *Fibers Polym.* **2021**, *22*, 2355–2367. [CrossRef]
26. Al-Sawarees, D.K.; Darwish, R.M.; Abu-Zurayk, R.; Masri, M.A. Assessing silver nanoparticle and antimicrobial combinations for antibacterial activity and biofilm prevention on surgical sutures. *J. Appl. Microbiol.* **2024**, *135*, lxae063. [CrossRef]
27. Jamshaid, H.; Mishra, R.; Hussain, U.; Rajput, A.W.; Tichy, M.; Muller, M. Natural Fiber Based Antibacterial, Wound Healing Surgical Sutures by the Application of Herbal Antimicrobial Compounds. *J. Nat. Fibers* **2022**, *19*, 9531–9546.
28. Chen, Y.-G.; Li, C.-X.; Zhang, Y.; Qi, Y.-D.; Liu, X.-H.; Feng, J.; Zhang, X.-Z. Hybrid suture coating for dual-staged control over antibacterial actions to match well wound healing progression. *Mater. Horiz.* **2022**, *9*, 2824–2834. [CrossRef]
29. Chua, R.; Lim, S.K.; Chee, C.F.; Chin, S.P.; Kiew, L.V.; Sim, K.S.; Tay, S.T. Surgical site infection and development of antimicrobial sutures: A review. *Eur. Rev. Med. Pharmacol. Sci.* **2022**, *26*, 828–845. [PubMed]
30. Matz, D.; Engelhardt, S.; Wiencierz, A.; Soysal, S.D.; Misteli, H.; Kirchhoff, P.; Heizmann, O. Do Antibacterial Skin Sutures Reduce Surgical Site Infections After Elective Open Abdominal Surgery?—A Prospective, Randomized Controlled Single-Center Trial. *J. Clin. Med.* **2024**, *13*, 6803. [CrossRef]
31. Zhang, J.; Li, X.; Cheng, M.; Wan, K.; Yan, S.; Peng, W.; Duan, G.; Wu, Y.; Wen, L. MoO3-X nanodots coated suture for combating surgical site infection via antibacterial and anti-inflammatory properties. *Nanomed. Nanotechnol. Biol. Med.* **2024**, *60*, 102757.
32. Liu, M.; Zhang, Y.; Liu, K.; Zhang, G.; Mao, Y.; Chen, L.; Peng, Y.; Tao, T.H. Biomimicking antibacterial opto-electro sensing sutures made of regenerated silk proteins. *Adv. Mater.* **2021**, *33*, 2004733.
33. Baygar, T.; Ugur, A.; Karaca, I.R.; Kilinc, Y.; Gultekin, S.E.; Sarac, N. Fabrication of a Biocompatible Nanoantimicrobial Suture for Rapid Wound Healing after Surgery. *ACS Omega* **2024**, *9*, 22573–22580. [PubMed]
34. Zlobina, O.; Bugaeva, I.; Glukhova, I.; Glukhova, A. Pichkhidze SYa. Experimental modification and investigation of antibacterial surgical suture material. Главный редактор **2023**, *1*, 51–56.
35. Bhouri, N.; Debbabi, F.; Lassoued, M.A.; Abderrahmen, M.; Abdessalem, S.B. Wound infections preventing using antibacterial chitosan/Laurus nobilis essential oil emulsion on PET braided surgical sutures. *Colloids Surf. A Physicochem. Eng. Asp.* **2022**, *643*, 128678. [CrossRef]
36. Schmitz, N.-D.; Ovington, L.; Berlin, J.; Zhang, S.; Collier, J. Optimal usage of antibacterial sutures for wound closure in clinical trials addressing SSI. *Lancet* **2023**, *401*, 1497–1498. [CrossRef] [PubMed]
37. Mathew, S.; Kumar, K.V.; Prabhu, A.; Shastry, R.P.; Rajesh, K. Braided silk sutures coated with photoreduced silver nanoparticles for eradicating Staphylococcus aureus and Streptococcus mutans infections. *J. Microbiol. Methods* **2024**, *220*, 106923. [CrossRef]
38. Basov, A.; Dzhimak, S.; Sokolov, M.; Malyshko, V.; Moiseev, A.; Butina, E.; Elkina, A.; Baryshev, M. Changes in number and antibacterial activity of silver nanoparticles on the surface of suture materials during cyclic freezing. *Nanomaterials* **2022**, *12*, 1164. [CrossRef]
39. Schäfer, S.; Aavani, F.; Köpf, M.; Drinic, A.; Stürmer, E.K.; Fuest, S.; Grust, A.L.C.; Gosau, M.; Smeets, R. Silk proteins in reconstructive surgery: Do they possess an inherent antibacterial activity? A systematic review. *Wound Repair Regen.* **2023**, *31*, 99–110. [CrossRef]
40. Edis, Z.; Bloukh, S.H. Thymol, a Monoterpenoid within Polymeric Iodophor Formulations and Their Antimicrobial Activities. *Int. J. Mol. Sci.* **2024**, *25*, 4949. [CrossRef]
41. Chen, Y.-G.; Li, C.-X.; Zhang, Y.; Qi, Y.-D.; Feng, J.; Zhang, X.-Z. Antibacterial sutures coated with smooth chitosan layer by gradient deposition. *Chin. J. Polym. Sci.* **2022**, *40*, 1050–1061. [CrossRef]
42. Kandathil, A.M.; Aslam, S.A.; Abidha, R.; Cherian, M.P.; Sudarsanan, M. Evaluation of Microbial Adherence on Antibacterial Suture Materials during Intraoral Wound Healing: A Prospective Comparative Study. *J. Contemp. Dent. Pract.* **2023**, *24*, 515–520. [PubMed]
43. Ranjbar-Mohammadi, M.; Sa'di, V.; Moezzi, M.; Saghafi, R. Fabrication and characterization of antibacterial suture yarns containing PLA/tetracycline hydrochloride-PVA/chitosan nanofibers. *Fibers Polym.* **2022**, *23*, 1538–1547. [CrossRef]
44. Deng, X.; Qasim, M.; Ali, A. Engineering and polymeric composition of drug-eluting suture: A review. *J. Biomed. Mater. Res. Part A* **2021**, *109*, 2065–2081. [CrossRef]

45. Lee, Y.; Kim, H.; Kim, Y.; Noh, S.; Chun, B.; Kim, J.; Park, C.; Choi, M.; Park, K.; Lee, J. A multifunctional electronic suture for continuous strain monitoring and on-demand drug release. *Nanoscale* **2021**, *13*, 18112–18124. [CrossRef] [PubMed]
46. Deng, X.; Gould, M.; Ali, M.A. Fabrication and characterisation of melt-extruded chitosan/keratin/PCL/PEG drug-eluting sutures designed for wound healing. *Mater. Sci. Eng. C* **2021**, *120*, 111696. [CrossRef]
47. Mendez, K.; Whyte, W.; Freedman, B.R.; Fan, Y.; Varela, C.E.; Singh, M.; Cintron-Cruz, J.C.; Rothenbücher, S.E.; Li, J.; Mooney, D.J. Mechanoresponsive Drug Release from a Flexible, Tissue-Adherent, Hybrid Hydrogel Actuator. *Adv. Mater.* **2024**, *36*, 2303301. [CrossRef]
48. Parikh, K.S.; Omiadze, R.; Josyula, A.; Shi, R.; Anders, N.M.; He, P.; Yazdi, Y.; McDonnell, P.J.; Ensign, L.M.; Hanes, J. Ultra-thin, high strength, antibiotic-eluting sutures for prevention of ophthalmic infection. *Bioeng. Transl. Med.* **2021**, *6*, e10204. [CrossRef]
49. Manoukian, O.S.; Rudraiah, S.; Arul, M.R.; Bartley, J.M.; Baker, J.T.; Yu, X.; Kumbar, S.G. Biopolymer-nanotube nerve guidance conduit drug delivery for peripheral nerve regeneration: In vivo structural and functional assessment. *Bioact. Mater.* **2021**, *6*, 2881–2893. [CrossRef] [PubMed]
50. Di, X.; Liang, X.; Shen, C.; Pei, Y.; Wu, B.; He, Z. Carbohydrates used in polymeric systems for drug delivery: From structures to applications. *Pharmaceutics* **2022**, *14*, 739. [CrossRef] [PubMed]
51. Anup, N.; Chavan, T.; Chavan, S.; Polaka, S.; Kalyane, D.; Abed, S.N.; Venugopala, K.N.; Kalia, K.; Tekade, R.K. Reinforced electrospun nanofiber composites for drug delivery applications. *J. Biomed. Mater. Res. Part A* **2021**, *109*, 2036–2064. [CrossRef]
52. Cazorla-Luna, R.; Martín-Illana, A.; Notario-Pérez, F.; Ruiz-Caro, R.; Veiga, M.-D. Naturally occurring polyelectrolytes and their use for the development of complex-based mucoadhesive drug delivery systems: An overview. *Polymers* **2021**, *13*, 2241. [CrossRef]
53. Mu, J.; Luo, D.; Li, W.; Ding, Y. Multiscale polymeric fibers for drug delivery and tissue engineering. *Biomed. Technol.* **2024**, *5*, 60–72. [CrossRef]
54. Arun, Y.; Ghosh, R.; Domb, A.J. Biodegradable hydrophobic injectable polymers for drug delivery and regenerative medicine. *Adv. Funct. Mater.* **2021**, *31*, 2010284.
55. Osi, B.; Khoder, M.; Al-Kinani, A.A.; Alany, R.G. Pharmaceutical, biomedical and ophthalmic applications of biodegradable polymers (BDPs): Literature and patent review. *Pharm. Dev. Technol.* **2022**, *27*, 341–356.
56. Lee, J.; Jang, E.H.; Kim, J.H.; Park, S.; Kang, Y.; Park, S.; Lee, K.; Kim, J.-H.; Youn, Y.-N.; Ryu, W. Highly flexible and porous silk fibroin microneedle wraps for perivascular drug delivery. *J. Control. Release* **2021**, *340*, 125–135. [PubMed]
57. Pires, P.C.; Mascarenhas-Melo, F.; Pedrosa, K.; Lopes, D.; Lopes, J.; Macário-Soares, A.; Peixoto, D.; Giram, P.S.; Veiga, F.; Paiva-Santos, A.C. Polymer-based biomaterials for pharmaceutical and biomedical applications: A focus on topical drug administration. *Eur. Polym. J.* **2023**, *187*, 111868.
58. Rostamitabar, M.; Abdelgawad, A.M.; Jockenhoevel, S.; Ghazanfari, S. Drug-eluting medical textiles: From fiber production and textile fabrication to drug loading and delivery. *Macromol. Biosci.* **2021**, *21*, 2100021. [CrossRef]
59. Wu, S.; Dong, T.; Li, Y.; Sun, M.; Qi, Y.; Liu, J.; Kuss, M.A.; Chen, S.; Duan, B. State-of-the-art review of advanced electrospun nanofiber yarn-based textiles for biomedical applications. *Appl. Mater. Today* **2022**, *27*, 101473. [PubMed]
60. Naser, M.A.; Sayed, A.M.; Abdelmoez, W.; El-Wakad, M.T.; Abdo, M.S. Biodegradable suture development-based albumin composites for tissue engineering applications. *Sci. Rep.* **2024**, *14*, 7912.
61. Antoniac, I.; Antoniac, A.; Gheorghita, D.; Gradinaru, S. In vitro study on biodegradation of absorbable suture Materials used for surgical applications. *Mater. Plast.* **2021**, *58*, 130–139. [CrossRef]
62. Nouri, A.; Shirvan, A.R.; Li, Y.; Wen, C. Biodegradable metallic suture anchors: A review. *Smart Mater. Manuf.* **2023**, *1*, 100005.
63. Szabelski, J.; Karpiński, R. Short-Term Hydrolytic Degradation of Mechanical Properties of Absorbable Surgical Sutures: A Comparative Study. *J. Funct. Biomater.* **2024**, *15*, 273. [CrossRef] [PubMed]
64. Wu, H.; Guo, T.; Zhou, F.; Bu, J.; Yang, S.; Dai, Z.; Teng, C.; Ouyang, H.; Wei, W. Surface coating prolongs the degradation and maintains the mechanical strength of surgical suture in vivo. *Colloids Surf. B Biointerfaces* **2022**, *209*, 112214.
65. Li, Y.; Meng, Q.; Chen, S.; Ling, P.; Kuss, M.A.; Duan, B.; Wu, S. Advances, challenges, and prospects for surgical suture materials. *Acta Biomater.* **2023**, *168*, 78–112.
66. Alhulaybi, Z.A. Fabrication and Characterization of Poly (lactic acid)-Based Biopolymer for Surgical Sutures. *ChemEngineering* **2023**, *7*, 98. [CrossRef]
67. Mukherjee, C.; Varghese, D.; Krishna, J.; Boominathan, T.; Rakeshkumar, R.; Dineshkumar, S.; Rao, C.B.; Sivaramakrishna, A. Recent advances in biodegradable polymers–properties, applications and future prospects. *Eur. Polym. J.* **2023**, *192*, 112068.
68. Al-Shalawi, F.D.; Hanim, M.A.; Ariffin, M.; Kim, C.L.S.; Brabazon, D.; Calin, R.; Al-Osaimi, M.O. Biodegradable synthetic polymer in orthopaedic application: A review. *Mater. Today Proc.* **2023**, *74*, 540–546. [CrossRef]
69. Sun, S.; Chen, C.; Zhang, J.; Hu, J. Biodegradable smart materials with self-healing and shape memory function for wound healing. *RSC Adv.* **2023**, *13*, 3155–3163. [CrossRef] [PubMed]

70. Zarinkolah, Z.; Hosseinkhani, S.; Nikkhah, M. Investigation on the mechanical, thermal, bio-degradation, and bio-compatibility properties of poly (lactic acid)/poly (ethylene glycol) blend. *IIUM Eng. J.* **2021**, *22*, 223–233.
71. Liu, W.-C.; Chang, C.-H.; Chen, C.-H.; Lu, C.-K.; Ma, C.-H.; Huang, S.-I.; Fan, W.-L.; Shen, H.-H.; Tsai, P.-I.; Yang, K.-Y. 3D-Printed double-helical biodegradable iron suture anchor: A rabbit rotator cuff tear model. *Materials* **2022**, *15*, 2801.
72. Chen, Y.; Sun, Y.; Wu, X.; Lou, J.; Zhang, X.; Peng, Z. Rotator cuff repair with biodegradable high-purity magnesium suture anchor in sheep model. *J. Orthop. Transl.* **2022**, *35*, 62–71.
73. Yang, N.; Venezuela, J.; Zhang, J.; Wang, A.; Almathami, S.; Dargusch, M. Evolution of degradation mechanism and fixation strength of biodegradable Zn–Cu wire as sternum closure suture: An in vitro study. *J. Mech. Behav. Biomed. Mater.* **2023**, *138*, 105658. [CrossRef]
74. Xu, L.; Liu, Y.; Zhou, W.; Yu, D. Electrospun Medical Sutures for Wound Healing: A Review. *Polymers* **2022**, *14*, 1637. [CrossRef] [PubMed]
75. Akombaetwa, N.; Bwanga, A.; Makoni, P.A.; Witika, B.A. Applications of electrospun drug-eluting nanofibers in wound healing: Current and future perspectives. *Polymers* **2022**, *14*, 2931. [CrossRef] [PubMed]
76. Han, H.; Tang, L.; Li, Y.; Li, Y.; Bi, M.; Wang, J.; Wang, F.; Wang, L.; Mao, J. A multifunctional surgical suture with electroactivity assisted by oligochitosan/gelatin-tannic acid for promoting skin wound healing and controlling scar proliferation. *Carbohydr. Polym.* **2023**, *320*, 121236. [CrossRef]
77. Zhu, J.; Jin, Q.; Zhao, H.; Zhu, W.; Liu, Z.; Chen, Q. Reactive oxygen species scavenging sutures for enhanced wound sealing and repair. *Small Struct.* **2021**, *2*, 2100002. [CrossRef]
78. Öksüz, K.E.; Kurt, B.m.; Şahin İnan, Z.D.; Hepokur, C. Novel bioactive glass/graphene oxide-coated surgical sutures for soft tissue regeneration. *ACS Omega* **2023**, *8*, 21628–21641. [CrossRef]
79. Kim, K.; Kim, H.; Do, S.; Kim, H. Potential of Aligned Electrospun PLGA/SIS Blended Nanofibrous Membrane for Tendon Tissue Engineering. *Polymers* **2023**, *15*, 2313. [CrossRef] [PubMed]
80. Teno, J.; Pardo-Figuerez, M.; Evtoski, Z.; Prieto, C.; Cabedo, L.; Lagaron, J.M. Development of Ciprofloxacin-Loaded Electrospun Yarns of Application Interest as Antimicrobial Surgical Suture Materials. *Pharmaceutics* **2024**, *16*, 220. [CrossRef] [PubMed]
81. Madheswaran, D.; Sivan, M.; Valtera, J.; Kostakova, E.K.; Egghe, T.; Asadian, M.; Novotny, V.; Nguyen, N.H.; Sevcu, A.; Morent, R. Composite yarns with antibacterial nanofibrous sheaths produced by collectorless alternating-current electrospinning for suture applications. *J. Appl. Polym. Sci.* **2022**, *139*, 51851. [CrossRef]
82. Rivera, P.; Villegas, C.; Cabezas, R.; Pérez, B.; Torres, A.; de Dicastillo, C.L.; Garrido, L.; Galvez, P.; Araya, C.; Romero, J. Development of PLA suture materials by extrusion, electrospinning and supercritical CO_2 impregnation of ibuprofen and naproxen. *J. Supercrit. Fluids* **2023**, *194*, 105854. [CrossRef]
83. Mohamadinooripoor, R.; Kashanian, S.; Omidfar, K. Poly-ε-Caprolactone/Propolis Electrospun Yarns as Suture. *Fibers Polym.* **2023**, *24*, 2641–2651. [CrossRef]
84. Li, Y.; Luo, H.; Li, Y.; Huang, P.; Xu, J.; Zhang, J.; Cai, P.; He, H.; Wu, J.; Li, X. Surface biofunctional bFGF-loaded electrospun suture accelerates incisional wound healing. *Mater. Des.* **2023**, *225*, 111451. [CrossRef]
85. Yang, Z.; Liu, S.; Li, J.; Wu, G.; Zhang, M.; Li, F.; Jia, L.; Zhang, Y.; Li, H.; Liu, X. Study on Preparation of Core-Spun Yarn Surgical Sutures by Compositing Drug-Loaded Nanofiber Membrane with PLA and Its Controllable Drug Release Performance. *Fibers Polym.* **2023**, *24*, 4181–4193. [CrossRef]
86. Madheswaran, D.; Sivan, M.; Hauzerova, S.; Kostakova, E.K.; Jencova, V.; Valtera, J.; Behalek, L.; Mullerova, J.; Nguyen, N.H.; Capek, L. Continuous fabrication of braided composite nanofibrous surgical yarns using advanced AC electrospinning and braiding technology. *Compos. Commun.* **2024**, *48*, 101932. [CrossRef]
87. Abhari, R.E.; Snelling, S.J.; Augustynak, E.; Davis, S.; Fischer, R.; Carr, A.J.; Mouthuy, P.-A. A hybrid electrospun-extruded polydioxanone suture for tendon tissue regeneration. *Tissue Eng. Part A* **2024**, *30*, 214–224. [CrossRef]
88. Jin, D.; Yang, S.; Wu, S.; Yin, M.; Kuang, H. A functional PVA aerogel-based membrane obtaining sutureability through modified electrospinning technology and achieving promising anti-adhesion effect after cardiac surgery. *Bioact. Mater.* **2022**, *10*, 355–366. [CrossRef]
89. Singh, P.; Pandey, P.; Arya, D.K.; Anjum, M.M.; Poonguzhali, S.; Kumar, A.; Gupta, R.; Rajamanickam, V.M.; Singh, S.; Chaurasia, S. Biomimicking dual drug eluting twisted electrospun nanofiber yarns for post-operative wound healing. *Biomed. Mater.* **2023**, *18*, 035006. [CrossRef]
90. Dang, J.; Huang, S.; Li, S.; Liu, J.; Chen, Z.; Wang, L.; Wang, J.; Chen, H.; Xu, S. Effects of the Biomimetic Microstructure in Electrospun Fiber Sutures and Mechanical Tension on Tissue Repair. *ACS Appl. Mater. Interfaces* **2024**, *16*, 29087–29097. [PubMed]
91. Pisani, S.; Croce, S.; Mauramati, S.; Marmonti, M.; Cobianchi, L.; Herman, I.; Dorati, R.; Avanzini, M.A.; Genta, I.; Benazzo, M.; et al. Engineered Full Thickness Electrospun Scaffold for Esophageal Tissue Regeneration: From In Vitro to In Vivo Approach. *Pharmaceutics* **2022**, *14*, 252. [CrossRef] [PubMed]

92. Sharma, D.; Dhingra, S.; Banerjee, A.; Saha, S.; Bhattacharyya, J.; Satapathy, B.K. Designing suture-proof cell-attachable copolymer-mediated and curcumin-β-cyclodextrin inclusion complex loaded aliphatic polyester-based electrospun antibacterial constructs. *Int. J. Biol. Macromol.* **2022**, *216*, 397–413.
93. Ye, Y.; Zhou, Y.; Jing, Z.; Xu, Y.; Yin, D. Electrospun heparin-loaded nano-fiber sutures for the amelioration of achilles tendon rupture regeneration: In vivo evaluation. *J. Mater. Chem. B* **2021**, *9*, 4154–4168. [CrossRef] [PubMed]
94. Sharma, D.; Satapathy, B.K. Understanding release kinetics and collapse proof suture retention response of curcumin loaded electrospun mats based on aliphatic polyesters and their blends. *J. Mech. Behav. Biomed. Mater.* **2021**, *120*, 104556.
95. Nezhentsev, A.; Abhari, R.E.; Baldwin, M.J.; Mimpen, J.Y.; Augustyniak, E.; Isaacs, M.; Mouthuy, P.A.; Carr, A.J.; Snelling, S.J. In vitro evaluation of the response of human tendon-derived stromal cells to a novel electrospun suture for tendon repair. *Transl. Sports Med.* **2021**, *4*, 409–418. [PubMed]
96. Li, Y.; Wang, J.; Wang, Y.; Cui, W. Advanced electrospun hydrogel fibers for wound healing. *Compos. Part B Eng.* **2021**, *223*, 109101. [CrossRef]
97. Reddy, B.; In, K.H.; Panigrahi, B.B.; Paturi, U.M.R.; Cho, K.; Reddy, N. Modeling tensile strength and suture retention of polycaprolactone electrospun nanofibrous scaffolds by artificial neural networks. *Mater. Today Commun.* **2021**, *26*, 102115.
98. Li, A.; Wang, L.; Qin, X. Manufacturing and Application of Electrospinning Nanofiber Yarn. *Electrospinning Fundam. Methods Appl.* **2024**, 45–69. [CrossRef]
99. Zhang, H.; Lan, D.; Wu, B.; Chen, X.; Li, X.; Li, Z.; Dai, F. Electrospun piezoelectric scaffold with external mechanical stimulation for promoting regeneration of peripheral nerve injury. *Biomacromolecules* **2023**, *24*, 3268–3282. [CrossRef] [PubMed]
100. Barcena, A.J.R.; Ravi, P.; Kundu, S.; Tappa, K. Emerging Biomedical and Clinical Applications of 3D-Printed Poly(Lactic Acid)-Based Devices and Delivery Systems. *Bioengineering* **2024**, *11*, 705. [CrossRef]
101. Wickramasinghe, S.; Peng, C.; Ladani, R.; Tran, P. Analysing fracture properties of bio-inspired 3D printed suture structures. *Thin-Walled Struct.* **2022**, *176*, 109317. [CrossRef]
102. Wickramasinghe, S.; Do, T.; Tran, P. Flexural behavior of 3D printed bio-inspired interlocking suture structures. *Mater. Sci. Addit. Manuf.* **2022**, *1*, 9. [CrossRef]
103. Nash, R.J.; Li, Y. Experimental and numerical analysis of 3D printed suture joints under shearing load. *Eng. Fract. Mech.* **2021**, *253*, 107912. [CrossRef]
104. Choi, D.J.; Choi, K.; Park, S.J.; Kim, Y.-J.; Chung, S.; Kim, C.-H. Suture fiber reinforcement of a 3D printed gelatin scaffold for its potential application in soft tissue engineering. *Int. J. Mol. Sci.* **2021**, *22*, 11600. [CrossRef] [PubMed]
105. Zhou, L.; Li, Y.; Tu, Q.; Wang, J. A 3D printing mold method for rapid fabrication of artificial blood vessels. *Colloids Surf. A Physicochem. Eng. Asp.* **2023**, *662*, 130952. [CrossRef]
106. Xu, X.; Yu, S.; Ma, L.; Mao, J.; Chen, H.; Zhu, Z.; Wang, L.; Lin, H.; Zhang, J.; Wang, Z. Multifunctional high-simulation 3D-printed hydrogel model manufacturing engineering for surgical training. *Int. J. Bioprinting* **2023**, *9*, 355–368. [CrossRef]
107. Altuntas, U.; Coker, D.; Yavas, D. Creating tougher interfaces via suture morphology in 3D-printed multi-material polymer composites by fused filament fabrication. *Addit. Manuf.* **2023**, *61*, 103359. [CrossRef]
108. Ye, T.; Chai, M.; Wang, Z.; Shao, T.; Liu, J.; Shi, X. 3D-Printed hydrogels with engineered nanocrystalline domains as functional vascular constructs. *ACS Nano* **2024**, *18*, 25765–25777. [CrossRef]
109. Ghazi, A. A call for change. Can 3D printing replace cadavers for surgical training? *Urol. Clin.* **2022**, *49*, 39–56. [CrossRef]
110. RG, A.P.; Bajaj, G.; John, A.E.; Chandran, S.; Kumar, V.V.; Ramakrishna, S. A review on the recent applications of synthetic biopolymers in 3D printing for biomedical applications. *J. Mater. Sci. Mater. Med.* **2023**, *34*, 62.
111. Marconi, S.; Mauri, V.; Negrello, E.; Pugliese, L.; Pietrabissa, A.; Auricchio, F. Quantitative Assessment of 3D printed blood vessels produced with J750™ digital anatomy™ for suture simulation. *bioRxiv* **2022**. [CrossRef]
112. Wu, W.; Dong, Y.; Liu, H.; Jiang, X.; Yang, L.; Luo, J.; Hu, Y.; Gou, M. 3D printed elastic hydrogel conduits with 7, 8-dihydroxyflavone release for peripheral nerve repair. *Mater. Today Bio* **2023**, *20*, 100652. [CrossRef] [PubMed]
113. Lalième, L.; Adam, J.; Nataf, P.; Khonsari, R.H. 3D-printed suture guide for thoracic and cardiovascular surgery produced during the COVID19 pandemic. *Ann. 3d Print. Med.* **2021**, *1*, 100005. [CrossRef] [PubMed]

pharmaceutics

MDPI

Review

Lipid Biomimetic Models as Simple Yet Complex Tools to Predict Skin Permeation and Drug–Membrane Biophysical Interactions

Eduarda Fernandes [1,*], Carla M. Lopes [2,3,4] and Marlene Lúcio [1,5,*]

1 CF-UM-UP—Centro de Física das Universidades do Minho e Porto, Departamento de Física, Universidade do Minho, 4710-057 Braga, Portugal
2 FFP-I3ID—Instituto de Investigação, Inovação e Desenvolvimento, FP-BHS—Biomedical and Health Sciences Research Unit, Faculdade de Ciências da Saúde, Universidade Fernando Pessoa, 4200-150 Porto, Portugal; cmlopes@ufp.edu.pt
3 UCIBIO—Applied Molecular Biosciences Unit, MedTech–Laboratory of Pharmaceutical Technology, Faculty of Pharmacy, University of Porto, 4050-313 Porto, Portugal
4 Associate Laboratory i4HB, Institute for Health and Bioeconomy, Faculty of Pharmacy, University of Porto, 4050-313 Porto, Portugal
5 CBMA—Centro de Biologia Molecular e Ambiental, Departamento de Biologia, Universidade do Minho, 4710-057 Braga, Portugal
* Correspondence: eduardabfer@gmail.com (E.F.); mlucio@fisica.uminho.pt (M.L.)

Abstract: The barrier function of the skin is primarily determined by its outermost layer, the *Stratum Corneum* (SC). The SC consists of corneocytes embedded in a lipid matrix composed mainly of ceramides, cholesterol, and free fatty acids in equimolar proportions and is organised in a complex lamellar structure with different periodicities and lateral packings. This matrix provides a diffusion pathway across the SC for bioactive compounds that are administered to the skin. In this regard, and as the skin administration route has grown in popularity, there has been an increase in the use of lipid mixtures that closely resemble the SC lipid matrix, either for a deeper biophysical understanding or for pharmaceutical and cosmetic purposes. This review focuses on a systematic analysis of the main outcomes of using lipid mixtures as SC lipid matrix models for pharmaceutical and cosmetic purposes. Thus, a methodical evaluation of the main outcomes based on the SC structure is performed, as well as the main recent developments in finding suitable new in vitro tools for permeation testing based on lipid models.

Keywords: *Stratum Corneum*; SC lipid model systems; SC surrogate; in vitro skin permeation

1. Introduction

The skin is the largest organ of the body, and its primary function is to protect the body against invasion by exogenous substances, including dust, pathogens, compounds, and ultraviolet radiation [1–3]. This protective function is pivotal in diverse fields of research, such as toxicology and risk assessment and pharmaceutical and cosmetic products development. The major barrier function of the skin is located in the *Stratum Corneum* (SC), the outermost layer of the human skin, which is composed of corneocytes embedded in a matrix of lipid multi-layered lamellar structures known as the intercellular lipid matrix (ILM). This ILM is rich in ceramides (Cers) (40–50%), cholesterol (Chol) (20–25%), free fatty acids (FFAs) (15–25%), and a lesser quantity of cholesterol sulfate (ChS) (5–10%) that are self-associated in multilayers [2,4,5]. Corneocytes are grouped in clusters separated by micron-sized channels [6]. The lipids in the intercorneocytes are assembled in parallel and as repeated bilayers, i.e., head-to-head and tail-to-tail [7]. The SC's lipid structure and composition differ from that of other biological membranes that comprise crystalline phospholipids. SC has two lamellar phases with repeated distances of 5–6 and 12–14 nm [7].

The chemical composition and the structural organisation of SC may be defined as a "brick and mortar" model, where the "bricks" are the cells and the "mortar" is the ILM [8], both conferring resistance to skin permeation. Most topically applied compounds are unable to permeate the SC due to its corneocyte-rich nature, and consequently, the transport through the skin predominantly occurs via passive diffusion through the ILM. The passive diffusion of bioactive compounds across this lipid lamellar structure implies a tortuous movement between corneocytes through interlamellar regions of the ILM (i.e., tail–tail region of intercorneocyte lipids) [6,7,9]. Additionally, the SC presents high density and low hydration (15–20%) compared to the body's average 70% hydration. Therefore, the SC acts as a barrier for the diffusion of the majority of compounds, specifically the ones that present high molecular weight (>500 Da) and inadequate solubility [2,10,11]. A non-linear pH gradient between the upper and lower regions of the SC, which may range between 4.5 on the surface and 7.0 at the lower regions, can also limit the diffusion through the skin. Additionally, other important physiological conditions, such as temperature, as well as the presence of enzymes, can influence the rate of skin permeation [12–15].

Despite the numerous benefits that the skin provides as a route of administration for several compounds, the permeation of molecules via skin layers remains a challenge because of the significant variability between people's skin and between skin sites on an individual [4,7,16], as well as the skin barrier's effectiveness in limiting the type and amount of compound that can permeate it (mainly its outermost layer of epidermis, the SC) [7,17]. Therefore, estimating skin permeation rates is crucial in pharmaceutical and cosmetic research to hasten the development of dermal or transdermal compounds and/or formulations. However, estimating the permeability of a specific compound through the skin is often difficult due to the very complex nature of the different structures and mechanisms that comprise the permeation pathway [17]. Skin permeation studies are carried out for two main reasons: to study the impermeability of the skin to toxic chemicals, allergens, and irritants or to evaluate the ability of a drug/compound, commonly carried in a delivery system, to permeate the skin barrier and be delivered to a particular skin layer or even directly into the bloodstream. In this context, different skin models have been developed, ranging from in vitro to in vivo [18], from the simplest to the most complex.

Conventional in vitro skin permeation studies typically employ the Franz diffusion cell, a device that sandwiches a skin surrogate between donor and receptor compartments. Ideally, human skin would be the gold standard for such investigations to ensure an accurate representation of the processes involved, but its complex nature and limited availability, as well as the inter- and intra-individual variability and ethical constraints, hinder its widespread use in the early stages of compounds and/or formulation research [19–22]. As a result, animal skin surrogates are often employed, albeit with significant anatomical disparities and non-standardised protocols leading to inconsistent permeability data [23–26]. Moreover, with the European Union (EU) Cosmetic Regulation (EC 1223/2009) [27] imposing constraints on products involving animal testing in the EU, as well as the EU REACH regulation (Registration, Evaluation, Authorization and Restriction of Chemicals) [28] recommending/requiring the use of alternatives to animal assays [29,30], there is a clear and urgent need for efficient, precise, and cost-effective models to be used in the early stages of drug/compound research. In this regard, artificial model membranes based on lipid mixtures are attracting increasing attention. Since the 1990s, these lipid mixtures that mimick the SC ILM have been employed for various purposes, including biophysical characterisation and, more recently, interaction and permeation profiling for cosmetic and pharmaceutical applications. While the literature includes reviews on the use of lipid models for biophysical SC ILM characterisation [31–35], none of them focus on the development of these models into a new generation of skin diffusion and permeation tools. This review aims to bridge this gap by providing an overview of the main outcomes of biophysical SC ILM characterisation through lipid models and delving into their transformation into innovative lipid-based SC surrogates (SCS). The described SCS hold the promise of revolutionising in vitro studies in pharmaceuticals and cosmetics, offering a powerful alternative to traditional skin models for diffusion and permeation investigations.

2. Composition and Molecular Assembly of the *Stratum Corneum* Lipid Matrix

The human ILM is mainly composed of mixtures of Cers, Chol, and FFAs. Cers are sphingolipids consisting of a fatty acyl chain (resultant from a fatty acid) amide-linked to a sphingoid base. According to the type of sphingoid base, different nomenclatures are attributed to Cers: sphingosine (S), dihydrosphingosine (dS), phytosphingosine (P), and 6-hydroxysphingosine (H). The fatty acyl chain has also different nomenclatures with two to four hydroxyl (OH) functional groups and a monosubstituted amide group (N–C=O), which behave as hydrogen bond donors and acceptors [36] and can be non-substituted (N), α-hydroxylated (A), ω-hydroxylated (O), or bearing the ω-linoleyloxy group [37]. The combination of these letters designates the type of fatty acyl and sphingoid base in the Cers nomenclature. The structural variation of the long acyl chain linked to the sphingoid base via an amide linkage leads to a broad diversity of Cers, with around 17 subclasses of Cers identified in human skin [38,39]. In the native SC lipid matrix, Cers have acyl chains that can range from long (C13 to C19), very long (C20 to C26), or ultralong (>C28) [40], and FFA's chain lengths can range from C16 to C30, with a predominance of C24 and C26 (33.7 and 25.2%, respectively) [41].

In Figure 1 are depicted the main structural differences between Cer headgroups.

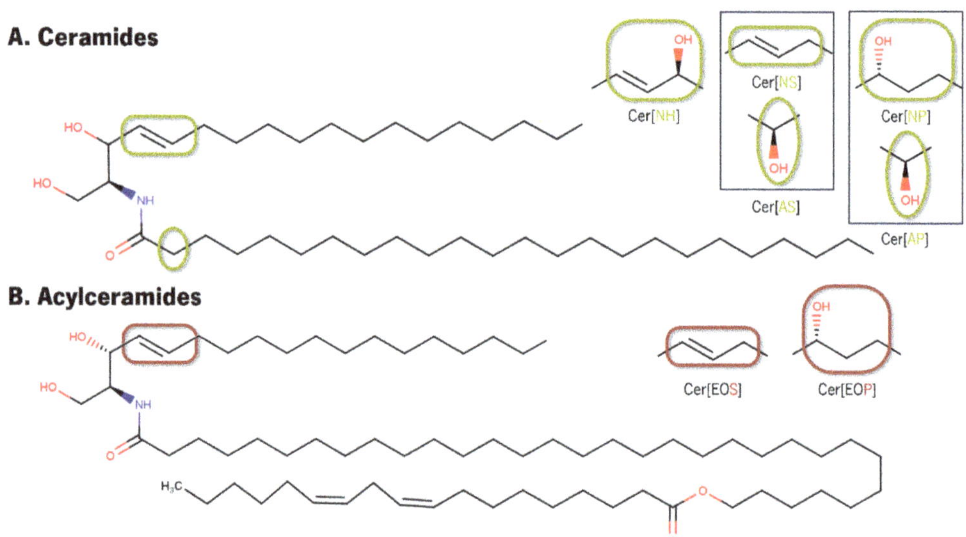

Figure 1. Schematic illustration of the main differences at the Cers headgroup level. Chemical structures were drawn in MarvinSketch® version 5.3.1.

Cer[AP] and Cer[NP] differ slightly at the headgroup level, with Cer[NP] featuring three OH groups and Cer[AP] having four OH groups [42].

Although the function of each Cer class has been the subject of intense debate and scrutiny, some consensus can be found in the literature. Considering the skin barrier function, while acylCers (Cer[EOS], Cer[EOP], and Cer[EOH]) are of pivotal importance due to their really long fatty acid chains [43,44], both classes of α-hydroxylated acyl chains (Cer[AS], Cer[AdS], Cer[AP], and Cer[AH]) and non-hydroxy fatty acid Cers (Cer[NdS], Cer[NH], Cer[NP], and Cer[NS]), with their involvement in intermolecular hydrogen bonding interactions, provide structural cohesion [45,46].

3. *Stratum Corneum* Lipid Model Membranes

Due to structural differences in Cers—acyl chain length and mobility—SC lipids are assembled in two lamellar phases: a short periodicity phase (SPP) with a repeated distance of ≈5–6 nm and a long periodicity phase (LLP) with a repeated distanced of ≈12–14 nm (Figure 2) [47]. Regarding lateral packing, at the human skin surface temperature (≈32 °C), the SC lipids are mainly assembled in an orthorhombic packing, while a low level of lipids follow hexagonal or liquid crystalline packing (Figure 2) [47].

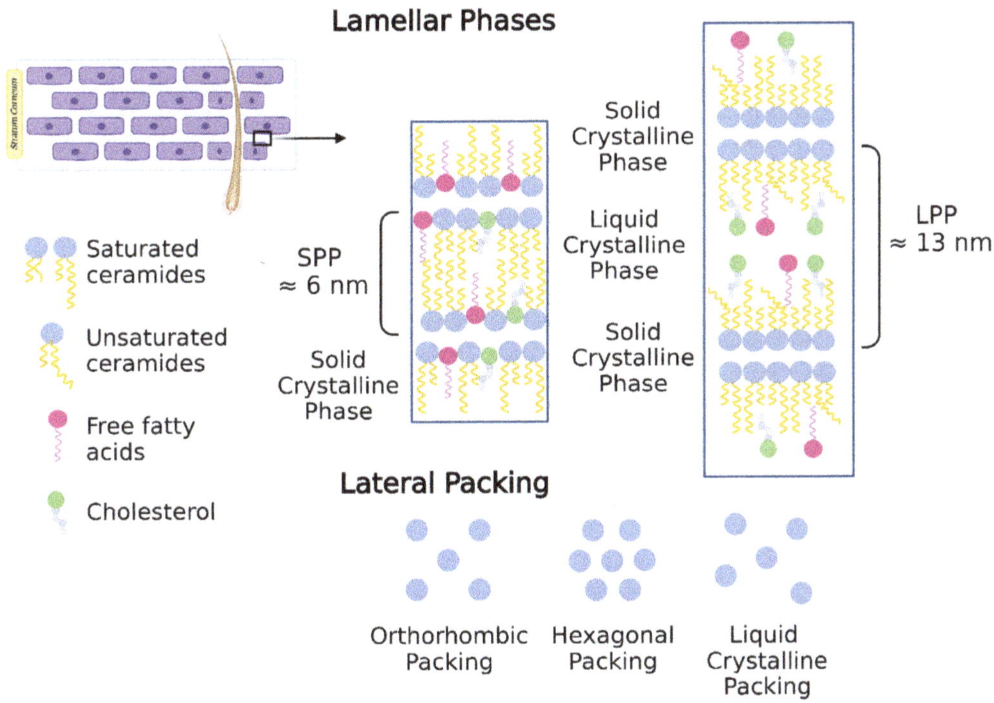

Figure 2. Lipids constituting the intercellular lipid matrix (ILM) of the *Stratum Corneum* (SC) are assembled in two crystalline coexistent lamellar phases: a short periodicity phase (SPP) and a long periodicity phase (LLP) and can be laterally packed in orthorhombic, hexagonal, or liquid-like unit cells (created with BioRender.com).

Despite their simplicity, SC lipid models can reproduce the characteristics of native SC lipid matrix assembly [48]. Numerous techniques, particularly X-ray diffraction, neutron diffraction, infrared (IR) spectroscopy, and nuclear magnetic resonance (NMR), have been employed to extensively decipher the structure of SC ILM through lipid mixtures of synthetic or natural lipids. Understanding the SC ILM structure is paramount to develop models that can mimick the SC barrier function in skin permeation studies. The most recent literature reporting examples of SC lipid model mixtures from 2010 onwards is presented in Table 1, with older studies [30,43–46,48–103] reviewed in the Supplementary Materials (Table S1).

Table 1. The most recent examples of *Stratum Corneum* (SC) lipid model mixtures reported in the literature, from 2010 onwards, for deciphering the SC structure or as in vitro platforms to study compound–SC lipid matrix interactions.

Lipid Model Composition	Main Objective	Characterisation Techniques	Temperature (T) and pH Conditions	Main Outcomes	Ref.
Cer[AP]:Chol:PA:ChS (55:25:15:5 wt%)	Study the OA impact on bilayer architecture	Neutron diffraction	T: 32 °C pH: 9.5	• OA had no effect on SC bilayer thickness, but it caused lipid disorder by interfering with alkyl chain alignment.	[104]
	Determine the electron density profile of SPP at full hydration and at different pH	Neutron diffraction	T: 25–60 °C pH: 5.0–9.0	• Repeated distances of ≈41 and ≈47 Å remain at full hydration and at different pH values. • Almost no water detected in the inter-bilayer spacing either in full or partial hydration SC states.	[105]
	Elucidate the IPM mechanism of action as a SC permeation enhancer	Neutron diffraction	T: 32 °C pH: ND	• IPM permeation results from bilayer perturbation and disordering effects. • IPM induces phase separation and disturbs the multilamellar lipid assembly.	[106]
Cer[AP]:Chol:SA:ChS (55:25:15:5 wt%)	Synthesis and application of a deuterated Cer[AP]-SA in a SC lipid model	Neutron diffraction	T: 32 °C pH: ND	• Deuterium label located in high-ordered lamellar structures, with a repeated distance of ≈44 Å, supporting the Cer hairpin conformation within bilayer structures previously proposed.	[107]
Cer[AP]:Chol:FFA:ChS (55:20:15:10 or 66:10:18:6 wt%) FFA = PA:SA:AA:BA:LA:CA (1.3:3.3:6.8:42.0:36.2:6.7 molar ratio)	Investigate the effects of different Chol–ChS ratios on the hydration and structure of a SC lipid model with a realistic FFA mixture	Neutron diffraction	T: 20 and 32 °C pH: ND	• SC systems structure with FFA mixture is close to that of PA containing systems, with characteristic small inter-bilayer space. • Ratio variation does not lead to any significant changes of membrane structure parameters. • Replacement of Chol by ChS increases membrane swelling, inhibits phase separation, and promotes a more homogeneous lipid system.	[108]

Table 1. *Cont.*

Lipid Model Composition	Main Objective	Characterisation Techniques	Temperature (T) and pH Conditions	Main Outcomes	Ref.
Cer[AP]:Chol:SA (1:0.7:1 molar ratio)	Explore the influence of hydrophilic permeation enhancer DMSO in SC model	Neutron diffraction, PCS, IR, and leakage studies	T: 32 °C pH: 7 and 10	• No specific interaction of 10% and 30% DMSO with SC model system. • It was hypothesised that the permeation enhancing effect of DMSO is foreign to the SC lipid matrix.	[109]
Cer[AP]:Chol:SA (1:0.7:1 molar ratio)	Develop and optimise an oligolamellar SC model	X-ray reflectivity and FTIR	ND	• New oligo-SC model has a similar total number of lipid membranes with no phase separation as native SC between corneocytes. • The preparation method is reproducible.	[110]
	Study the influence of hydrophilic (urea and taurine) permeation enhancers on SC model	Neutron reflectivity, IR, SWAXS, carboxyfluorescein leakage, PCS, Langmuir isotherms, BAM, and IRRAS	T: 21 °C pH: 5.5 and 7.4	• No specific interaction when 10% of urea or 5% taurine are added to SC monolayers. • Permeation enhancing effect of both substances might not be related with the SC lipid matrix.	[111]
	Elucidate the IPM mechanism of action as SC permeation enhancer	Langmuir isotherms, IRRAS, BAM, and GIXD	T: 21 °C pH: 5.5	• IPM showed a limited fluidising effect on the alkyl chains of SC lipids through a slight decrease in packing density of the system, irrespective of chain deuteration.	[112]
Cer[AP]:Chol:FFA:ChS (55:20:15:10 or 66:10:18:15 wt%) FFA 1 = PA:SA:AA:BA:LA:CA (1.3:3.3:6.7:41.7:36:6.7 molar ratio) FFA 2 = BA:LA:CA (8.8:7.7:1.4 molar ratio)	Study the influence of hydration and composition in the temperature behaviour of SC model	Neutron diffraction	T: 20–72 °C pH: ND	• SC model with 20% of Chol and 60% humidity undergoes a phase transition in the 63–67 °C range, which, upon Chol decreasing content, shifts to larger temperatures. • The hydration process increases repeated distance in 1 Å at 25 °C and in 1.6 Å at 57 °C, • Phase separation occurs upon hydration at 57 °C, and the FFA-rich phase changed the repeated distance.	[113]

Table 1. *Cont.*

Lipid Model Composition	Main Objective	Characterisation Techniques	Temperature (T) and pH Conditions	Main Outcomes	Ref.
Cer[AS]:Chol:PA (Variable ratio)	Characterise the fluidity and microstructure of a SC lipid model	Fluorescence anisotropy and SAXS	T: 25 to 80 °C pH: 5	• SC bilayers formed three kinds of lamellar structures and two kinds of hydrocarbon chain packing. • Hexagonal/orthorhombic packing ratio was close to the value found in human isolated SC.	[114]
Cer[AS]:Chol:PA (26.5:13.9:59.6 mol%)	Investigate the effects of l- and d-menthols as permeation enhancers	Fluorescence anisotropy, DSC, and WAXS	T: 25 to 80 °C pH: 5	• Permeation enhancing effects of optically active menthols, with d-menthol provoking stronger effects than l-menthol.	[115]
Cer[NP]:Chol:LA (1:1:1 molar ratio)	Investigate the SC model in a temperature range	Neutron diffraction and ^{2}H solid-state NMR spectroscopy	T: 25–80 °C pH: ND	• Coexistence of two lipid phases at 32 °C, with repeated distances of 54.2 Å and 43.0 Å. • Model assembly of SC lamellar phases is proposed: the thicker phase consists of Cer[NP] in a hairpin conformation mixed with Chol and LA, while the thinner phase contains Cer[NP] in hairpin conformation.	[116]
	Explore the effect of ethanol on SC model	DSC, FTIR, and ^{2}H-NMR	T: 25–75 °C pH: 5.4	• Ethanol had two effects: (i) a concentration-dependent effect on the reduction of the SC phase transition temperature, and (ii) at high concentrations, it selectively extracted the FFA component from the SC lipid assemby.	[117]
Cer[NS]:Chol:BA (1:0:0 or 1:0.3:1 molar ratio)	Study the interactions between NIPAM nanogels cross-linked with MBA and SC lipid models	Neutron reflectivity and TEM	T: 24–42 °C pH: ND	• Nanogels interact with the SC lipid model mainly through hydrophobic interactions, and these interactions increase with the MBA content.	[118]

Table 1. *Cont.*

Lipid Model Composition	Main Objective	Characterisation Techniques	Temperature (T) and pH Conditions	Main Outcomes	Ref.
Cer[NS]:Chol:(BA:LA) (1:1:1 molar ratio)	Explore the effect of various surfactants on the structure of SC model	Neutron scattering and MD	T: 32 °C pH: ND	• A new mechanism for surfactant as a permeation enhancer is proposed: surfactant increases the barrier function by incorporating phase-separated Chol molecules into the SPP structure of SC membrane, while the hydrophilic surfactant headgroup remarkably increases hydration and fluidity, which eventually results in better permeability.	[119]
Cer[NS]:Chol:FFA (1:1:1 molar ratio) FFA = LA, PA or PA:SA:AA:BA:LA (1.8:4:7.6:47.8:38.9 mol%)	Examine the effect of chain length of FFAs on thermotropic phase behaviour and mixing properties of SC mixtures	FTIR and Raman imaging spectroscopy	T: 20–90 °C pH: 5	• Thermotropic phase behaviour and mixing properties of lipid mixtures strongly depend on chain length and its hydration state. • While a complex mixture of FFA and single LA did not significantly change the SC mixture phase behaviour, in mixtures containing only PA, phase separation was observed after hydration and long-term storage. • Evidence that matching the FFA chain length and Cer acyl chain promotes mixing.	[120]
Cer[NdS]:Chol:PA (1:1:1 molar ratio) D-*erythro*- or L-*threo*-[NdS]	Investigate the contribution of stereochemistry of Cer[NdS] in SC model	DSC and SWAXS	T: 20–180 °C pH: ND	• Inter- and intramolecular interaction was greater in the D-*erythro*-based system than in the L-*threo*-based ones, resulting in appreciable differences in lamellar periodicities. • Stereostructure of Cer molecules can be determinant in lamellar periodicities.	[121]

Table 1. *Cont.*

Lipid Model Composition	Main Objective	Characterisation Techniques	Temperature (T) and pH Conditions	Main Outcomes	Ref.
Cer[AP]/Cer[NS]:Chol:SA (1:1:1 molar ratio)	Develop an IR spectroscopy approach for studying lateral phase separation and lamellar structure formation in SC models	FTIR	T: 31 °C pH: 5.5	• It was possible to detect the reorganisation of Cer amide H-bonds from disordered to ordered forms, appearance of ordered Cer chain orthorhombically packed, and the phase separation of orthorhombically packed domains of SA with the H-bonded dimeric headgroup. • Kinetic IR spectroscopy measurements showed to be promising in monitoring the temporal sequence for phase separation and lamellar structure formation.	[122]
Cer:Chol:PA (1:1:1 molar ratio) Cer = [AS], [NP], [NP]-SA or [NS]	Investigate the structure of hydrated SC lipid model	Mid-IR spectroscopy	T: 30–35 °C pH: 5.5	• Between 30 and 35 °C, orthorhombic and hexagonal chain packing are observed, which favoured the formation of orthorhombic packing with decreased hydration. • Hydrated Cers exhibited high solid crystalline- to liquid crystalline-phase transition temperatures. • At low temperature, phase separation is observed: a Cer-PA domain and a PA-Chol domain. Cer[NS]-SA improved the system miscibility. • Formation of distinct crystalline-like domains embedded in a more fluid phase supported the domain mosaic model. • Headgroup architecture of Cer mainly determines the structure of hydrocarbon chains, favouring the orthorhombic perpendicular subcell lattice.	[123]

Table 1. *Cont.*

Lipid Model Composition	Main Objective	Characterisation Techniques	Temperature (T) and pH Conditions	Main Outcomes	Ref.
Cer[AP]/Cer[NP]:Chol:SA (55:25:20 wt%)	Investigate the lamellar structure of SC model	Neutron diffraction and NMR	T: 32 °C pH: 7.4	• Absence of only one OH group caused drastic structural changes, supporting the great influence of the Cer headgroup in the structure of the SC membrane. • Cer[NP]-based system exhibited higher lamellar order than the Cer[AP] system, with completely different temperature-dependent behaviour and hydration characteristics. • Cer[NP] was able to prevent the model membrane from significant hydration and swelling, due to intense intra- and intermolecular headgroup interactions.	[42]
Cer[EOS]:Chol:FFA (Variable ratio) FFA = PA:SA:AA:BA:TA:LA:CA (1.8:4:7.7:42.6:5.2:34.7: 4.1 molar ratio)	Investigate the molecular assembly and lipid packing in SC model	SAXS	T: 20–100 °C pH: ND	• Equimolar ratio forms only one lamellar phase with a long repeated distance of 14.7 nm, with FFA and Cer chains undergoing phase transition in different temperature ranges, while part of their chains are mixing in orthorhombic lattice. • Although deviations in the FFA content are less critical for formation of the 14.7 nm lamellar phase than the Chol content, a certain degree of FFA chain length is important.	[124]
Cer[AdS]:Chol:PA (Variable ratio)	Examine the quantitative effect of Chol reduction in SC model	Raman, DSC, PXRD, and SAXS	T: 20–120 °C pH: ND	• Amount of Chol in the SC model was inversely correlated with the hexagonal structure ratio (RHex/Ort) in the packing structure. • RHex/Ort values are highly correlated with the amount of lipid elution effect of surfactants.	[125]

Table 1. *Cont.*

Lipid Model Composition	Main Objective	Characterisation Techniques	Temperature (T) and pH Conditions	Main Outcomes	Ref.
Cer[NS16]/Cer[NS24]:Chol:FFA:ChS (1:1:1 molar ratio + 5 wt%) FFA = LA or PA:SA:AA:BA:LA (1.8:4:7.6:47.8:38.8 wt%)	Investigate the membrane behaviour of long and very long Cer[NS] in SC model	FTIR	T: 28–100 °C pH: 5.5	• While very long Cer[NS24] preferred an extended conformation in which FFAs (either LA or heterogeneous mixture) were associated with Cer chains, shorter Cer[NS16] were less FFA miscible, being mostly phase separated.	[126]
Cer[NS16]/Cer[NS24]:Chol:FFA (1:1:1 molar ratio) FFA = PA, LA, AA or LA:AA:SA:PA (84.6:8.75:4.57:2.05 mol%)	Study the influence of acyl chain length on lipid mixing properties in SC model	Raman and IR	T: 33 °C pH: 5	• The combination of Cer and FFA bearing very long chains is required for the formation of homogeneous mixtures, while shorter chained FFAs lead to micro-sized domains.	[127]
$_{BB}$Cer[NP]:Chol:SA (1:1:1 or 0.5:1:1 molar ratio)	Investigate the thermotropic and kinetics of lipid dynamics and domain formation in normal and Cer deficient SC model	FTIR	T: 19–85 °C pH: 5.5	• Kinetic and thermotropic studies displayed differences in both Cer and FFA chain fluidity and reduction in Cer levels had an ordering effect. • In a Cer-deficient model, the formation of Cer orthorhombic domains is retarded, affecting the lipid barrier structure.	[128]
$_{BB}$Cer:Chol:SA (1:1:1 molar ratio)	Study the water vapor uptake and surfactant sorption onto SC model	QCM	T: 22 °C pH: ND	• Surfactant sorption was found to be concentration-dependent even beyond its critical micelle concentration. • QCM results are in accordance to reported clinical data, which suggest that this approach is promising to monitor the barrier properties of the SC model.	[129]

Table 1. *Cont.*

Lipid Model Composition	Main Objective	Characterisation Techniques	Temperature (T) and pH Conditions	Main Outcomes	Ref.
DPPC:Chol (7:3 molar ratio) Cer[AP]:Chol:SA (14:10:14 molar ratio)	Explore the effect of a natural and four synthetic surfactants on two SC monolayer models	Langmuir isotherms, surface dilatational rheology, and fluorescence microscopy	T: 21 °C pH: ND	• The four synthetic surfactants were capable of solubilising lipids from the SC monolayer. • Natural saponin-rich extracts increased surface pressure greater than synthetic ones and enhanced SC elastic properties. • No evidence of SC lipids removal by saponin-rich extract suggests a surfactant action through changes in the physical state of the monolayer.	[130]
Cer[EOS]:Cer[AP]: Chol:BA (23:10:33:33 wt%)	Study the effect of IPM in the SC lipid model assembly	Neutron diffraction	T: 32, 50 and 70 °C pH: ND	• IPM had a disordering effect in the rigid lamellar structure and influenced the phase behaviour of SC model possibly by interaction with lipid headgroups. • IPM prevented the formation of LPP by favouring the SPP formation and decreasing SC barrier properties.	[131]
Cer[EOS]:Cer[AP]: Chol:PA:ChS (30:30:20:15:5 wt%)	Characterise the LPP SC lipid model in excess water	SAXS	T: ND pH: 7.2 or 9	• In excess water at pH 7.2, SC lipid membrane revealed two SPP with repeated distances of 47 and 35.7 Å and a LPP with a repeated distance of 127 Å. Increasing the pH to 9 resulted in the destruction of LPP and in a single SPP with repeated distance of 48.3 Å.	[132]

Table 1. *Cont.*

Lipid Model Composition	Main Objective	Characterisation Techniques	Temperature (T) and pH Conditions	Main Outcomes	Ref.
Cer:Chol:SA+ChS (1:1:1 molar ratio – 5 wt%) Cer = [NP]:[AP], [NP]:[NS], [AP]:[NS] or [NP]:[AP]:[NS] (1:1 or 1:1:1 molar ratio)	Characterise each Cer role in the SC membrane: effects on assembly, miscibility, and thermotropic behaviour	SWAXS and IR spectroscopy	T: 20–80 °C pH: ND	• Cer[NP] showed the ability to establish very strong H-bonds within its solid crystalline structure, which results in a phase separation effect and orthorhombic lattice that weakens when Cer[NP] is combined with the other two Cer types. • The importance of acyl chain length was demonstrated, and it was clear that even a small variation in the Cer headgroup structure or minor change in the membrane composition can lead to great differences in the phase behaviour of the SC model system.	[133]
Cer[NS]:Cer[EOS]:FFA:Chol:ChS (Variable ratio) FFA = PA:SA:AA:BA LA (1.8:4:7.6:47.8:38.8 mol%)	Investigate the lamellar structure in SC model	SAXS	ND	• In the SC model, medium and very long lamellar phases may be formed in the absence of Cer[EOS], whereas these Cer likely stabilise the final lipid assembly of LPP by riveting the adjacent lipid layers. • An increased ChS:Chol molar ratio modulated the membrane polymorphism.	[134]
Cer[NP]:Cer[AP]:Chol:FFA (Variable ratio)	Understand the SC lipid matrix assembly	Neutron Diffraction and MD	ND	• Except for Chol, all lipids showed a strong H-bonding network, which is correlated with tighter lipid packing. A strong H-bonding network is determinant in SC packing, but the location of -OH groups is equally important.	[135]
Cer[EOS]:Cer[NS]: Chol:FFA (variable ratio) FFA = PA:SA:AA:BA:LA (4:8:48:40 mol%)	Investigate the effects of replacing Cer[NS] by Cer[EOS]	SAXS and solid-state and diffusion NMR	T: 32 °C pH: ND	• SC model is assembled in the LPP phase, where the rigid headgroup of Cer[EOS] is anchored at the lamellar interfatial region, and the isotropic fluid middle and terminal segments of linoleate chains of Cer[EOS] show slow translational diffusion. • Unusual self-assembled structure demonstrated a strong impact on barrier function.	[136]

Table 1. *Cont.*

Lipid Model Composition	Main Objective	Characterisation Techniques	Temperature (T) and pH Conditions	Main Outcomes	Ref.
Cer[NP]:*D-/L-*Cer[AP]:Chol:LA (0.66:0.34:0.7:1 or 0:1:0.7:1 molar ratio)	Study the influence of the Cer[AP] conformation on the lamellar structure of SC model	Neutron diffraction and SAXS	T: 32 °C pH: ND	• Strong influence of Cer[AP] conformation on lamellar and nanostructure of the multilamellar SC system. • While D- alone showed a crystalline-like behaviour, L- presented no lamellar phase. • When combined with Cer[NP], the D-based system adopted a more native-like structure, while L-promoted a very crystalline-like behaviour.	[137]
Cer[EOS]:Cer[NS]:Chol:PA (Variable ratio)	Investigate the effects of vaporised *L*-menthol on SC lipid matrix	DSC, SAXS, and ATR-FTIR	T: 20–120 °C pH 5	• SC mixture formed characteristic orthorhombic and hexagonal packing structures. • *L*-menthol showed the ability to act as a permeation enhancer through a liquid crystalline lamellar structure by affecting the Cer carbon chain and dissociating the H-bond hydrophilic groups.	[138]
Cer[EOS]:Cer[NS]:Chol:FFA (0.4:0.6:1:1 molar ratio) FFA = LA:BA:SA:PA (84.6:8.75:4.57:2.05 molar ratio)	Characterise the chain order of Cer[EOS] in SC model forming LPP	SAXS, H-NMR, and IR	T: 25–70 °C pH: 5	• LPP structure is formed by a solid ordered phase in which highly disordered liquid domains formed by the oleate chain of Cer[EOS] are embedded. • Cer[EOS] acts as a strong modulator of SC fluidity balance.	[139]
Cer[NS]:Cer[AP]:Chol:LA (variable ratio)	Explore the impact of Cer[NS] on SC model structure	Neutron diffraction and MD	T: 32 °C pH: ND	• Both Cers showed a similar influence on lamellar structure and promoted lipid miscibility with Chol; • The SC model composed of [NS]/[NP] 2:1 and 1:2 presented repeated distances similar to those of natural-like SPP.	[140]
Cer[NP]:Cer[AP]:Chol:ChS:LA:FA:SA (33:22:25:5:7.5:3.75:3.75 mol%)	Investigate the effect of model CPEs on SC liposomes	Sodium fluorescein leakage and fluorescence anisotropy	T: 25 °C pH: 9	• SC liposomes as promising models for high-throughput screening of permeation enhancers effectivity.	[141]

Table 1. *Cont.*

Lipid Model Composition	Main Objective	Characterisation Techniques	Temperature (T) and pH Conditions	Main Outcomes	Ref.
Cer[NP]:Cer[AP]:Chol:ChS: LA:PA:SA (33:22:25:5:7.5:3.75:3.75 mol%)	Investigate the potential of SC liposomes for CPEs permeation study (ethanol as model) and comparison with animal skin permeation studies	Fluorescein leakage, fluorescence anisotropy, animal skin impedance, and hairless rat skin permeation	T: 25 °C pH: 9	• Good correlation between permeation in SC liposomes and in hairless rat skin. • Ethanol action on SC lipids and permeation-enhancing effect mainly occurs through polar headgroup perturbation.	[37]
Cer:Chol:LA (1:1:1 molar ratio) Cer = [NS]:[NH] or [NS]:[AP] (1:1 molar ratio)	Understand the basic structure of SC membranes	NMR, DSC, and WAXS	T: 20–120 °C pH: 5.5	• Coexistence of several phases at all temperatures tested: at skin temperature, molecules formed a solid crystalline phase with orthorhombic lipid packing and only a small fraction of lipids is in a liquid crystalline state. • While the Cer[AP]-based SC system formed a lamellar phase with SPP, the Cer[NH]-containing system formed a lamellar phase with 10.7 nm periodicity.	[142]
Cer[AP]:$_{br}$Cer[EOS]: Chol:BA (10:23:33:33 wt%)	Explore the localisation of $_{br}$Cer[EOS] species within LPP of SC model	Neutron diffraction	ND	• SC lipid mixture formed a LPP-sized by 114 or 118 Å with two coexisting SPP of 48 Å and 45 Å. • Cer[EOS] was found in LPP and absent in SPP, while CER[AP] was found in both short phases but not within the LPP.	[143]
$_{br}$Cer[EOS]:Cer[AP]: Chol:BA (23:10:33:33 wt%)	Synthesis of new artificial Cer[EOS] species and implication to SC model structure	DSC, FTIR, Raman, Neutron diffraction, and MD	T: 20–120 °C pH: ND	• Synthesised $_{br}$Cer[EOS] showed to be promising in substituting native form in terms of the formation of well-ordered lipid bilayers with a lipid assembly equivalent to the model matrix comprising native Cer[EOS].	[144]
Cer[NS]:Cer[AS]:Chol:PA:ChS (25:15:25:25:10 wt%)	Understand the nature of Posintro^{TM} permeation into SC lipid matrix	FRET, ITC, AFM, Cryo-TEM, and EIS	T: 5, 15, 25 or 37 °C pH: 7.4	• Posintro^{TM} nanoparticles disturbed the SC bilayers in a DC-Chol content-dependent manner. • Nanoparticle permeation into skin may be involving their fusion with the SC lipid matrix.	[145]

Table 1. *Cont.*

Lipid Model Composition	Main Objective	Characterisation Techniques	Temperature (T) and pH Conditions	Main Outcomes	Ref.
Cer[EOS]:Cer[NS]: Chol:FFA (0.4:0.6:1:1 molar ratio) FFA = PA:SA:AA:BA:LA (1.8:4:7.6:47.8:38.8 mol%)	Study the interaction between Cer[NS] and FFA chains in LPP	FTIR and SAXS	T: 0–90 °C pH: ND	• Cer[NS] acyl chain and FFAs are mixed and colocalised within central region of LPP. • Cer[NS] adopted an extended conformation that increases the packing density, inhibits membrane swelling, and allows flexibility of the SC system.	[146]
Cer[NP]:Cer[AP]:Chol:LA (0.66:0.34:0.7:1 or 0.34:0.66:0.7:1 molar ratio)	Investigate the influence of Cer[NP] and Cer[AP] on SC nanostructure	Neutron diffraction	T: 32–37 °C pH: ND	• SC mixture showed similar thickness to the SPP native SC lipid matrix, with a repeated distance of 5.45 nm. • Cer[AP] are assembled in such a way that its C24 chain is somehow shorter than the C24 chain of Cer[NP], not overlapping as much with the opposite lamellar leaflet. • In the [NP]:[AP] 2:1 ratio system, chains are mostly straight-packed, and C24 chains show a broad overlapping region in the lamellar midplane.	[147]
Cer[NS]:Cer[EOS]: Chol:FFA (0.6:0.4:1:1 molar ratio) FFA = LA:AA:SA:PA (84.6:8.75:4.57:2.05 mol%)	Study the lipid spatial distribution of SC model	Raman spectroscopy and AFM-IR	T: 33 °C pH: 5.0	• SC model showed to be overall homogeneous with small, slightly enriched, and depleted regions in a lipid component. • Fluid nature of the Cer[EOS] chain was confirmed, while the rest of the lipid matrix was highly ordered, supporting the three-layer model for the SC structure.	[148]
Cer[EOS]:Cer[NS]:SP: SA:Chol:LA (Variable ratio)	Determine the effect of Cer reduced concentration on LPP structure	Neutron diffraction, SAXS, and FTIR	T: 23 or 37 °C pH: 5.0	• LPP structure remained undisturbed at least until the replacement of 25% of Cer[NS]. • An increased ceramidase activity or reduced Cer synthase activity does not lead to dramatic changes in the lipid phase behaviour of Cers. • With increased FFA concentration, Cers in LPP tend to redistribute.	[149]

Table 1. *Cont.*

Lipid Model Composition	Main Objective	Characterisation Techniques	Temperature (T) and pH Conditions	Main Outcomes	Ref.
Cer[NP]:Cer[AP]:Cer[EOS]:Chol:LA (0.6:0.3:0.1:0.7:1 molar ratio)	Examine the lamellar and nanostructure of SC model	Neutron diffraction	T: 32 °C pH: ND	• The SC mixture was able to mimick natural SPP with a lamellar repeated distance of 5.47 nm, in which LPP formation was not detected. • Cer[EOS] was integrated into SPP spanning multiple layers.	[150]
Cer[EOS]:Cer[NS]:Cer[NP]:Chol:LA (0.4:0.3:0.3:1:1 molar ratio)	Examine the location of Cer[NP] and Cer[NS] in LPP unit cell	Neutron diffraction	T: 37 °C pH: 5.0	• In LPP of a SC model system presenting a repeated distance of 12.6 nm, the acyl chain of Cer[NP] was predominantly located in the central part of the trilayer structure, with a minor fraction at the unit cell boundary, similar to Cer[NS]. • Both Cers adopted a linear conformation in the LPP unit.	[151]
Cer:Chol:FFA (1:1:1 molar ratio) Cer = [NS]:[NP24]:[AS]:[NP16]:[AP] (60:19:5:11:6 molar ratio) FFA = PA:SA:AA:BA:TA:LA:CA (1.8:4:7.7:42.6:5.2:34.7:4.1 molar ratio)	Study the molecular assembly of SPP	Neutron diffraction	T: 25 °C pH: 5.0	• SPP were characterised with a repeated distance of 5.4 nm in a typical bilayer structure. • Cers, with interdigitating acyl chains, were symmetrically assembled in the centre of the unit cell.	[152]
	Determine the FFA and Chol location into SPP of SC model	Neutron diffraction	T: 25 °C pH: ND	• SC model was assembled in the lamellar phase with a repeated distance of 53.9 Å. • Chol headgroup was found to be somewhat inward from the unit cell boundary, with its tail located 6.2 Å from the unit cell centre. • FFA headgroups were located at the unit cell boundary, with their acyl chains encircling the core of the unit cell.	[153]

Table 1. *Cont.*

Lipid Model Composition	Main Objective	Characterisation Techniques	Temperature (T) and pH Conditions	Main Outcomes	Ref.
Cer:Chol:FFA (1:1:1 molar ratio) Cer = [EOS]:[NS]:[NP]:[AS]:[NP16]:[AP] or [EOS]:[NS] (39:38:10:3:6:4 or 40:60 mol%) FFA = PA:SA:AA:BA:TA:LA:CA (1.8:4:7.6:47.8:38.8 mol%)	Explore both assembly and conformation of Cer[NS] within LPP unit cell	Neutron diffraction and SAXS	T: 25 and 32 °C pH: ND	• In the repeating trilayer LPP unit, the acyl chain of Cer[NS] was located in the central and outer layers, while sphingosine chain was located exclusively in the middle of the outer layers. • Cer[NS] showed the same behaviour regardless the Cers mixture studied.	[154]
Cer:Chol:FFA (1:0–1:0/1 molar ratio) Cer = [EOS]:[NS]:[NP24]:[AS]:[NP16]:[AP] (0:60:18:5:10:7 or 40:36:11:3:6:4 mol%) FFA = PA:SA:AA:BA:TA:LA:CA (1.8:4:7.7:42.6:5.2:34.7:4.1 mol%)	Explore the effect of Chol on lipid assembly in each lamellar phase	FTIR and SAXS	T: 20–60 °C pH: 5.0	• A minimum of 0.2 Chol in 1:0.2:1 molar ratio mixture was required for the formation of each lamellar phase, and its gradual increment increased the fraction of lipids forming very dense orthorhombic lateral packing. • Chol revealed to be an important building block for the formation of orthorhombic lateral packing of lipid chains and to induce the formation of both SPP and LPP.	[155]
Cer:Chol:FFA (1:1:1 molar ratio) Cer = [EOS]:[NS]:[NP24]:[AS]:[NP16]:[AP] (40:60:0:0:0:0 or 40:36:11:3:6:4 mol%) FFA = PA:SA:AA:BA:TA:LA:CA:MA (1.8:4:7.7:42.6:5.2:34.7:4.19 mol% with increased amount of PA)	Investigate the effects of FFA chain lengths (C16–C28) on lamellar phase and lateral packing of SC model	SAXS, FTIR, and transepidermal water loss measurements	T: 0–90 °C pH: 5.0	• Alterations in LPP behaviour are not strictly correlated with chain length, and shorter FFA chain length models exhibit a different behaviour compared to their longer counterparts. • Only models FFA C20 and longer have orthorhombic phase transition temperatures that increase with the increasing chain length. • LPP is a robust structure that can withstand a certain threshold of change before its barrier function is affected.	[156]

Table 1. *Cont.*

Lipid Model Composition	Main Objective	Characterisation Techniques	Temperature (T) and pH Conditions	Main Outcomes	Ref.
Cer:Chol:FFA (1:1:1 molar ratio) Cer = *L*-/*O*-/*S*-[EOS]:[NP]:[NS24]:[AS]:[NS16]:[AP] (30:42:13:3.4:7.5:4.1 mol%) FFA = PA:SA:AA:BA:TA:LA:CA (1.3:3.2:6.9:42:5.3:37:4.7 mol%)	Study the influence of Cer[EOS] moiety on SC lipid assembly	FTIR and SAXS	T: −9.5–90 °C pH: 5.0	• While Cer *S* stabilised the orthorhombic phase, *L*- and *O*-forms promoted the presence of LPP. • The copresence of the three Cer[EOS] may contribute to a proper skin barrier function.	[157]
Cer:Chol:FFA (1:1:1 molar ratio) Cer = [EOS]:[NS]:[NP24]:[AS]:[NP16]:[AP] (13.3:12:3.7:1:1:1.3 mol%) FFA = PA:SA:AA:BA:TA:LA:CA (0.6:1.3:2.6:14.2:1.7:11.5:1.4 mol%)	Study the molecular assembly of LPP in the SC	Neutron diffraction	T: 25 °C pH: 5.0	• SC lipid mixtures with high content of Cer[EOS] formed LPP with a repeated distance of ≈130 Å in a trilayer structure with Cer[NS] and LA interdigitated mostly at the unit cell.	[158]
Cer:Chol:FFA (1:1:1 molar ratio) Cer = [EOS]:[NS]:[NP24]:[AS]:[NP16]:[AP] (13.3:12:3.7:1:2:1.3 mol%) FFA = PA:SA:AA:BA:TA:LA:CA (0.6:1.3:2.6:14.2:1.7:11.5:1.4 mol%)	Determine the lipid components' location in the LPP unit cell	Neutron diffraction	T: 25 °C pH: ND	• A molecular model is proposed for the LPP unit cell with repeated distance of ≈129 Å: (i) Chol is located in two outer layers of LPP (at ≈26 Å from the cell unit centre), (ii) Cer[EOS] (at ≈21.4 Å from the cell unit centre) is linking two outer layers with the central lipid layers, and (iii) the two other lipid subclasses are predominantly located in the central layer of LPP.	[159]
Cer[NS18]/Cer[NP18]:SA:Chol (1:1:1 molar ratio)	Explore the chain-matched and headgroup influence on SC assembly	^{2}H-NMR	T: 32, 50 and 75 °C pH: ND	• Mixtures containing Cer[NP] tend to create more fluid phases than Cer[NS]: the extra OH of Cer[NP] generates a steric conflict when mixed with chain-matched SA and Chol, which destabilises the orthorhombic lattice of SC lipid mixtures and favours the production of liquid crystalline phases.	[160]

Table 1. *Cont.*

Lipid Model Composition	Main Objective	Characterisation Techniques	Temperature (T) and pH Conditions	Main Outcomes	Ref.
Cer[EOS]:Cer[NS]:Chol:FFA (0.3:0.7:0.45:1 molar ratio) FFA = PA:SA:AA:BA:LA (1.8:3.9:7.5:47.8:39 wt%)	Study the Cer[EOS] molecular behaviour in a SC lipid model	SAXS, WAXS, ^{2}H-NMR, and Neutron diffraction	T: 32 °C pH: 5.4	• LPP structure with alternating fluid (sphingosine chain-rich), rigid (acyl chain-rich), isotropic (linoleate-rich), rigid (acyl-chain rich), and fluid layers (sphingosine chain-rich).	[161]

AA—arachidonic acid; AFM—atomic force microscopy; ATR-FTIR—attenuated total reflection Fourier-transform infrared spectroscopy; BA—behenic acid; BAM—Brewster angle microscopy; $_{BB}$Cer—bovine brain ceramide; $_{br}$Cer—branched ceramide; CA—cerotic acid; Cer—ceramide; Cer[AdS]—α-hydroxy fatty acid/dihydrosphingosine base ceramide; Cer[AP]—α-hydroxy fatty acid/phytosphingosine base ceramide; Cer[EOS]—ester-linked ω-hydroxy fatty acid/sphingosine base ceramide; Cer[NdS]—non-hydroxy fatty acid/dihydrosphingosine base ceramide; Cer[NH]—non-hydroxy fatty acid/6-hydroxy-sphingosine base ceramide; Cer[NP]—non-hydroxy fatty acid/phytosphingosine base ceramide; Cer[NS]—non-hydroxy fatty acid/sphingosine base ceramide; Chol—cholesterol; ChS—cholesteryl sulfate; CPE—chemical permeation enhancer; DMSO—dimethyl sulfoxide; DPPC—dipalmitoylphosphatidylcholine; DSC—differential scanning calorimetry; EIS—electrochemical impedance spectroscopy; FFA—free fatty acid; FRET—Förster resonance energy transfer; FTIR—Fourier-transform infrared spectroscopy; GIXD—Grazing incidence X-ray diffraction; IPM—isopropyl myristate; IR—infrared; IRRAS—infrared reflection absorption spectroscopy; ITC—isothermal titration calorimetry; LA—lignoceric acid; LPP—long periodicity phase; MA—montanic acid; MBA—N′-methylenebisacrylamide; MD—molecular dynamics; ND—not discriminated; NIPAM—n-isopropylacrylamide; NMR—nuclear magnetic resonance; OA—oleic acid; PA—palmitic acid; PCS—photon correlation spectroscopy; QCM—quartz–crystal microbalance; SA—stearic acid; SAXS—small-angle X-ray scattering, SWAXS—small and wide angle X-ray scattering; SC—*Stratum Corneum*; SP—Sphingosine; SPP—short periodicity phase; TA—tricosylic acid; TEM—transmission electron microscopy; UV—ultraviolet; WAXS—wide-angle X-ray scattering; wt—weight.

As detailed in Table 1, both natural and/or synthetic Cers can be employed in SC mimetic mixtures. Natural Cers can be isolated from different human/animal sources, which is indicated by the subscript prior to Cer nomenclature: pig ($_{pig}$Cer), human isolated ($_{h}$Cer), or bovine brain ($_{BB}$Cer). While acyl chain lengths vary greatly in natural Cers, synthetic Cers have well-defined chain lengths. Synthetic forms are more prone to be affected by the FFA composition, and its use in general leads to reduced repeated distances [49]. Multiple attempts have been dedicated to unravel the minutiae of the SC ILM structure, as evidenced by the extensive literature on this subject. Although SC lipid models for studying compounds interactions with the SC lipid matrix are also presented in Table 1, this review will mainly focus on the structural characteristics of such models.

The lamellar coexistence of two different periodicities, i.e., SPP and LPP, is consensual; however, their spatial assembly remains unclear, and several models have been proposed (see [34] for a detailed review of the structural organisation of SPP and LPP). Notwithstanding, the most widely accepted models in the literature can be assigned to the sandwich model for LPP and the armature reinforcement model as an extension of the sandwich model for the SPP. Therefore, both periodicity phases and their most widely accepted assembly model will be detailed in the next sections.

3.1. The Long Periodicity Phase (LPP)

The sandwich model was originally proposed for LPP based on X-ray diffraction data obtained from isolated human skin. This model suggests a ternary lamellar structure with two solid crystalline layers surrounding a more fluid liquid crystalline core (Figure 3A) [43,50,162,163]. The unique backbone of Cer[EOS], which displays both rigid and mobile regions, has a pivotal role in the formation of this ternary lamellar structure due to its flexible segments at terminal regions, rigid middle segments near the carboxyl groups, and linoleate segments demonstrating rapid isotropic reorientation but slow self-diffusion (Figure 3B) [50,134,136,153,157,161]. These isotropic fluid chains are a fundamental and essential feature of the LPP structure in the context of the sandwich model [136].

Considering that the Cer[EOS] headgroup is anchored in the lamellar interface of the LPP structure and that linoleate chains are covalently bonded to the remaining rigid regions of the molecule preventing translational diffusion, a less dense packing at the end of the molecule or in the middle of the lipid layer would allow for isotropic reorientation [136]. The inner liquid crystalline phase layer has been observed even in the absence of water, and it is attributed to the steric confinement of the linoleate chains of Cer[EOS], which results in local hydrocarbon nanodroplets (Figure 3C) [51,139,161]. Indeed, the unsaturation level has an active role on the LPP formation, as observed by the lack of LPP as a consequence of replacing Cer[EOS]-linoleate by the -stearate form [157]. Similarly, the proportion of lipids forming LPP was reduced to a greater extent in $_{h}$Cer:Chol mixtures in which $_{h}$Cer[EOS] was replaced by a synthetic form of Cer[EOS]-linoleate as opposed to $_{h}$Cer[EOS]-oleate mixtures [50]. De Sousa Neto et al. studied the role of different unsaturation degrees and found that, in the presence of Cer[EOS], even at high temperatures, the stearate moiety stabilised the interchain coupling, and the LPP formation was only possible in the presence of a specific level of unsaturation, such as Cer[EOS]-linoleate and -oleate moieties [157]. Moreover, the high degree of mobility at C18, along with conformational disordering and/or folding of the linoleate moiety in the inner headgroup regions, compensates the long chains of Cers and FFAs from opposite directions [159]. Therefore, as longer chain lengths lead to enhanced van der Waals interactions and reduced interchain distances [49], Cer[EOS] with its elongated acyl chains acts as a molecular rivet, sticking together the opposing lipid sheets and reinforcing the LPP structure [134,162]. Notwithstanding, the lack of a multilamellar structure in a Cer[EOS]:Chol:PA mixture [52] leads to investigations on other pivotal factors affecting the LPP formation. Several evidence suggesst that the presence of other Cers alongside Cer[EOS] contributes to LPP formation. In a mixture containing 15 mol% Cer[EOS] and 85 mol% Cer[NS], along with Chol and FFA, the coexistence of LPP and SPP predominantly displaying an orthorhombic packing was observed.

However, increasing the Cer[EOS] concentration to 40 mol% resulted in exclusive LPP formation [146]. Consequently, the optimal molar ratio of Cer[EOS]:Cer[NP]:$_{BB}$Cer[EOH] as the Cer component for the formation of LPP in a lipid mixture was investigated, and the authors found that, aside from a low sensitivity towards changes in the component ratio, the optimal fraction on the Cer component was indeed dependent on the presence or absence of FFAs [53]. Although Cer[EOS] seems to contribute to stabilise the lateral orthorhombic packing in the centre of the lamellae by increasing the lipids forming this packing lattice on the LPP [157], there is also a pivotal role of FFA in governing the lateral packing of the lipid mixtures, since both orthorhombic and hexagonal lateral packing are required for LPP formation [157,164]. The addition of long-chain FFA into a mixture of $_{h}$Cer:Chol induced a transition from a hexagonal to an orthorhombic packing [50], whereas the FFA presence, even at low concentrations, lead to the formation of orthorhombic lateral packing, and in its absence, only hexagonal packing was observed [54]. However, although LPP formation occurred, low FFA levels led to a considerable amount of Chol-phase separating, while a higher FFA content led to decreased phase separation without affecting the LPP formation [159]. This was also observed in a mixture of $_{h}$Cer:Chol:FFA, in which varying the FFA proportions revealed a distribution of Cer and FFA into two lamellar phases, with a minor portion of Chol separated into solid crystalline domains [54]. Furthermore, other mixtures employing $_{h}$Cer with Chol, FFA, and ChS in varying ratios were also investigated [51]. In the absence of FFA and ChS, the mixture predominantly displayed a LPP phase with hexagonal lateral packing, while a small fraction of the lipids formed a fluid phase. However, in the absence of $_{h}$Cer[EOS], only a minor portion of the lipids formed LPP, with SPP predominating [51]. When FFA was added to the mixture containing the whole $_{h}$Cer, it revealed a SPP supremacy, which was predominantly orthorhombically packed [51]. This is in line with another study that reported that, although an initial increase in the relative amount of FFA or Cer[EOH] promoted LPP formation, when an optimal amount of Cer:FFA components was exceeded, it led to reduced LPP and a shift toward SPP dominance [49].

Investigations using $_{pig}$Cer have revealed both similarities and differences between $_{h}$Cer- and $_{pig}$Cer-containing systems. Common aspects include LPP dominance and hexagonal packing in $_{h}$Cer:Chol mixtures, consistent phase behaviour despite Cer:Chol ratio adjustments, a transition from hexagonal to orthorhombic packing upon FFA addition, increased Chol miscibility due to a ChS presence, and a significant reduction in the LPP phase when Cer[EOS] was absent. On the other hand, the primary differences were observed in the liquid lateral packing of $_{h}$Cer:Chol:FFA only upon ChS addition; the addition of FFA not only induced an orthorhombic transition but also decreased the proportion of lipids forming LPP in $_{h}$Cer mixtures, and larger fractions of Chol could be intercalated into lamellar phases of $_{h}$Cer compared to $_{pig}$Cer mixtures [51]. These differences were assigned to the documented association between the proportion of lipids forming fluid phases [50], with the fraction of Cer containing linoleic acid linked to a ω-hydroxy fatty acid being higher in the $_{h}$Cer than in $_{pig}$Cer [55].

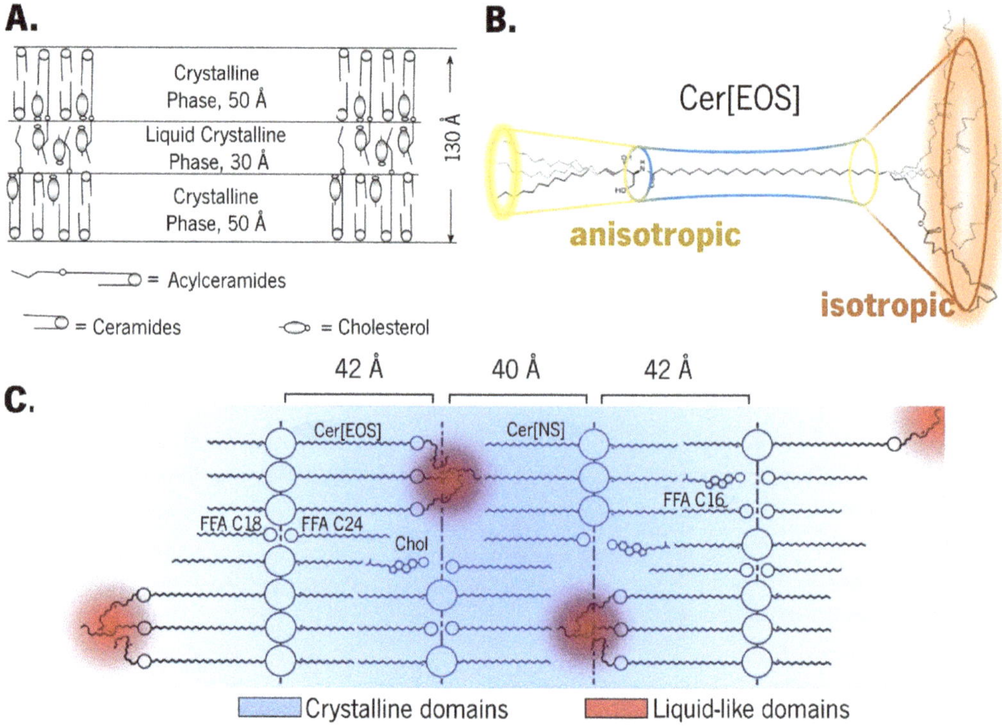

Figure 3. (**A**) Schematic illustration of the sandwich model purposed for a long periodicity phase (LPP) structure. Adapted with permission from [162]. Copyright 2007, Springer Nature. (**B**) Representation of the shape of a Cer[EOS] molecule. Adapted with permission from [161]. Copyright 2023, Elsevier. (**C**) Depiction of the suggested locations of the liquid-like domains in the LPP structure. Adapted with permission from [139]. Copyright 2018, Elsevier.

These studies lead to other critical factors affecting the LPP formation, in which all the mixture components have specific roles that must be balanced. The presence of Chol has been also demonstrated to be required for both LPP formation and to increase the packing density within the LPP unit cell [43,56,155]. A Chol:Cer[EOS] ratio of 1:2 has been identified as the minimum Chol content required for LPP formation, allowing for a certain degree of LPP flexibility [56,155]. The precise location of Chol in the LPP is unknown; nevertheless, it is unlikely to be predominantly situated in the middle lipid layer of the LPP unit cell [158]. Therefore, Chol potentially bridges the gap between the inner headgroup and the chains of the sphingoid base from Cers and FFAs extending from the unit cell border by locating them in the outer layer of the LPP and contributing to the increased density in the structure [155,159]. Although variations in the FFA content appear to exert a less critical influence on LPP formation compared to the Chol content, a specific range of FFA chain lengths remains essential to promote Chol solubilisation [124]. This aspect was previously mentioned, in which FFA decreased the Chol-phase separation. Moreover, for a proper lipid packing, it is crucial to maintain an optimal chain length variation within the lipid mixture, whether in the FFAs or the Cer fractions [49]. Assuming the Cer[EOS] headgroup is at the unit cell boundary, its C30 acyl chain extends towards the unit cell centre, terminating at the ester bond linking the linoleate [159]. Furthermore, McIntosh et al. demonstrated that Chol may be preferentially located at the outer sides of the LPP cell, which is composed of two bilayers with Chol asymmetrically distributed in each bilayer [57]. The hydrophobic part of Chol aligns closely with the saturated acyl chain of Cers, resulting in strong van der Waals interactions [57,159]. Assuming a perpendicular orientation to the basal plane of Cer[EOS], its ester group must be located very close to the Chol headgroup [159]. While the presence of the unsaturated linoleate moiety in the hydrophilic headgroup region is uncommon, it can be supported by Chol stabilisation. This support may arise from favouring hydrogen bonding with the carbonyl group, entropic stabilisation due to the high conformational disordering of the linoleate, and the limited headgroup hydrophilicity compared to phospholipid bilayers because the SC lipid matrix typically contains only one to two bound water molecules per lipid molecule [152,159], supported by the limited sensitivity of LPP to hydration levels [56,58]. Notably, variations in the headgroup assembly of Cers did not appear to be a prerequisite for LPP formation [155].

In a composition comprising Cer[EOS]:Cer[AP]:Chol (33:22:45 wt%), a lamellar phase with repeated distances resembling SPP was formed, accompanied by a Chol phase separated. This suggests that Cer[EOS] can also be accommodated within the SPP phase, possibly by spanning a layer and extending its acyl chains into neighbouring layers [59]. Interestingly, the addition of PA did not lead to LPP formation, probably due to differences in the chain lengths between short-chained PA and long-chained Cer[EOS], which hindered proper lipid mixing [59]. Similar findings were obtained when PA was replaced with longer-chained BA, TA, or CA [60]. Hence, it is important to acknowledge that the presence of Cer[EOS] is not the only prerequisite for LPP formation, and other additional parameters must be balanced for LPP formation with proper lipid miscibility.

3.2. The Short Periodicity Phase (SPP)

The suggestion that Cer[EOS] can be accommodated within the SPP phase [59,60] (Figure 4A), along with other evidence of Cers adopting different conformations when present in either LPP or SPP, has led to the armature reinforcement model.

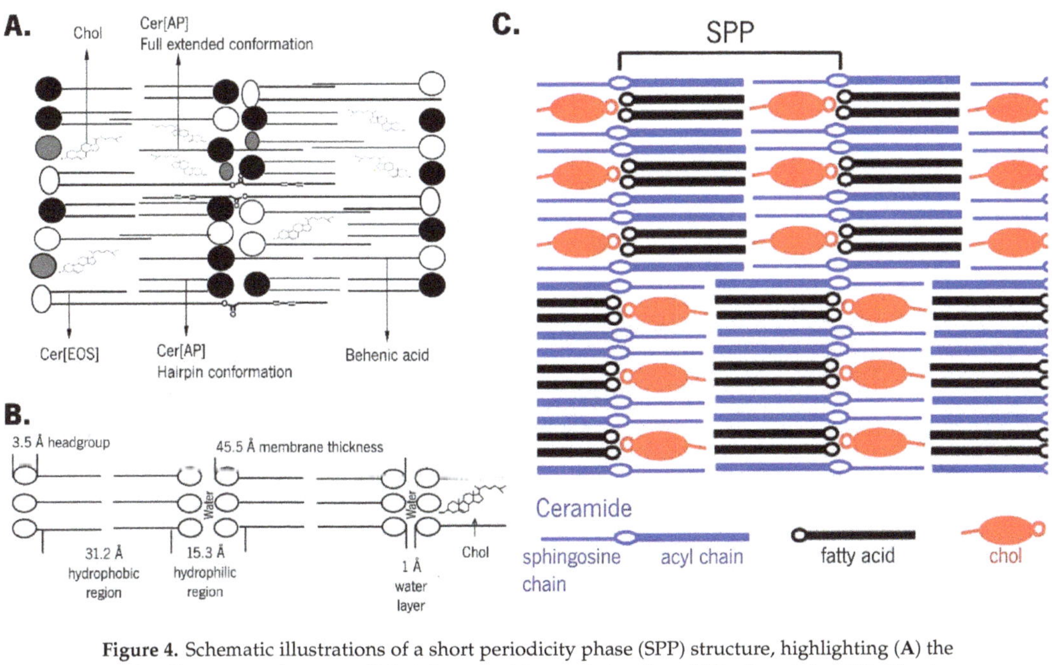

Figure 4. Schematic illustrations of a short periodicity phase (SPP) structure, highlighting (**A**) the Cer[EOS] accommodation on SPP. Adapted with permission from [60]. Copyright 2009, Elsevier. (**B**) The reinforcement model. Adapted with permission from [61]. Copyright 2005, Springer Nature. (**C**) The asymmetric lamellae with alternating directions. Adapted with permission from [126]. Copyright 2014, American Chemical Society.

This model is an extension of the sandwich model but specifically applies to the SPP [34]. Cers, as double-chain amphiphiles, can adopt either a hairpin conformation with both chains pointing in the same direction or a fully extended conformation with chains pointing in opposite directions [126]. The adhesion effect promoted by an extremely small intermembrane space is critical for the sandwich model, as the conformational features of Cers are associated with small-sized intermembrane spaces [162]. While several models for the orientation of the hydrocarbon chains have been proposed, the key distinction between the sandwich and armature reinforcement models lies in the postulation of a limited number of Cers in a fully extended conformation while the remaining Cers adopt a hairpin conformation [62,165]. Therefore, the fully extended conformation of Cers serves to strengthen the structural framework [162], which is symbolised by the headgroups linking the opposing headgroup regions of two adjacent leaflets.

Cer[AP] typically arranges symmetrically between the two leaflets of the bilayer, with its polar group serving as a bridge between the leaflets. The polar groups of opposite bilayers form the interface between two membrane leaflets, which is described as the polar headgroup adhesion effect [61]. The almost nonexistent intermembrane space is a consequence of this effect and imposes that the absence of a water layer between subunits creates optimal conditions for the permeation of Cer[EOS] between two layers [61]. Due to the strong lateral hydrogen bonding provided by Cer[AP], Cer[EOS] is obligated to arrange itself in the highly stable SPP [45,59]. This Cer[AP] imposition is also true for various long-chain FFAs arranged within SPP through chain interdigitation in the centre of the membrane or the formation of FFA-rich phases [59]. A model system containing Cer[NP]:Cer[AP] has been found to form lamellar structures with a thickness resembling the native SPP, wherein the long tails overlap within the lamellar centre, as observed in native SC. Interestingly, differences were reported depending on the dominant Cer present in the mixture. A higher content of Cer[AP] resulted in acyl chains with a more tilted packing, reducing their overlap within the midplane but having no effect on the lamellar phase. In contrast, a predominance of native-like Cer[NP] led to straight chains with a broad overlapping region in the lamellar midplane [147]. When used as the sole Cer in the SC model system, a Cer[AP] with 18 carbon atoms induced the formation of SPP [107]. Phase separation imposed by different Cer[AP] conformations was observed in a mixture of Cer[AP]:Chol:SA at a 1:0.7:1 molar ratio, which persisted even after the heating/cooling cycle [110,111]. Although a fully extended conformation of Cer[AP] is highly stable and energetically preferred [105,110,111], the introduction of large amounts of water and/or energy (e.g., heat) can result in a chain-flip of Cer[AP] molecules [61]. This supports the reinforcement model (Figure 4B) [63,162]: after hydration in excess water, the short chains of Cer[AP] undergo a chain-flip transition from a fully extended to hairpin conformation, resulting in increased intermembrane space, allowing water to enter and the disappearance of the narrow contact between adjacent layers [63]. The lipid mixtures of Cer[AP]:Chol:PA:ChS (55:25:15:5 wt%) have been characterised [61,64,65,105] and employed to test the permeation enhancement ability of various chemical compounds [66,104,106]. While vesicular preparation demonstrated phase separation [66,105], which was generally attributed to hairpin and fully extended conformations [166], when the mixture is prepared in multiple lipid bilayer stacks, no phase separation occurs, and a high degree of lamellar order is described [61,104,106]. The only phase in this system is assigned to a fully extended conformation of Cer[AP], which is known to promote an extremely strong intermembrane attraction, resulting in closer neighbouring bilayers, dense contact, and reduced water diffusion in the lateral direction [61] (Figure 4C).

The acyl chains of Cers and FFA cause interdigitation in the SPP, with Chol compensating for the long FFA chains that extend beyond the bilayer's core [126,158]. Cer[NS] and FFA C24 acyl chains exhibited distinct assemblies within the cores of the LPP and SPP [57,152]. In a mixture with Cer, Chol, and FFA, the long Cer[NS16] or very long-chained Cer[NS24] exhibited a fully extended conformation, which persisted with or without ChS and under hydration [126]. Models containing Cer[NS16] showed lower ordering and miscibility of this Cer with tested FFAs when compared to its C24 homologue. Although the results suggest either a hairpin or extended conformation of the used Cer with a random distribution of its symmetric chains, the hairpin conformation would result in a cross-sectional area very close to that of Chol. This proximity may facilitate miscibility and explain the higher sensitivity of Cer[NS16]-containing mixtures compared to Cer[NS24]-containing mixtures. Moreover, although LA mixed well with the Cer[NS24] in a SPP lamellar phase, some phase separation of Chol occurred, either with or without ChS [126]. These results contradict an asymmetric assembly of SPP in which Cer and FFA chains were partially interdigitated with Chol symmetrically located in both leaflets of the bilayer, slightly inward from the SPP unit boundary [152,153]. Therefore, Školová et al. purposed an explanation for these observations based on two hypotheses: (1) the use of a heterogeneous mixture of Cer that would affect lipid mixing; (2) it can be consistent at a macroscopic scale if the asymmetric assembly with extended Cer is considered, wherein Chol is associated with its sphingosine chain and FFA with its acyl chain, alternating in orientation rather than aligning in the same direction [126].

3.3. Cholesterol Influence on Stratum Corneum Lipid Assembly

The influence of Chol on lipid assembly was explored in a SC membrane model comprising Cer[EOS]:Cer[NS]:Cer[NP24]:Cer[AS]:Cer[NP16]:Cer[AP] as the Cer mixture [155]. The study revealed that, in addition to increasing lipid density, Chol played a crucial role in the formation of both LPP and orthorhombic lateral packing in SPP [155]. The proportion of lipids engaging in orthorhombical packing increased gradually with the increasing Chol content. This phenomenon could be attributed to the positioning of Chol within the gaps between the very long-chained FFAs, resulting in tighter orthorhombic packing (Figure 5) [125]. Actually, an inverse relationship could be observed between the Chol content and the occurrence of hexagonal lipid packing [125]. A maximum Chol level of 50% relative to the level of FFA was identified, as further increments led to the absence of a liquid-ordered phase [155]. Although Chol exhibited similar effects on the lateral packing of both LPP and SPP, some differences were observed: a larger fraction of lipid adopted orthorhombic packing in LPP compared to SPP, suggesting a distinct distribution of Chol in the LPP and SPP [155]. Additionally, a linear trend of decreasing repeated distances with the increasing Chol content in the membrane after heating/cooling cycles was noticed. The authors proposed two possible interpretations, neither of which precludes their coexistence: Chol increased the tilt angle with respect to the base plane of the membrane and/or Chol reduced the order of the hydrocarbon chains of other membrane components [67,111].

Where various studies have highlighted Chol as a significant factor influencing the miscibility of Cers and FFA, with substantial impacts in the thermotropic behaviour of SC lipid mixtures [68,155], phase interdigitation of Chol-rich crystalline domains has been frequently reported not only in SC lipid model systems [67,111] but also in native SC [167]. Pursuing a deeper comprehension of the Chol influence on the SC lipid membrane structure, Zbytovskáa et al. conducted a study combining different ratios of Cer[AP]:PA:Chol:ChS [67]. Their findings revealed that, for all the tested ratios, the fluidising effect of Chol was evident when the system was below the main phase transition temperature (T_m). Conversely, when the lipid system was above the T_m, Chol increased the chain order [67].

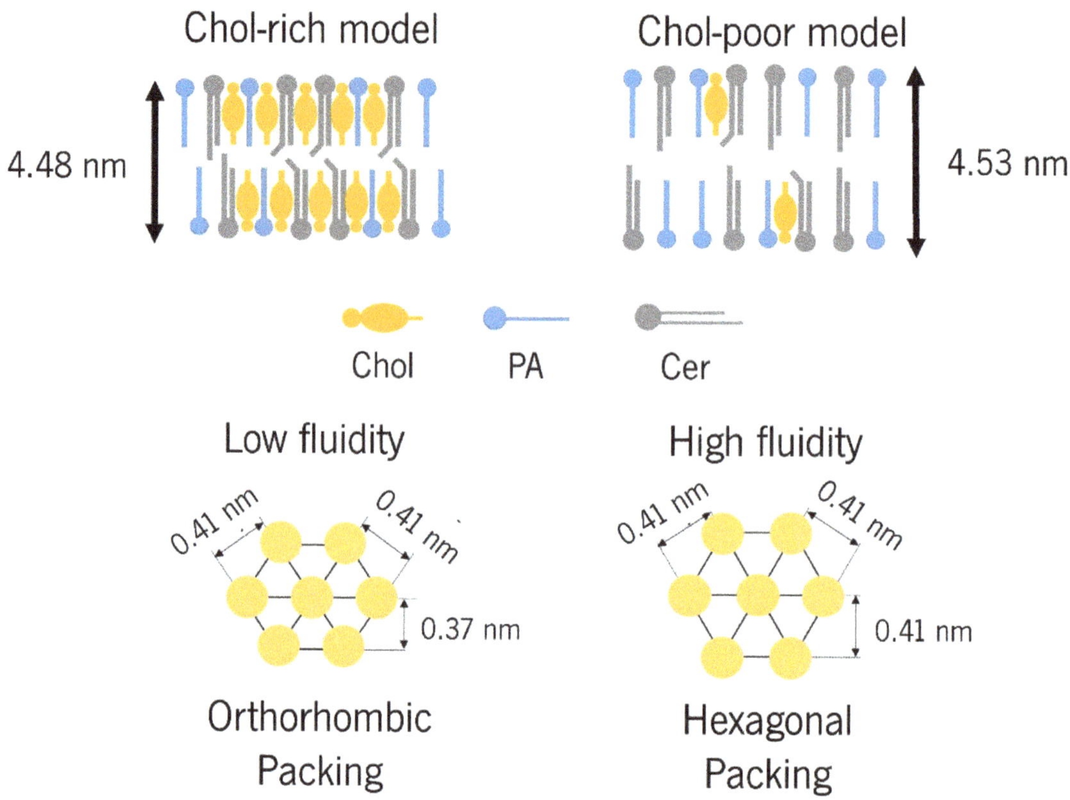

Figure 5. Schematic diagram of the cholesterol (Chol) influence on intercellular lipid matrix organisation. Adapted with permission from [125]. Copyright 2022, Elsevier.

3.4. Phase Separation and Promoters of Miscibility on Stratum Corneum Lipid Models

Pure Cer[NP] is characterised by its high melting temperature, forming crystalline orthorhombic structures [69,116,133]. Both Cer[NS] and Cer[NP] occupy the same location within the LPP unit, adopting predominantly an extended conformation with the (phyto)sphingosine and acyl chains on either side of the headgroup [151,168]. This linear conformation benefits the skin's barrier function by reducing the cross-sectional area per lipid molecule compared to the hairpin conformation, which enables a tighter packing of hydrocarbon chains [168]. Additionally, it enhances the connection between adjacent lipid layers, reducing permeability in the SC model system and discouraging lipid lamellae swelling during hydration [139]. Indeed, at room temperature, SC lipid mixtures typically exhibit a significant portion of their lipids in a solid crystalline phase [55]. The presence of phase-separated domains of Cers and FFAs arranged in an orthorhombic chain packing contributes to the low permeability of skin [69–72]. Subsequently, further homogenisation and the formation of a liquid crystalline phase occur at high temperatures [68,70,73]. Ordered-disordered phase transition has also been observed at temperatures similar to those found in native SC lipids [55,70,72,74].

In a model using $_{BB}$Cer[NP] in combination with Chol and PA at variable ratios, a common trend is observed: increased miscibility with the rising temperature. However, a high Chol content results in a delayed attainment of complete miscibility (from the 40–50 °C range to 75 °C) [73]. In Cer[NS] mixed with Chol or SA, Chol exhibited good miscibility with Cer at physiological temperatures and across a wide range of temperatures. In contrast, systems containing SA tend to phase separate into a SA-rich phase, achieving only moderate miscibility at high temperatures [68]. The kinetics of orthorhombic domain formation in SC lipid models were monitored via infrared spectroscopy by Mendelsohn et al. [122]. Conformational ordering and orthorhombic packing formation within the Cer[NS] chains occurred on a similar timescale as Cer headgroup region changes. In contrast, the formation of hydrogen-bonded dimers of SA started with a time lag of 3 to 4 h. The formation of FFA domains, along with conformational ordering of the lipid chains, likely required physical separation between hydrophobic regions, enabling their chains to segregate into orthorhombic domains [122]. The phenomenon of increased homogeneity with the temperature is contingent on several factors. Swelling in the LPP can be induced by elevated pH values or a high content of ChS, which promotes fluid phase formation due to its negatively charged sulphate group [57]. The sulphate group increases the molecular area per lipid, reduces the lipid packing density, enhances lipid miscibility, and strengthens its ability to form more hydrogen bonds with water compared to the OH group of Chol [51,108].

In the case of Cer[NP]:Chol:SA (55:25:20 w/w), phase separation occurs, with one phase exclusively composed of hairpin Cer[NP] molecules and the other phase consisting of a mixture of all components from the system, including fully extended Cer[NP] molecules that bridge adjacent bilayers [42]. In the hairpin conformation, the polar headgroup of Cer[NP18] is positioned between the two acyl chains, which are oriented in opposite directions [42]. Interestingly, this latter phase did not respond to applied contrast variation, as the headgroup of Cer[NP24] effectively shields against intrusion by water or deuterated water through the outstretched acyl chains and inter- and intramolecular hydrogen networks [42,157]. In equimolar Cer[NP]:Chol:LA, two phases coexist, with one phase being rich in fully extended Cer[NP24] and the other of reduced thickness, suggesting a distinct membrane structure and/or composition, possibly involving intercalated and/or tilted chains [116]. In systems involving Cer[NP] alone or combined with Cer[NS] or Cer[AP]:[NS] mixed with Chol:SA+ChS at a molar ratio of 1:1:1 + 5 wt%, three different phases were identified, corresponding to a phase rich in Cer[NP] with a minor incorporation of Chol and SA, a phase with high miscibility degree among all the components, and another phase resembling a pure SA phase with a minor presence of other components [133].

Although increasing the hydration or temperature induced phase separation in Cer[NS] mixtures, the demixing extent was highly dependent on the FFA chain length [120]. The presence of shorter FFAs reduced the temperature of the orthorhombic to hexagonal packing transition [120]. In equimolar mixtures of Cers, Chol, and FFA containing Cer[NS], Cer[NP], Cer[NP-18:1], or Cer[AS], the lipid assembly was dependent on the employed Cer [123]. Differences in the hydrocarbon chains were also investigated by comparing Cer[NP]- and Cer[NP-18:1]-based mixtures. Phase segregation was observed in the Cer[NP]-based system, whereas the use of Cer[NP18] promoted the formation of a more homogeneous phase [123]. The formation of solid crystalline phases results in ordering all the methylene segments into an all-*trans* configuration to maximise the van der Waals interaction, and maybe the double bond present at the middle of hydrocarbon chain leads to a strong reduction in the ability of Cer[NP-18:1] to form solid crystalline phases [123]. Depending on the used Cer, variations in the mixing properties with PA were noticed, including the coexistence of orthorhombic and hexagonal packing, along with the phase segregation of PA enriched with Chol from the phases of Cer[NP] with PA [123]. However, when SA is used instead of PA, both lipids adopt hexagonal chain packing, making mixing more likely [75]. The phase separation observed between FFAs and Cer has been attributed to a hydrophobic mismatch between their chain lengths [67,133,146]. The effect of FFA on LPP was notably dependent on their chain length (ranging from C16 to C28) [146]. Longer FFAs increased the repeat distances of the lamellar phase, whereas shorter FFAs had the opposite effect. Notwithstanding, whereas the FFA chain length affected the lateral packing, the orthorhombic to hexagonal packing occuring in phase transition was not affected. When shorter FFAs were neighbouring longer Cer C24 acyl chains, it caused hydrophobic mismatching and elastic stress on the lipid lamellae. When the amount of stress exceeds the level that the lamellae can accommodate, a lipid reassembly is required to relieve it, resulting in phase separation or the formation of more energetically favourable structures [156]. Although the presence of FFAs with varying acyl chain lengths is regarded as a promoter of lipid miscibility [52], it is only evidenced when matching FFA chain lengths and Cer acyl chains occur [120]. For example, a mixture containing six FFAs with different chain lengths in a SC model system inhibited the separation of individual FFAs from the multilamellar structure, likely due to a partial interdigitation that resulted in free spaces within the system, which were filled by FFA through tilted chains towards free volume minimisation inside the system [108]. Contrastingly, Engstrom et al. observed a tendency for polar lipids to segregate in the crystalline state, even when the lipids belonged to the same family [76]. This phase separation is attributed to packing constraints imposed by differences in chain length, particularly when the disparity in hydrocarbon chain length exceeds four carbon atoms [76]. Furthermore, there is a general propensity for the presence of very long-chained FFA to favour the mixing of lipid components, as reported in other studies involving SC lipid systems containing Cer[NS24] [120,127]. Paz Ramos et al. found that Cer[NS] combined with C24 FFAs led to mixtures with significantly greater homogeneity than those incorporating chains containing C16 or C20. Remarkably, after a week of incubation, the C24 system preserved its homogeneity, whereas systems containing FFA with C20 and C16 showed notable variations in local composition over time [127]. In the context of phase separation, it appears that the intermolecular interactions are the primary driving force, which suggests that very long acyl chains contribute to a relative increase in van der Waals interactions, thereby promoting lipid mixing in the SC lipid matrix [127]. Notwithstanding, other studies have proposed that chain heterogeneity might play a more crucial role in enhancing lipid miscibility compared to hydrophobic chain matching [77]. This was exemplified by cases where even with optimal hydrophobic matching between components, such as in the Cer[NS16]:Chol:FFA C16 mixture, extensive phase separation occurred [77]. It is worth noting that FFAs do not form hydrogen bonds with themselves due to the unavailability of undissociated hydrogen atoms in their headgroups; instead, their oxygen atoms are available for hydrogen bonding with other lipids [135]. Considering that FFA preferentially forms hydrogen bonds with dominant Cer molecules [135], it becomes apparent that the polarity of the headgroup also influences the miscibility characteristics of SC lipid mixtures. Indeed, the influence of the headgroup can outweigh the influence of the chain length in terms of the membrane assembly [45,59,78]. In experiments involving monolayers composed of variable combinations of $_{\mathrm{pig}}$Cer, $_{\mathrm{BB}}$Cer, Cer[NP16], or Cer[NP24], along with Chol and PA (C16) or LA (C24), phase separation into distinct lipid domains was observed. For instance, short-chain Cer[NP16] mixed with Chol and PA, while long-chain Cer[NP24] formed domains independently without mixing with LA [78]. Furthermore, the internal nanostructure of the Cer[AP] systems was not linearly affected by an increase in the FFA chain length in the system (from C18 to C26) [45]. Initially, the increase in FFA chain length led to a decrease in the membrane repeated distance through a partial interdigitation of the FFAs chains. This interdigitation resulted from the tendency of FFAs to 'fit' into the membrane size created by Cer[AP]. The authors explained that the formation of FFA-rich phases in the presence of longer-chained FFAs with C24 and C26 chains was due to the lower solubility of long-chained FFA in the Cer[AP]-based SC system [45]. This propensity of longer-chain FFA to form separated FFA-rich phases in Cer[AP]-based SC systems, even in presence of Cer[EOS], was also reported by Kessner et al. [59]. Therefore, interactions mediated by Cer[AP] molecules appear to be the main forces determining the stability and dictating the main phase of the system [45,59].

Consequently, the presence of FFA exerts a notable influence on the packing of Cer molecules. When Cer[NS]:Cer[AP] at 2:1 and 1:2 molar ratios are mixed with Chol and LA, a single lamellar phase is formed with the formation of FFA- and Cer-enriched nanosized domains, yet without complete phase separation. Moreover, Chol was uniformly distributed throughout the system [140]. This reinforces the limited miscibility of Cer[NS] with other lipids [68,77,120,126]. Interestingly, a smaller amount of Cer[AP] exhibited good miscibility with a higher quantity of Cer[NS], but the reverse scenario was not observed. In these systems, the lipid chains exhibited a slight tilt, and an assembly of Cer molecules with long overhanging tail ends of Cer C24 acyl chains and the long LA C24 chains overlapping in the lamellar middle was proposed. This was counteracted by the shorter C18 chain of the sphingoid base and Chol, allowing the opposing longer chains to fill these gaps [140]. On other hand, Chol is described as 'line active', since it tends to promote the miscibility of lipids [76,78], which is possible through the maximisation of hydrophobic interactions with alkyl chains from Cers and FFA [64].

3.5. Ceramide Headgroup Influence on Stratum Corneum Assembly

The effects of the Cer headgroup on the assembly of SC membrane lipids have been extensively explored [42,70,122,123,133,149,160]. Cer headgroups play a crucial role in governing the behaviour of LPP and SPP, impacting parameters like packing density, lipid miscibility, and hydrogen bonding strength [75,169]. Engelbrecht et al. conducted a study focusing on the replacement of Cer[AP] with Cer[NP] within SC lipid membranes containing SA and Chol. This substitution, which involved the removal of a single OH group from the Cer molecule, led to drastic structural alterations [42]. Although both were important in conferring high lamellar order to SC models, the Cer[AP]-containing system exhibited a more homogeneous mixture with a lower lamellar order than the one containing Cer[NP], which showed the coexistence of two lamellar phases [42,75]. Both Cers induced the formation of a solid crystalline phase with tightly packed and aligned hydrocarbon chains. Cer[AS]m, which has an extra OH group and can adopt different chain conformations, can also arrange its chains in a hexagonal lattice [123]. Moreover, the absence of an OH group in the headgroup region appeared to prevent hydration or swelling of the SC model membrane, possibly due to strong intra- and intermolecular headgroup interactions [42]. Badhe et al. demonstrated that differences between Cer[AP] and Cer[NP] could be attributed to variations in the hydrogen bonding dynamics. In a study with both Cer[AP] and Cer[NP] at 2:1 or 1:2 ratios, they found that the dynamics of hydrogen bonding with FFA and Chol changed significantly depending on the concentration of Cers [135]. Surprisingly, even though the additional OH group in Cer[AP] should facilitate greater hydrogen bonding, a dominant presence of Cer[AP] resulted in less ordered packing. This suggests that hydrogen bonding and the localisation of OH groups both play important roles in determining the packing order [133,135]. The additional OH group in Cer[AP] caused a tilt in the hydrocarbon chains, affecting their mobility and leading to a disrupted packing [135]. Packing defects due to the extra OH group were also reported, with a Cer[NP] system exhibiting disturbed packing compared to a Cer[NS] [160]. The extra OH group generated a steric conflict, reducing the lateral packing and, consequently, shifted the phase transition of the liquid crystalline phase to lower temperatures [160]. The impact of a single OH group on phase separation kinetics and subsequent barrier function was also observed when Cer[NS] was replaced by Cer[AS]. The extra OH group in Cer[AS] altered the propensity and quantity of the hydrogen bonds formed by both the Cers and FFA headgroups [122]. The removal of an OH group and the introduction of a trans-double bond in position 4 of an C18 Cer[NP] increased the solid crystalline phase compared to the C24 variant [160,170]. Longer C24 chains compensated for the packing defect induced by the additional OH group of Cer[NP], likely due to more methylene segments contributing to van der Waals interactions, compensating for the entropic losses caused by steric conflicts at the headgroup region [160]. Contrastingly, decreasing the chain length to C16 led to less ordered phases [160]. Another study demonstrated that, while adding Cer[NH] with three OH groups or Cer[AP] with four OH groups to a ternary mixture containing Cer[NS] did not interfere with the formation of the solid crystalline phase, the incorporation of tri- or tretrahydroxylated Cers resulted in stronger forces holding the lipid system together [142]. The most significant difference between Cer[AP] and Cer[NH] is the ability of the latter to promote the formation of a lamellar phase close to LPP, which appears to be related to the presence of OH at position 6 of the sphingoid backbone [142]. In another approach, the two stereometric forms of Cer[AP]: *D*- and *L*-isomers [166] were incorporated into ternary or quaternary mixtures [137]. While ternary mixtures containing only *D*-isomer as Cer exhibited crystalline-like behaviour, quaternary mixtures incorporating Cer[NP] exhibited a more SC lipid matrix-like behaviour [137]. This mixture was arranged with mostly straight chains and overlapping long chains in the lamellar midplane [137]. In the case of the *L*-isomer, no lamellar phase was found when it was the only Cer in the mixture, and when combined with Cer[NP], a highly crystalline-like behaviour was reported, classifying it as not safe [137]. These distinctions can be attributed to the reported hairpin conformation associated with the *L*-enantiomer of Cer[AP], whereas the *D*-enantiomer is associated with the non-tilted form, which is eventually associated with longer repeat distances [110]. Sphingomyelin (SPM) is a precursor of skin Cers, and its presence in a mixture with Chol and PA instead of Cer demonstrated how different headgroup structures have a significant influence on the SC structure [70]. While the SPM-based mixture formed a homogeneous liquid ordered mixture, replacing it with a Cer introduced a rich polymorphism into the structure. Additionally, while the Cer-based system showed a highly ordered structure with Cer and FFA residing in an orthorhombic subcell at low temperatures, a homogenisation with the formation of a fluid phase was reported at high temperatures [70]. Cer[EOP] and Cer[EOS], which differ in terms of the presence of an additional OH group and 4,5-desaturation, showed distinct abilities to promote the formation of LPP, with Cer[EOS] being more efficient in this regard [49].

Nonetheless, there are also examples of lamellar phases that are adaptable to differences in lipid headgroup assembly. While Cer[NP] and Cer[NS] occupy similar positions in the LPP unit, water permeation can influence their crystalline phases due to an increased opening angle of the hairpin conformation with higher hydration levels [123,137,151]. A 2–3 °C shift in the orthorhombic to hexagonal transition temperature was associated with increased hydration conditions, which was attributed to weakened hydrogen bonds between opposing headgroups of lipid layers induced by the incorporation of small amounts of water into neighbouring layers of headgroups [120]. The optimal hydrogen bonding of the hydrated phytosphingosine headgroup was found to be incompatible with orthorhombic chain packing [165]. This change in orthorhombic packing induced by Cer[NP18] was not related to the missing *trans* double bond in the phytosphingosine backbone [160]. The extent of orthorhombic chain packing of Cers depends on factors such as their chemical structure, temperature, and hydration degree [123]. Higher-phase transition temperatures are associated with stronger interactions in the headgroup region, primarily between OH groups and the amide group. Phytosphingosine-based Cers exhibit extremely low amide I wavenumbers, indicating a strong involvement of these amide groups in a tight hydrogen bond network. Conversely, sphingosine-based mixtures show higher amide I frequencies, suggesting weaker hydrogen bonding interactions between headgroups [123]. Monolayer measurements have revealed that the headgroup area of phytosphingosines is larger when compared to sphingosine Cers, and a more open polar interface allows for a stronger hydrogen bond network [123].

3.6. Main Considerations for Stratum Corneum Barrier Function

Lipids in the LPP are arranged in two layers sandwiching a narrow central lipid layer containing a subpopulation of fluid lipids. In SC lipid systems, the formation of the LPP is dependent on achieving an optimal fraction of lipids that form a liquid-like phase. When this fraction deviates from the optimal range, either by being too low or too high, it leads to an increase in SPP formation at the expense of LPP formation [50]. Specifically, the linoleic acid moiety of Cer[EOS] is found in this central layer [50].

Fluid phases exhibit significantly higher permeability compared to crystalline phases. Given that the fluid phase is situated in the central layer of the LPP, permeation parallel to the basal plane of lamellae is faster than permeation perpendicular to the lamellae (across headgroup regions) [50]. An increased fraction of unsaturated acyl chains, particularly those linked to the long base, enhances the fluidity and, consequently, permeability. Membrane permeability is directly affected by the relative complexity of the SC lipid system used [133]. Maintaining a balanced relationship between the three key properties—orthorhombic packing, hydrogen bonding, and miscibility—is critical for establishing an effective barrier.

To summarise, the most critical characteristics of SC lipid systems include (1) a wide distribution of acyl chain lengths in Cers; (2) the coexistence of Cers based either on phytosphingosine, sphingosine, or 6-hydroxysphingosine bases in a single mixture; (3) the presence of acylCers; and (4) a range of FFA chain lengths varying between C16 and C26. These characteristics play a pivotal role in influencing lipid phase behaviour, with acylCers being crucial for LPP formation and the need for a broad distribution of FFA chain lengths to form both LPP and SPP without additional coexisting phases rich in FFA or Cers, replicating the lipid assembly found in human SC.

4. *Stratum Corneum* Lipid Models as Surrogates for Permeation Studies

In recent years, significant effort has been directed toward the development of skin surrogates that better mimick the complexity of the SC ILM. This endeavour arises from both the simplicity and intricacy of existing SC lipid model systems, along with the limitation associated with the current skin surrogates. Therefore, Table 2 provides a summary of recent developments in innovative SC surrogates (SCS) based on SC lipid models.

From a general point of view, a SCS can be described as a synthetic model of SC ILM that consists of depositing an appropriate mixture of lipids mimicking a SC ILM composition on a porous substrate. This porous substrate covered by the lipid mixture intends to be mounted in a typical diffusion cell between donor and acceptor compartments (Figure 6A). What differentiates the type of SCS is the approach followed to cover the filter membranes with lipids. Therefore, the approaches can be divided into spraying by airbrush (Figure 6A), the Skin-PAMPA™ (Figure 6B), and the Phospholipid Vesicle-based Permeation Assay (PVPA) (Figure 6C).

Table 2. Published approaches of *Stratum Corneum* surrogates (SCS) incorporating SC lipid biomimetic models for permeation studies.

SC Lipid Model Composition	Porous Substrate	Characterisation Techniques	Model Compounds	Main Outcomes	Ref.
Airbrush approaches					
Cer:Chol:FFA (1:1:1 mol ratio) Cer = [EOS]:[NS]:[NP24]: [AS]:[NP16]:[AP] (15:51:16:4:9:5 molar ratio) FFA = PA:SA:AA:BA:TA:LA:CA (1.3:3.3:6.7:41.7:5.4:36.8:4.7 molar ratio)	Polycarbonate 50 nm filter	SWAXS and CryoEM	n.a.	• Airbrush is able to homogeneously spray a SC mixture onto a polycarbonate filter. • Synthetic lipids in the SCS resembled the orientation of the ILM of SC: the width of the arcs of SCS were close to the ones in the SC. • SCS characteristics showed great potential for use as a skin barrier diffusion model.	[58]
		Diffusion studies	PABA, ethyl-PABA, and butyl-PABA	• Barrier properties of SCS prepared with a 12-μm-thick lipid layer are similar to those of isolated human SC. • SCS model can be easily tuned by adjusting the lipid mixture to mimick different healthy and diseased skin.	[30]
Cer:Chol:FFA (1:1:1 molar ratio) Cer = [EOS]:[NS]:[NP24]: [AS]:[NP16]:[AP] (15:51:16:4:9:5 molar ratio) FFA = PA:SA:AA:BA:TA:LA:CA (1.8:4:7.7:42.6:5.2:34.7:4.1 molar ratio)	Polycarbonate 50 nm filter	Different airbrush methods, SAXS, and diffusion studies	Benzoic acid	• Irrespective of the preparation method, benzoic acid permeation profiles were very similar to human SC. • While the rotor method is claimed to increase efficiency and reproducibility, the Linomat method reduced the lipid loss during lipid airbrush and was thus selected as the best method to spray lipids on porous substrate.	[79]
Cer:Chol:FFA:$ChSO_4$ (Variable ratio) Cer = [EOS]:[NS]:[NP24]: [AS]:[NP16]:[AP] (15:51:16:4:9:5 molar ratio) FFA = PA:SA:AA:BA:TA:LA:CA (1.8:4:7.7:42.6:5.2:34.7:4.1 mol%)	Polycarbonate 50 nm filter	FTIR, SAXS, and permeability studies	Benzoic acid	• Two SCS models showed significant changes in the benzoic acid steady-state flux: an excess of Chol decreased flux, while an excess of $ChSO_4$ increased the flux. • Lipid composition disorders in psoriasis and dry skin may not be responsible for impaired skin barrier function in vivo.	[171]

Table 2. *Cont.*

SC Lipid Model Composition	Porous Substrate	Characterisation Techniques	Model Compounds	Main Outcomes	Ref.
Cer:Chol:FFA (1:1:1 molar ratio) Cer = [EOS]:[NS]:[NP]:[AP] (15:51:16:4:9:5 molar ratio) FFA = PA:SA:AA:BA:TA:LA:CA or MyA:PA:SA:AA:BA (Variable ratio)	Polycarbonate 50 nm filter	SAXS, EM, and permeation studies	Benzoic acid	• Permeability of the SC system to benzoic acid closely follows that of human SC at temperatures ranging from 31 to 43 °C.	[172]
Cer[AP]:Chol:PA:ChS (55:25:15:5 wt%)	Polycarbonate 50 nm filter	SAXS, polarisation, confocal and ESEM microscopies, Raman imaging, FTIR, and diffusion studies	Urea	• A continuous uniformly distributed lipid layer on polycarbonate filters was obtained. • Membranes showed higher barrier function than isolated human SC.	[173]
Cer:Chol:FFA (1:1:1 molar ratio) Cer = [EOS]:[NS]:[NP24]:[NP16] (16:56:18:10 mol%)	Polycarbonate 50 nm filter	FTIR and interaction studies	γ-undecalactone, DDA, and diethyl 1,4-cyclohexanedicarboxylate	• The dose-dependent effects of the chemicals on the lateral molecular packing in the SCS were qualitatively identical to those observed by IR spectroscopy in human SC.	[174]
Cer[NS]:Chol:PA (1:1:1 molar ratio)	Polycarbonate 50 nm filter	Laurdan fluorescence, SEM, Raman scattering, and permeability studies	Benzoic acid and CAF	• SCS showed continuous coverage of gel-phase lipids, allowing it to be used in permeability studies. • Barrier properties in the same order as 'real' SC.	[175]
Cer:Chol:FFA (1:1:1 molar ratio) $_{pig}$Cer or $_{syn}$Cer = [EOS]:[EOP]:[NS]:[NP24]:[AS]:[NP16]:[AP] (Variable ratio) FFA = PA:SA:AA:BA:TA:LA:CA (1.8:4:7.7:42.6:5.2:34.7:4.1 mol%)	Polycarbonate 50 nm filter	SAXS, FTIR, and permeation studies	HC	• Reduced barrier in SCS prepared with $_{pig}$Cers compared to synthetic Cers. • Chain length distribution affects the lipid barrier.	[164]
Cer:Chol:FFA (1:1:1 molar ratio) Cer = [EOS]:[NS]:[NP24]:[AS]:[NP16]:[AP] (15:51:16:4:9:5 molar ratio) FFA = PA:SA:AA:BA:TA:LA:CA (1.8:4:7.7:42.6:5.2:34.7:4.1 mol%)	Polycarbonate 50 nm filter	SAXS, FTIR, TEWL, and permeation studies	HC	• Presence of MUFAs in the FFA fraction enhanced SCS permeability by inducing the formation of hexagonal lateral packing. • Lipid barrier function decreased with the increased unsaturation degree.	[176]

Table 2. *Cont.*

SC Lipid Model Composition	Porous Substrate	Characterisation Techniques	Model Compounds	Main Outcomes	Ref.
$_a$Cer:Chol:LA:ChS (1:1:1 molar ratio + 5 wt%)	Polycarbonate 15 nm filter	IR, SWAXS, and permeation studies	TH and IND	• Strong permeabilisation effects of pentadecasphingosine Cer with short acyls. • SCS permeability to relatively small and hydrophilic compounds was more sensitive to lamellar assembly than to the overall lipid chain order and packing.	[177]
Cer:Chol:FFA (1:1:1 molar ratio) Cer = [NS] or [NS]:[NP24]:[AS]:[NP16]:[AP] (60:19:5:11:6 mol%) FFA = PA:SA:AA:BA:LA(1.8:4:7.6:47.8:38.9 mol%)	Polycarbonate 50 nm filter	FTIR, XRD, and permeability studies	ethyl-PABA	• Ethyl-PABA flux was affected by the SCS FFA composition: flux was higher when a mixture of FFA was used than when a single FFA C24 was present. • Ethyl-PABA flux was not affected by the Cer composition.	[178]
Cer:Chol:SA:ChS (1:1:1 molar ratio + 5 wt%) Cer = [NP]:[AP] (Variable ratio)	Polycarbonate 15 nm filter	IR, EI, EM, and permeation studies	TH and IND	• SCS based on one Cer showed greater permeability for both chemicals (Cer[AP] more than Cer[NP]) than those based on their mixture, without significant difference from the flux on porcine skin.	[179]
Cer:Chol:FFA (1:1:1 molar ratio) Cer = [EOS]:[NS], [EOS]:[AS], [EOS]:[NP] or [EOS]:[AP] (40:60 mol%) FFA = PA:SA:AA:BA:LA (1.8:4:7.6:47.8:38.8 mol%)	Polycarbonate 50 nm filter	FTIR, X-ray, and permeability studies	ethyl-PABA	• Due to distinct in-phase behaviour, lipid chain packing, and headgroup interactions, sphingosine-based Cer SCS resulted in a higher permeability than the phytosphingosine SCS.	[169]
Cer:Chol:FFA (1:1:1 molar ratio) Cer = [EOS]:[NS]:[NS24]:[NS16]:[NP]:[AS24]:[AS16]:[AP24] (Variable ratio) FFA = PA:SA:AA:BA:LA (Variable ratio)	Polycarbonate 50 nm filter	FTIR, SWAXS, permeability studies, and TEWL	ethyl-PABA	• Changes in the SCS lipid compositions mimicking atopic dermatitis patients resulted in decreased barrier capability; FFA compositional changes showed a greater impact on barrier function than changes in the Cer profile.	[180]

Table 2. *Cont.*

SC Lipid Model Composition	Porous Substrate	Characterisation Techniques	Model Compounds	Main Outcomes	Ref.
$_{h}$Cer:Chol:FFA:ChS (1:1:1 molar ratio + 5 wt%) FFA = PA:SA:AA:BA:LA (1.8:4:7.6:47.8:38.8 mol%) Cer[NS]:Cer[EOS]:Chol:LA:ChS (Variable ratio)	Polycarbonate 15 nm filter	WAXS, FTIR, TEWL, electrical impedance, and permeation studies	TH and IND	• While 40% of the Chol SCS was less permeable than the control, Chol depletion did not compromise the barrier function to water, TH, and IND.	[181]
$_{a}$Cer[NS]:Chol:LA:ChS (1:1:1 molar ratio + 5 wt%)	Polycarbonate 15 nm filter	XRD, IR spectroscopy, permeability studies, water loss, and electrical impedance	TH and IND	• SCS containing 3-deoxy-Cer, as well as 1-deoxy-Cer, had similar permeabilities. • Methylation of the Cer[NS] nitrogen improved its miscibility with Chol, disordered lipid chains, and increased the permeability compared to the physiological Cer[NS] control.	[36]
Cer:Chol:LA (1:1:1 molar ratio) Cer = [EOS]:[NS24] or [EOS NS16] (Variable ratio)	Polycarbonate 50 nm filter	FTIR, SAXS, and permeation studies	ethyl-PABA	• Ethyl-PABA permeability increased significantly when ≥50% of Cer[NS24] was substituted with Cer[NS16].	[182]
Cer:Chol:FFA+ChS (1:1:1 molar ratio + 5 wt%) Cer = [NS], *R*-/*S*-[AS], [NdS], *R*-/*S*-[AdS], [NP] or *R*-/*S*-[AP]	Polycarbonate 15 nm filters	XRD, FTIR, and permeability studies	TH and IND	• Mixed effects highly dependent on the sphingoid base chain, configuration, and permeability marker used.	[37]
Cer:Chol:FFA (1:1:1 molar ratio) Cer = [EOS]:[NS] (10–90:90–10 mol%) FFA = PA:SA:AA:BA:LA (1.8:4:7.6:47.8:38.8 mol%)	Polycarbonate 50 nm filters	SAXS, FTIR, permeation studies, and TEWL	ethyl-PABA	• Steady-state flux only increased when the Cer[EOS] content was raised to 50%, becoming significantly higher when it was raised to ≥70%.	[183]

Table 2. *Cont.*

SC Lipid Model Composition	Porous Substrate	Characterisation Techniques	Model Compounds	Main Outcomes	Ref.
Cer:Chol:FFA:ChS (1:1:1 molar ratio + 5 wt%) Cer = [EOP]:[EOdS]:[EOS]:[AP]:[AS]:[NP]:[NS]:[NdS]:[AdS] (Variable ratio) FFA = BA:LA:AA:SA:PA (47.1:41.4:6.9:3.3:1.3 mol%)	Polycarbonate 15 nm filter	XRD, FTIR, electrical impedance, and permeability studies	TH and IND	• SCS with 10–20 mol% of acylCers mimick the nanostructure and permeability of healthy skin barrier lipids, while 0–5 mol% mimick some hallmarks of a diseased skin lipid barrier.	[184]
				• Permeability studies indicated that ω-OHCers could not be substituted by ω-O-acylCers. • Absence of ω-O-acylCers resulted in diminished barrier properties.	[185]
Cer[NS]:Chol:FFA:ChS (Variable ratio) FFA = PA:SA:AA:BA:LA (1.8:3.9:7.5:47.8:39 mol%)	Polycarbonate 15 nm filters	H-NMR, SAXS, WAXS, zeta potential, and permeability studies	TH and IND	• High ChS:Chol ratio increased the SCS permeability.	[186]
Cer[NS]:Chol:PA (1:1:1 molar ratio)	Polycarbonate 50 nm filters	SEM, DLS, and permeation studies	CAF	• CAF showed similar and comparable permeability profiles in SCS using two different approaches and in pig skin.	[187]
Skin-PAMPA™ approaches					
Certramides:Chol:SA (0 to 1:1:1 molar ratio)	Multiscreen-IP 450 nm pores	Permeability studies	CIP, NFD, and VER	• A relationship between the SCS concentration of certramides (from 0 to 100%) and the permeability of the compounds was found. • Strongest effect on permeability was caused by the certramide C12–C16.	[80]
Certramides:Chol:SA: Si oil	Stirwell™ PAMPA	Permeability studies	APAP, DCF, FUR, NAP, PEF, TH, and VER	• By comparison to three different databases, skin–PAMPA demonstrated high prediction capability.	[188]
PVPA-based approaches					
E80:$_{BSC}$Cer:Chol:PA:ChS (77:23:0:0:0 or 50:27.5:12.5:7.5:2.5 wt%)	Mixed cellulose ester 650 nm filters	Permeation studies	Flu, IBP, IND, SAL, FITC-dextran, and CAL	• PVPA approaches were able to distinguish between substances with different degrees of transdermal absorption.	[189]

Table 2. *Cont.*

SC Lipid Model Composition	Porous Substrate	Characterisation Techniques	Model Compounds	Main Outcomes	Ref.
E80:$_{BSC}$Cer:Chol:PA:ChS (50:27.5:12.5:2.5:7.5 wt%)	Mixed cellulose ester 650 nm filters	Permeation and stability studies	Liposomal formulations of DCF	• Adaptations to the original systems were performed towards nanoformulation testing. • All liposomal formulations of DCF exhibited significantly higher permeabilities than DCF solution. • PVPA has shown to be valuable for screening and the optimisation of lipid nanoformulations at the preformulation stage.	[190]
E80:$_{BSC}$Cer:Chol:PA:ChS (77:23:0:0:0 or 50:27.5:12.5:7.5:2.5 wt%)	Mixed cellulose ester 650 nm filters	Stability and permeation studies	CAF, ACV, CPL, and CAL (liposomal formulation and/or aqueous solution)	• PVPA systems were more promising in detecting permeability differences between aqueous solution and liposomal formulations than EpiSkin®. • PVPA showed greater stability than EpiSkin® (up to 2 weeks against 3-day testing window).	[191]
		Diffusion studies	CAF, DCF, CAP and CAL	• SCS showed good correlation by ranking model drugs similarly to the ranks obtained using a pig ear skin model and were comparable to the literature on permeation through healthy human skin.	[192]
EPC/SPC:$_{BSC}$Cer:Chol:FA:ChS (25–80:41–11:18–5: 3.75–1:11.25–3 wt%)	Nylon 450 nm filters	Menthol enhancer permeation studies, comparison with porcine ear skin and ATR-FTIR	FCA, PF, ALB, THC, and THP	• PVPA model was adapted towards employment in a permeation-enhancing effect evaluation. • The enhancement ratios obtained were in accordance with data from porcine ear skin. • ATR-FTIR analysis detected a similar mechanism of menthol in both models (PVPA and porcine ear skin).	[193]
EPC:Chol (77:23 wt%)	Nylon 450 nm filters	SEM, electrical resistance, ATR-FTIR, and permeation studies	FCA, PF, ALB, THC, and THP	• P_{app} values obtained through the PVPA-based system were well correlated with the values obtained through porcine ear skin. • Better relevance between porcine ear skin and the PVPA-based system was found compared to that between porcine ear skin and a commercial Strat-M® membrane.	[194]

Table 2. *Cont.*

SC Lipid Model Composition	Porous Substrate	Characterisation Techniques	Model Compounds	Main Outcomes	Ref.
Cer:EPC:Chol:SA:ChS (50:25:12.5:10:2.5 wt%)	Polycarbonate 400 nm filters	SEM, electrical resistance permeation, and stability studies	CAL, CAF, CSP, DCF, MTX, and NAP	• Modified PVPA presented low permeability of hydrophilic markers when compared to the original PVPA (mimicking the intestinal barrier). • Good data correlation with permeability studies in a pig ear model.	[195]
		Stability studies and permeability studies	CAL	• Modified PVPA kept stable for 12 weeks when stored at −20 °C. • Barrier integrity was not affected by the presence of a set of cosolvents, but ethanol altered its integrity.	[196]

$_{a}$Cer—analog ceramides; AA—arachidic acid; ALB—albiflorin; APAP—acetaminophen; ATR-FTIR—attenuated total teflection Fourier-transform infrared; BA—behenic acid; BSC—bovine spinal cord; CA—cerotic acid; CAF—caffeine; CAL—calcein; CAP—chloramphenicol; Cer—ceramide; Cer[AdS]—α-hydroxy fatty acid/dihydrosphingosine base ceramide; Cer[AP]—α-hydroxy fatty acid/phytosphingosine base ceramide; Cer[EOdS]—ω-hydroxy fatty acid/dihydrosphingosine base ceramide; Cer[EOS]—ester-linked ω-hydroxy fatty acid/sphingosine base ceramide; Cer[NdS]—non-hydroxy fatty acid/dihydrosphingosine base ceramide; Cer[NP]—non-hydroxy fatty acid/phytosphingosine base ceramide; Cer[NS]—non-hydroxy fatty acid/sphingosine base ceramide; Chol—cholesterol; ChS—cholesteryl sulfate; CIP—ciprofloxacin; CSP—cyclosporine; DCF—diclofenac; DDA—dodecyl acetate; DLS—dynamic light scattering; E-80—egg phospholipid lipoid E-80; EM—electron microscopy; EPC—egg phosphatidylcholine; ESEM—environmental scanning electron microscopy; FCA—ferulic acid; FFA—free fatty acid; Flu—flufenamic acid; FTIR—Fourier-transform infrared; FUR—furosemide; HC—hydrocortisone; $_{h}$Cer—human isolated ceramide; IBP—ibuprofen; IND—indomethacin; IR—infrared; LA—lignoceric acid; MTX—methotrexate; NAP—naproxen; NFD—nifedipine; PA—palmitic acid; PABA—*p*-aminobenzoic acid; PAMPA—parallel artificial membrane permeability assay; PEF—pefloxacin; PF—paeoniflorin; PVPA—phospholipid vesicle-based permeation assay; $_{pig}$Cer—pig isolated ceramide; SA—stearic acid; SAL—salicylic acid; SAXS—small-angle X-ray scattering; $_{syn}$Cer—synthetic ceramide; SEM—scanning electron microscopy; SPC—soybean phosphatidylcholine; SWAXS—small- and wide-angle X-ray scattering; TA—tricosylic acid; TEWL—transepidermal water loss; TH—theophylline; THC—tetrahydrocolumbamine; THP—tetrahydropalmatine; WAXS—wide-angle X-ray scattering; VER—verapamil; XRPD—X-ray powder diffraction.

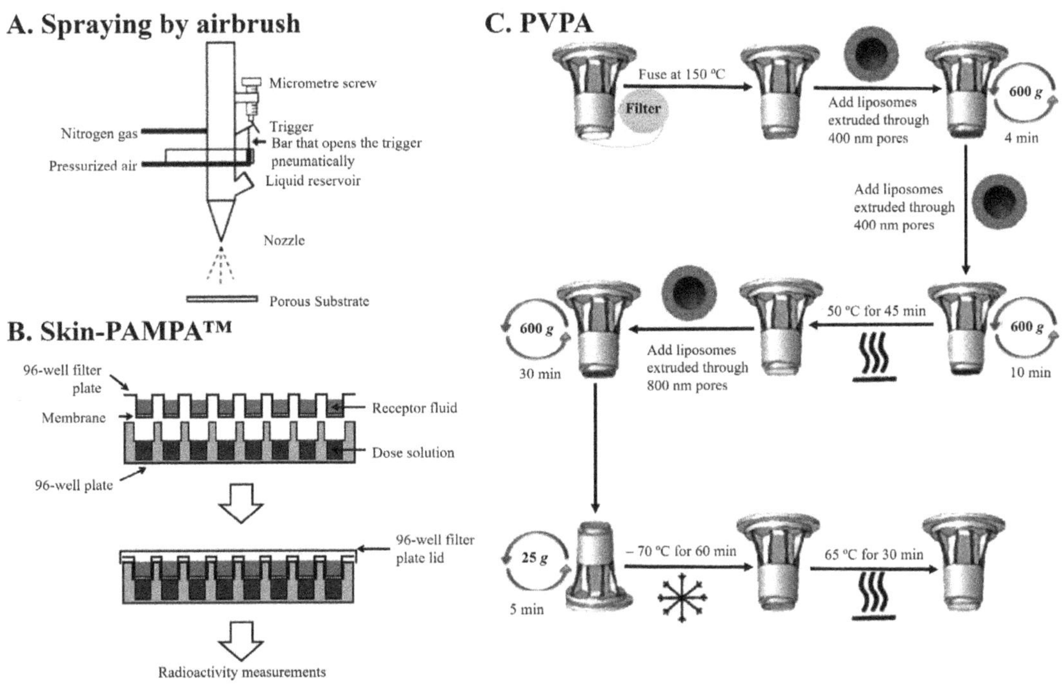

Figure 6. Schematic representation of each approach described to obtain *Stratum Corneum* lipid mixture-based surrogates for in vitro permeation evaluation. (**A**) Spraying by airbrush. Adapted with permission from [58]. Copyright 2006, Elsevier. (**B**) Skin-Parallel Artificial Membrane Permeability Assay (Skin-PAMPA™). Adapted with permission from [197]. Copyright 2013, Elsevier. (**C**) Phospholipid Vesicle-based Permeation Assay (PVPA). Adapted with permission from [189]. Copyright 2013, Wiley.

4.1. Spraying by Airbrush

The deposition of the SC lipid mixture onto a porous substrate via the spraying method entails a series of steps that have been optimised through time. Initially, lipids are dissolved in a hexane:96% ethanol (2:1 v/v) solution to a final desired concentration of 4.5 mg·mL^{-1}. Subsequently, this solution is sprayed onto the substrate under a nitrogen flow, typically employing a Linomat system with an extended y-arm (Figure 6A). Following that, an incubation period is employed, which can vary in temperature (from 70 to 90 °C, depending on the Tm of the used mixture) and duration (from 10 to 30 min), culminating in a cooling step to room temperature that can last 30 min, 3 h, or overnight.

In studies comparing three different methods for lipid airbrushing—manual, using a rotor, or the Linomat—the Linomat apparatus was found to be the most suitable for preparing SCS due to its efficiency and the uniformity of the membrane thickness compared to the other two methods. Although using benzoic acid as model compound, the steady-state flux was similar for the three SCS, and the values were very similar to human SC, so the authors selected the Linomat as the most appropriate apparatus for preparing the SCS [185]. Furthermore, the most critical factors in the spraying protocol to ensure the suitability of the SCS for permeability tests, in which it is imperative to obtain a densely packed and uniform layer, were investigated [58]. The research findings suggest that these critical factors include (i) the total lipid concentration; (ii) the composition of the organic solvent mixture; and (iii) the spraying equipment parameters, like the distance between the airbrush and the filter, the nitrogen pressure during both spraying and drying stages, and the amount of lipid reaching the porous substrate [58]. Deviation in the process conditions can lead in suboptimal outcomes, such as organic solvent evaporation occurring from the inside to the periphery and potential overspray of lipid solution onto the periphery of the filters, which typically occurs when nitrogen pressure is too high or the evaporation rates for the lipid solution are inadequate. During solvent evaporation, the lipid concentration increases, leading to lipid crystallisation based on their solubility. In diffusion experiments, non-hydrated lipid membranes must maintain their integrity when exposed to donor and acceptor solutions for an extended duration. Preliminary studies indicated that the SCS remained intact for at least 20 h during a typical diffusion experiment [58]. The lipid membranes cooled to room temperature before hydration exhibited minimal pore formation, with no water pools within the membranes at room temperature [58].

The use of this approach is prevalent in the scientific literature, and it is used not exclusively for permeation experiments but also for the biophysical and dynamical characterisation of SC permeability. An equimolar mixture of Cer[NS]:Chol:PA airbrushed in polycarbonate membranes formed a gel-phase structure capable of maintaining robustness even at high temperatures [175]. The permeability of caffeine (CAF) and benzoic acid in this SCS was comparable to that of real SC. However, there were variations in the membrane thickness (ranging from 16 to 26 μm) due to some sprayed droplets not coalescing into the continuous layer [175]. SCS composed solely of phytosphingosine Cers (Cer[NP] and Cer[AP]) demonstrated permeability characteristics similar to human skin when theophylline (TH) and indomethacin (IND) were used as model compounds [179]. However, the steady-state time for both permeants in SC membranes was shorter than in human skin. Variable ratios of the Cers content were tested, and while Cer[AP] showed the highest TH permeability, there was no significant difference from the TH and IND flux in porcine skin [179]. The influence of sphingosine- and phytosphingosine-based Cers on SCS permeability was compared. The steady-state flux of ethyl-PABA was significantly lower in phytosphingosine-based SCS compared to their sphingosine counterparts. There was no significant difference in the steady-state flux of ethyl-PABA between SCS with non-hydroxy fatty acid-based Cers and their α-hydroxy fatty acid counterparts [169]. Kovacik et al. investigated the effects of Cer α-hydroxylation and stereochemistry on the barrier properties. Both saturated sphingosine double bonds of Cer[NdS] and Cer[NP] and α-hydroxylation increased the water loss compared to the control Cer[NS], except for Cer(R)-[AdS], where water loss did not change. TH, a small molecule with balanced lipophilicity, showed decreased permeability upon Cer α-hydroxylation, with a significant difference between Cer(R)-[AdS] and Cer(S)-[AdS] diastereomers. IND, being larger and more lipophilic, exhibited similar permeability across α-hydroxylated Cers, except for Cer(R)-[AdS] and Cer(S)-[AdS] diastereomers, which showed twice the permeability of Cer[NdS] [37]. Furthermore, Kovácik et al. synthesised several unnatural Cer[NS]: 1-deoxy-, 3-deoxy-, and N-Me- and combined them with LA, Chol, and ChS for permeability testing [36]. Modifications at the polar headgroup had a significant effect on the SC model properties. The presence of an OH group at the C1 position was crucial for a proper lipid mixture, as the 1-deoxy-Cer[NS]-based system exhibited significantly higher permeability compared to physiological Cer[NS]. The N-methylation of Cer[NS] increased the water loss threefold and TH and IND permeability tenfold compared to the control [36]. Moreover, the influence of increasing the Cer[EOS] content in SCS barrier function was investigated [183]. Increasing the Cer[EOS] content from 10 to 30% of the total Cer fraction did not significantly affect the ethyl-PABA permeability. However, a further increase to 50, 70, and 90% of the Cer[EOS] content resulted in enhanced permeability, accompanied by increased water loss at higher Cer[EOS] contents (70 and 90%), attributed to the formation of a liquid-like phase.

The SCS were also examined for their utility in testing permeation enhancers, particularly those affecting the SC lipid matrix, such as Azone® and L-Pro2 [183]. Groen et al. examined the suitability of a SCS to mimick human SC by studying the interactions between the SC lipid system and three models from a class of chemicals frequently used as fragrance raw materials: γ-undecalactone, dodecyl acetate, and diethyl 1,4-cyclohexanedicarboxylate [174]. Between the several advantages, including the possibility to investigate the hydrogen bonding between headgroups, one limitation was assigned to SCS models over SC by tape-stripping: the impossibility so far to determine the in-depth profile of compounds [174].

4.2. Skin-Parallel Artificial Membrane Permeability Assay (Skin-PAMPA™)

Originally, the PAMPA method was developed for the rapid assessment of the passive membrane transport of molecules. Due to its potential cost-effectiveness and high-throughput capabilities, PAMPA has been widely adopted in the pharmaceutical industry [80]. The PAMPA method involves a 96-well plate with a filter that separates two compartments: the donor compartment containing a buffer solution of the compound under investigation and the acceptor compartment containing fresh buffer solution (Figure 6B). Over time, adaptations of PAMPA have been made to evaluate the epithelial intestinal barrier and blood–brain barrier (BBB) permeation. Recently, Sinko et al. introduced Skin-PAMPA™. In Skin-PAMPA™, the SC ILM lipid model consists of a lipid mixture containing certramides (instead of Cers), Chol, SA, and silicon oil [80,188]. Certramides, which are analogous of Cers, differ from natural Cers in that they lack unsaturated bonds in the side chain but share structural similarities, such as similar molecular size, hydrogen bond/acceptor capabilities, and high lipophilicity [80,188]. Skin-PAMPA showed a limited correlation with the epidermis, but a stronger correlation was observed with full-thickness skin [188]. When comparing Skin-PAMPA™ with other artificial models and using porcine skin as the reference, permeability data for a selection of drugs solubilised in different vehicles were examined. The results indicated that, among several artificial models, including Strat-M®, Skin-PAMPA™ provided the highest correlation by accurately ranking four out of six vehicles [197].

Despite claims of structural similarity between certramides and Cers and good correlations with animal models, the absence of unsaturated bonds and the presence of silicone oil in the mixture may limit the ability of the Skin-PAMPA™ tool to accurately replicate the complex structure of SC ILM. As discussed in the previous section, the composition of the lipid mixture employed has a direct impact on membrane permeability, and the presence of unsaturated acyl chains is critical for the formation of either LPP and liquid-like domains crucial for the proper elasticity of SC lipid models [50,157,159]. Furthermore, analysing the statements taken before as the most critical characteristics of SC lipid systems, Skin-PAMPA™ fails in the coexistence of different bases of Cers in a single mixture and in the presence of acylCers and also fails to guarantee a range of FFA chain lengths varying between C16 and C26.

4.3. Phospholipid Vesicle-Based Permeation Assay (PVPA)

The other proposal to obtain SCS is based on a lipid covering barrier created by a tight layer of liposomes on a filter support—PVPA [189,191]. The PVPA model was initially developed to mimick the intestinal barrier based on a membrane filter support with deposited liposomes of a defined composition [198,199]. A schematic representation of the full method to fabricate PVPA membranes is depicted in Figure 6C. The original PVPA model waas prepared by placing the liposomes through centrifugation onto a mixed cellulose filter, followed by solvent evaporation and freeze–thaw cycling to promote liposome fusion, resulting in a tight barrier. This vesicle-based barrier was characterised by unrequired agitation during the experiments, which was explained by the presence of aqueous compartments immobilised within the vesicle's matrix. Later, a modified version of the PVPA model with phospholipids composed by a SC lipid matrix lipid mixture was designed [189], in which the permeability of different compounds was evaluated and validated by comparison with data obtained from animal skin models and from in silico [189]. The model was then compared to a commercially available reconstructed human skin model—EpiSkin®—and the authors reported that the PVPA model was more effective, less expensive, and had longer storage stability than EpiSkin® [191]. Other permeation experiments were conducted based on this approach, including testing the PVPA relevance in preformulation studies and evaluating permeation enhancers [190,192,193,195,196]. Shakel et al. then presented a modified PVPA. The authors sonicated the multilamellar vesicles (MLVs) colloidal dispersion to decrease and homogenise the liposome size towards the adaptation to polycarbonate filter support with 400 nm. Then, the modified PVPA briefly consisted of retaining/passing of a portion of large unilamellar vesicles (LUVs) through a membrane filter with successive centrifugation, followed by freeze–thawing until the pores were filled by the LUVs and a layer on top began to form by the addition of MLVs [195]. While the PVPA from Engesland et al. was stable up to 2 weeks of storage at −70 °C, Moniz et al. showed that their modified PVPA [195] could maintain its integrity and lipid content over 12 weeks when stored at −20 °C, without changes in the calcein (CAL) permeability [196]. In the same investigation, while the modified PVPA model was not affected by the presence of a set of cosolvents that were usually employed as skin permeation enhancers (Tween® 20, Tween® 80, and PEG 400), ethanol at 10% (*v*/*v*) altered the integrity of the barrier [196].

The main disadvantage of the PVPA approach lies in the presence of phosphatidylcholine in the SC mimetic mixture, which does not translate the subclasses characteristically found in SC.

5. Commercially Available and Patented Skin Models

The use of skin, whether human or animal, has long been regarded as the gold standard for assessing (trans)dermal permeability. The commercial products like Transderm-Scop®, Androderm®, and Alora® were excised human skin used as a surrogate for in vivo studies [11,200]. Yet, ethical, technical, and economic challenges have stalled their widespread use. Obtaining human skin poses hurdles due to ethical concerns, high variability, low throughput, and technical challenges, limiting its use. Similarly, obtaining animal skin is time-consuming, technically challenging, and costly, and excised skins do not accurately represent human skin physiology and structure. In addition, both in vivo animal and human studies and ex vivo studies using excised skin models contribute significantly to the carbon footprint of pharmaceutical industries, raising sustainability concerns. The pharmaceutical and industry/research sectors are increasingly inclined to adopt other in vitro testing methods to lessen the reliance on in vivo studies or ex vivo surrogates. This trend is further emphasised by the EU prohibition on animal testing in cosmetic and toxicology evaluations by the directive 76/768/EEC [201], demanding alternative skin permeation testing methods.

In response to this, models equivalent to living skin have been developed by culturing human skin components such as keratinocytes and fibroblasts, and several human skin cell cultures are commercially available as skin models: EpiSkin™ model (Episkin, L'Oréal, Lyon, France), EpiDerm™model (MatTek Corporation, Ashland, MA, USA), SkinEthic™ (Episkin, L'Oréal, Lyon, France), Labskin™ (Labskin, Lyon, France), EpiCS® model (CellSystems, Troisdorf, Germany), Straticell model (Straticell, Les Isnes, Belgium), and Labcyte model (Gamagori, Japan); Full-thickness models such as the StrataTest® model (Stratatech, Madison, WI, USA), Phenion Full-Thickness Skin Model (Phenion, Düsseldorf, Germany), GraftSkin® (Apligraf, Organogenesis, MI, USA), EpiDermFT® (MatTek Corporation, Ashland, MA, USA), and Vitrolife-Skin™ model (Kyoto, Japan) [11,202]. These models have been employed in Franz diffusion cells for permeation assessments or incorporated into microfluidic chips (known as 'skin-on-chip'). However, they still fail to address several issues related to excised tissue, proving expensive, requiring specialised technical skills, and inadequately replicating human skin permeation. Primarily utilised for skin irritation and toxicity assessments, these cellular models present challenges that surpass the conveniences offered by microfluidic analysis. Despite the potential benefits of microfluidic technology in enhancing screening processes, the integration of skin cell-based models presents obstacles to achieving high-throughput screening. These hindrances include limitations in storage capacity of the cell-based models, lack of reproducibility, technical handling requirements, and elevated screening costs [11].

Finally, non-cellular skin models have been developed to access skin permeation. membranes. Most models are synthetic membranes for transdermal delivery composed of silicone-based membranes such as Silatos™ (LMA, Better by design, London, UK), Silastic® (Dow Corning Corporation, Midland, MI, USA), and multiple layers of polyester sulfone Strat-M® (Merck Millipore, Burlington, MA, USA). While synthetic membranes offer certain advantages over human or animal skins, establishing a definitive correlation with human SC barrier function, especially in finite dose applications, remains a challenge. Additionally, these models, predominantly composed of hydrophobic polymers, exhibit varying permeation efficiencies. While they provide reasonable estimations for hydrophilic compounds, they tend to overestimate the permeation of lipophilic compounds [11].

Recently, there have been advancements in synthetic barriers with compositions that more closely mimick natural skin. For instance, Permeapad® (PHABIOC GmbH, Karlsruhe, Germany, patent WO 2016/078667 A1) [203] features two cellulose membranes with a phospholipid layer in between, while Skin-PAMPA™ (Pion Inc., Billerica, MA, USA), already detailed in the previous section, utilises a model containing a synthetic amide-based compound containing certramide, free fatty acid, and cholesterol. Permeapad® offers a membrane model for use in Franz diffusion cells or as part of the PermeaPad® Plate consisting of two 96-well plates. The upper plate serves as the donor, with an insert plate integrating the membrane model, and the bottom plate acts as the acceptor. The bottom PermeaPad® Plate can be conveniently placed in standard fixtures of 96-well plates from HPLC equipment to quantify permeated compounds. On the other hand, Skin-PAMPA™ is available as the SuperQuick® Skin-PAMPA Kit (Creative bioarray, New York, NY, USA) comprising two 96-well plates, with the upper plate serving as the donor and the bottom plate as the acceptor. The skin model is provided as an organic solution to be added to the donor plate and allowed to dry. Like Permeapad®, the bottom plates of Skin-PAMPA™ can be placed in standard fixtures of 96-well plates from HPLC equipment for the quantification of permeated compounds.

Besides Permeapad® and the SuperQuick® Skin-PAMPA Kit, there are no competing high-throughput screening commercially available solutions incorporating cell-free mimetic models of skin. However, some patented devices can be applied for cells-on-chip (US10870823B2 [204], WO2021168511A1 [205], and US2022265176A1 [206]). Few other cell-free skin models have been patented:

CN114577678A—This Chinese national patent presents liposomes hydrated with 10% keratin arranged between two sealed polycarbonate membranes forming a mimetic model of skin. Filed by Nanjing Traditional Chinese Medicine University in 2022, the model offers the advantage of containing keratin, representing a protein component of the skin. However, polycarbonate membranes lack physiological relevance and do not mimick any skin component. There is no associated high-throughput device for a permeation assessment [207].

WO2008011812A1—This international patent describes synthetic phospholipid polymers containing amphipathic regions. Filed by the Shanghai Cancer Institute in 2008, these polymers mimick the dual affinity of the lipid matrix (hydrophilic and hydrophobic portions). However, they do not possess physiological relevance, failing to mimick the lamellar and lateral organisation crucial for the skin's permeation barrier functions [208].

6. Conclusions and Prospects

Over the last 30 years, numerous research groups have contributed to a better understanding of the structure of skin and the mechanisms of its permeation, and due to increasingly sophisticated biophysical techniques, our knowledge on skin permeation has grown markedly. The promising outcomes obtained from the analysis of model drugs and other permeability markers underscore the clear potential of SCS to be employed in a more standardised, less expensive, and more ethical way when investigating drugs or cosmeceuticals, especially those intended for compromised skin conditions, compared to other current models. SCS can potentially reduce the costs and the use of animal testing in the early stages of drug candidates, drugs, and cosmeceuticals development. While each discussed approach has its own set of advantages and disadvantages, a significant shared advantage of all SCS models is their versatility in allowing the use of different lipid compositions. These lipid compositions can be tailored to mimick specific skin disorders that are challenging to reproduce in clinical studies. SCS models are appealing, because the lipid composition can be precisely defined and modified as needed to simulate the desired conditions. This flexibility enables researchers to investigate how various chemicals impact both healthy and diseased skin through subtle changes in lipid systems.

SCS systems are well suited for testing a wide range of chemical classes that may come into contact with SC lipids and potentially affect skin barrier properties negatively. Therefore, SCS can find extensive applications in the fields of skin biophysics, dermatology, transdermal drug delivery, and risk assessment.

In conclusion, future directions of the use of skin models should encopass the following main features:

(i) an increasingly mimetic composition and structure of SC ILM essential to reproduce its permeation barrier function. To this end, not only the lipid composition should be considered but also the protein fraction (e.g., keratin).
(ii) adapting skin models to high-throughput screening. The integration of skin cell-free models into microfluidic chips that mimick the skin laminar flow can boost throughput and reduce cost, cut human intervention and errors, and set a new standard for skin permeation studies.
(iii) adding a new dimension to skin permeation evaluations by providing molecular/biophysical insights into permeation, vital to predicting potential skin toxicity.
(iv) investing in models that not only mimick healthy skin but also injured skin conditions that enable researchers to delve into the intricate dynamics of how various skin conditions influence compound permeation, offering invaluable insights into treatment efficacy and safety. Mimicking permeation behaviour in diseased skin aligns with the principles of personalised medicine, emphasising tailored treatments based on individual patient profiles and needs. This would empower researchers to deepen their understanding of how specific skin conditions impact barrier function and permeability, thereby facilitating the development of precise therapeutic interventions.

Supplementary Materials: The following supporting information can be downloaded at https://www.mdpi.com/article/10.3390/pharmaceutics16060807/s1: Table S1. Examples of *Stratum Corneum* (SC) lipid model mixtures reported in the literature for either deciphering SC structure or as in vitro platforms to study compound-SC lipid matrix interaction (previous to 2010).

Author Contributions: Conceptualisation and writing—E.F., C.M.L. and M.L.; Visualisation, E.F.; funding acquisition, M.L. and C.M.L. All authors have read and agreed to the published version of the manuscript.

Funding: This work was supported by the Portuguese Foundation for Science and Technology (FCT) in the framework of the Strategic Funding UIDB/04650/2020 and UIDB/04050/2020 (CF-UM-UP and CBMA), UIDP/04378/2020 and UIDB/04378/2020 (Research Unit on Applied Molecular Biosciences, UCIBIO), and the project LA/P/0140/2020 (Associate Laboratory Institute for Health and Bioeconomy, i4HB). E. F. is funded by FCT, POPH, and FEDER/COMPETE through grant SFRH/BD/147938/2019.

Institutional Review Board Statement: Not applicable.

Informed Consent Statement: Not applicable.

Data Availability Statement: Not applicable.

Conflicts of Interest: The authors declare no conflicts of interest.

References

1. Kolarsick, P.; Kolarsick, M.A.; Goodwin, C. Anatomy and Physiology of the Skin. *J. Dermatol. Nurses' Assoc.* **2011**, *3*, 203–213. [CrossRef]
2. Lopes, C.M.; Silva, J.; Real Oliveira, M.E.C.D.; Lúcio, M. Chapter 14—Lipid-based colloidal carriers for topical application of antiviral drugs. In *Design of Nanostructures for Versatile Therapeutic Applications*; Grumezescu, A.M., Ed.; William Andrew Publishing: Norwich, NY, USA, 2018; pp. 565–622. [CrossRef]
3. Zaidi, Z.; Lanigan, S.W. Skin: Structure and Function. In *Dermatology in Clinical Practice*; Lanigan, S.W., Zaidi, Z., Eds.; Springer: London, UK, 2010; pp. 1–15. [CrossRef]
4. Elias, P.M. Stratum Corneum Defensive Functions: An Integrated View. *J. Investig. Dermatol.* **2005**, *125*, 183–200. [CrossRef] [PubMed]
5. Elias, P.M. Structure and function of the stratum corneum extracellular matrix. *J. Investig. Dermatol.* **2012**, *132*, 2131–2133. [CrossRef] [PubMed]
6. Richard, C.; Cassel, S.; Blanzat, M. Vesicular systems for dermal and transdermal drug delivery. *RSC Adv.* **2020**, *11*, 442–451. [CrossRef] [PubMed]
7. Souto, E.B.; Fangueiro, J.F.; Fernandes, A.R.; Cano, A.; Sanchez-Lopez, E.; Garcia, M.L.; Severino, P.; Paganelli, M.O.; Chaud, M.V.; Silva, A.M. Physicochemical and biopharmaceutical aspects influencing skin permeation and role of SLN and NLC for skin drug delivery. *Heliyon* **2022**, *8*, e08938. [CrossRef] [PubMed]
8. Michaels, A.S.; Chandrasekaran, S.K.; Shaw, J.E. Drug permeation through human skin: Theory and invitro experimental measurement. *AIChE J.* **1975**, *21*, 985–996. [CrossRef]
9. Baroli, B. Penetration of nanoparticles and nanomaterials in the skin: Fiction or reality? *J. Pharm. Sci.* **2010**, *99*, 21–50. [CrossRef] [PubMed]
10. Walters, K.A.; Roberts, M.S. The structure and function of skin. In *Dermatological and Transdermal Formulations*; Walters, K.A., Ed.; Marcel Dekker: New York, NY, USA, 2002; pp. 1–41.
11. Neupane, R.; Boddu, S.H.S.; Renukuntla, J.; Babu, R.J.; Tiwari, A.K. Alternatives to Biological Skin in Permeation Studies: Current Trends and Possibilities. *Pharmaceutics* **2020**, *12*, 152. [CrossRef] [PubMed]
12. Schmid-Wendtner, M.H.; Korting, H.C. The pH of the Skin Surface and Its Impact on the Barrier Function. *Ski. Pharmacol. Physiol.* **2006**, *19*, 296–302. [CrossRef] [PubMed]
13. Wagner, H.; Kostka, K.-H.; Lehr, C.-M.; Schaefer, U.F. pH profiles in human skin: Influence of two in vitro test systems for drug delivery testing. *Eur. J. Pharm. Biopharm.* **2003**, *55*, 57–65. [CrossRef]
14. Lee, S.H.; Jeong, S.K.; Ahn, S.K. An Update of the Defensive Barrier Function of Skin. *Yonsei Med. J.* **2006**, *47*, 293–306. [CrossRef] [PubMed]
15. Blank, H.I. Measurement of pH of the skin surface. *J. Investig. Dermatol.* **1939**, *2*, 67–79. [CrossRef]
16. Schaefer, H.; Redelmeier, T.E.; Lademann, J. Skin Penetration. In *Contact Dermatitis*; Johansen, J.D., Frosch, P.J., Lepoittevin, J.-P., Eds.; Springer: Berlin/Heidelberg, Germany, 2011; pp. 215–227. [CrossRef]
17. Jepps, O.G.; Dancik, Y.; Anissimov, Y.G.; Roberts, M.S. Modeling the human skin barrier—Towards a better understanding of dermal absorption. *Adv. Drug Deliv. Rev.* **2013**, *65*, 152–168. [CrossRef]
18. Supe, S.; Takudage, P. Methods for evaluating penetration of drug into the skin: A review. *Ski. Res. Technol. Off. J. Int. Soc. Bioeng. Ski.* **2021**, *27*, 299–308. [CrossRef] [PubMed]
19. Nategholeslam, M.; Gray, C.G.; Tomberli, B. Stiff Spring Approximation Revisited: Inertial Effects in Nonequilibrium Trajectories. *J. Phys. Chem. B* **2017**, *121*, 391–403. [CrossRef] [PubMed]
20. Lundborg, M.; Wennberg, C.L.; Narangifard, A.; Lindahl, E.; Norlen, L. Predicting drug permeability through skin using molecular dynamics simulation. *J. Control. Release* **2018**, *283*, 269–279. [CrossRef] [PubMed]
21. Abraham, M.H.; Martins, F. Human skin permeation and partition: General linear free-energy relationship analyses. *J. Pharm. Sci.* **2004**, *93*, 1508–1523. [CrossRef] [PubMed]
22. Mitragotri, S.; Edwards, D.A.; Blankschtein, D.; Langer, R. A mechanistic study of ultrasonically-enhanced transdermal drug delivery. *J. Pharm. Sci.* **1995**, *84*, 697–706. [CrossRef] [PubMed]
23. Schmook, F.P.; Meingassner, J.G.; Billich, A. Comparison of human skin or epidermis models with human and animal skin in in-vitro percutaneous absorption. *Int. J. Pharm.* **2001**, *215*, 51–56. [CrossRef]
24. Catz, P.; Friend, D.R. Transdermal delivery of levonorgestrel. VIII. Effect of enhancers on rat skin, hairless mouse skin, hairless guinea pig skin, and human skin. *Int. J. Pharm.* **1990**, *58*, 93–102. [CrossRef]
25. Chilcott, R.P.; Barai, N.; Beezer, A.E.; Brain, S.I.; Brown, M.B.; Bunge, A.L.; Burgess, S.E.; Cross, S.; Dalton, C.H.; Dias, M.; et al. Inter- and intralaboratory variation of in vitro diffusion cell measurements: An international multicenter study using quasi-standardized methods and materials. *J. Pharm. Sci.* **2005**, *94*, 632–638. [CrossRef] [PubMed]
26. Haq, A.; Goodyear, B.; Ameen, D.; Joshi, V.; Michniak-Kohn, B. Strat-M® synthetic membrane: Permeability comparison to human cadaver skin. *Int. J. Pharm.* **2018**, *547*, 432–437. [CrossRef]
27. *EC1223/2009*; Cosmetic Products Regulation. European Parliament, Council of the European Union: Brussels, Belgium, 2009.

28. *EC1907/2006;* Registration, Evaluation, Authorization and Restriction of Chemicals (REACH). European Parliament, Council of the European Union: Brussels, Belgium, 2006.
29. Silva, R.J.; Tamburic, S. A State-of-the-Art Review on the Alternatives to Animal Testing for the Safety Assessment of Cosmetics. *Cosmetics* **2022**, *9*, 90. [CrossRef]
30. de Jager, M.; Groenink, W.; Bielsa i Guivernau, R.; Andersson, E.; Angelova, N.; Ponec, M.; Bouwstra, J. A novel in vitro percutaneous penetration model: Evaluation of barrier properties with p-aminobenzoic acid and two of its derivatives. *Pharm. Res.* **2006**, *23*, 951–960. [CrossRef] [PubMed]
31. Kessner, D.; Ruettinger, A.; Kiselev, M.A.; Wartewig, S.; Neubert, R.H. Properties of ceramides and their impact on the stratum corneum structure. Part 2: Stratum corneum lipid model systems. *Ski. Pharmacol. Physiol.* **2008**, *21*, 58–74. [CrossRef] [PubMed]
32. Bouwstra, J.A.; Gooris, G.S. The Lipid Organisation in Human Stratum Corneum and Model Systems. *Open Dermatol. J.* **2010**, *4*, 10–13. [CrossRef]
33. Bouwstra, J.; Pilgram, G.; Gooris, G.; Koerten, H.; Ponec, M. New aspects of the skin barrier organization. *Ski. Pharmacol. Appl. Ski. Physiol.* **2001**, *14* (Suppl. 1), 52–62. [CrossRef]
34. Schmitt, T.; Neubert, R.H.H. State of the Art in Stratum Corneum Research. Part II: Hypothetical Stratum Corneum Lipid Matrix Models. *Ski. Pharmacol. Physiol.* **2020**, *33*, 213–230. [CrossRef]
35. Schroeter, A.; Eichner, A.; Mueller, J.; Neubert, R.H.H. The Importance of Stratum Corneum Lipid Organization for Proper Barrier Function. In *Percutaneous Penetration Enhancers Chemical Methods in Penetration Enhancement*; Dragicevic, N., Maibach, H.I., Eds.; Springer-Verlag: Berlin/Heidelberg, Germany, 2015; pp. 19–38. [CrossRef]
36. Kovacik, A.; Pullmannova, P.; Pavlikova, L.; Maixner, J.; Vavrova, K. Behavior of 1-Deoxy-, 3-Deoxy- and N-Methyl-Ceramides in Skin Barrier Lipid Models. *Sci. Rep.* **2020**, *10*, 3832. [CrossRef]
37. Kovacik, A.; Pullmannova, P.; Opalka, L.; Silarova, M.; Maixner, J.; Vavrova, K. Effects of (R)- and (S)-alpha-Hydroxylation of Acyl Chains in Sphingosine, Dihydrosphingosine, and Phytosphingosine Ceramides on Phase Behavior and Permeability of Skin Lipid Models. *Int. J. Mol. Sci.* **2021**, *22*, 7468. [CrossRef]
38. van Smeden, J.; Hoppel, L.; van der Heijden, R.; Hankemeier, T.; Vreeken, R.J.; Bouwstra, J.A. LC/MS analysis of stratum corneum lipids: Ceramide profiling and discovery. *J. Lipid Res.* **2011**, *52*, 1211–1221. [CrossRef]
39. Masukawa, Y.; Narita, H.; Shimizu, E.; Kondo, N.; Sugai, Y.; Oba, T.; Homma, R.; Ishikawa, J.; Takagi, Y.; Kitahara, T.; et al. Characterization of overall ceramide species in human stratum corneum. *J. Lipid Res.* **2008**, *49*, 1466–1476. [CrossRef] [PubMed]
40. Rabionet, M.; Gorgas, K.; Sandhoff, R. Ceramide synthesis in the epidermis. *Biochim. Biophys. Acta* **2014**, *1841*, 422–434. [CrossRef]
41. van Smeden, J.; Boiten, W.A.; Hankemeier, T.; Rissmann, R.; Bouwstra, J.A.; Vreeken, R.J. Combined LC/MS-platform for analysis of all major stratum corneum lipids, and the profiling of skin substitutes. *Biochim. Biophys. Acta* **2014**, *1841*, 70–79. [CrossRef]
42. Engelbrecht, T.N.; Schroeter, A.; Hauß, T.; Demé, B.; Scheidt, H.A.; Huster, D.; Neubert, R.H.H. The impact of ceramides NP and AP on the nanostructure of stratum corneum lipid bilayer. Part I: Neutron diffraction and 2H NMR studies on multilamellar models based on ceramides with symmetric alkyl chain length distribution. *Soft Matter* **2012**, *8*, 6599–6607. [CrossRef]
43. Bouwstra, J.A.; Gooris, G.S.; Dubbelaar, F.E.R.; Weerheim, A.M.; Ijzerman, A.P.; Ponec, M. Role of ceramide 1 in the molecular organization of the stratum corneum lipids. *J. Lipid Res.* **1998**, *39*, 186–196. [CrossRef] [PubMed]
44. de Jager, M.; Gooris, G.; Ponec, M.; Bouwstra, J. Acylceramide head group architecture affects lipid organization in synthetic ceramide mixtures. *J. Investig. Dermatol.* **2004**, *123*, 911–916. [CrossRef]
45. Ruettinger, A.; Kiselev, M.A.; Hauss, T.; Dante, S.; Balagurov, A.M.; Neubert, R.H. Fatty acid interdigitation in stratum corneum model membranes: A neutron diffraction study. *Eur. Biophys. J.* **2008**, *37*, 759–771. [CrossRef] [PubMed]
46. Flach, C.R.; Mendelsohn, R.; Rerek, M.E.; Moore, D.J. Biophysical studies of model stratum corneum lipid monolayers by infrared reflection-absorption spectroscopy and Brewster angle microscopy. *J. Phys. Chem. B* **2000**, *104*, 2159–2165. [CrossRef]
47. Bouwstra, J.A.; Gooris, G.S.; van der Spek, J.A.; Bras, W. Structural investigations of human stratum corneum by small-angle X-ray scattering. *J. Investig. Dermatol.* **1991**, *97*, 1005–1012. [CrossRef]
48. Percot, A.; Lafleur, M. Direct observation of domains in model stratum corneum lipid mixtures by Raman microspectroscopy. *Biophys. J.* **2001**, *81*, 2144–2153. [CrossRef] [PubMed]
49. de Jager, M.W.; Gooris, G.S.; Ponec, M.; Bouwstra, J.A. Lipid mixtures prepared with well-defined synthetic ceramides closely mimic the unique stratum corneum lipid phase behavior. *J. Lipid Res.* **2005**, *46*, 2649–2656. [CrossRef] [PubMed]
50. Bouwstra, J.A.; Gooris, G.S.; Dubbelaar, F.E.; Ponec, M. Phase behavior of stratum corneum lipid mixtures based on human ceramides: The role of natural and synthetic ceramide 1. *J. Investig. Dermatol.* **2002**, *118*, 606–617. [CrossRef] [PubMed]
51. Bouwstra, J.A.; Gooris, G.S.; Dubbelaar, F.E.R.; Ponec, M. Phase behavior of lipid mixtures based on human ceramides: Coexistence of crystalline and liquid phases. *J. Lipid Res.* **2001**, *42*, 1759–1770. [CrossRef] [PubMed]
52. de Jager, M.W.; Gooris, G.S.; Dolbnya, I.P.; Bras, W.; Ponec, M.; Bouwstra, J.A. The phase behaviour of skin lipid mixtures based on synthetic ceramides. *Chem. Phys. Lipids* **2003**, *124*, 123–134. [CrossRef] [PubMed]
53. de Jager, M.W.; Gooris, G.S.; Dolbnya, I.P.; Ponec, M.; Bouwstra, J.A. Modelling the stratum corneum lipid organisation with synthetic lipid mixtures: The importance of synthetic ceramide composition. *Biochim. Biophys. Acta* **2004**, *1664*, 132–140. [CrossRef] [PubMed]

54. Gooris, G.S.; Bouwstra, J.A. Infrared spectroscopic study of stratum corneum model membranes prepared from human ceramides, cholesterol, and fatty acids. *Biophys. J.* **2007**, *92*, 2785–2795. [CrossRef] [PubMed]
55. Bouwstra, J.A.; Gooris, G.S.; Cheng, K.; Weerheim, A.; Bras, W.; Ponec, M. Phase behavior of isolated skin lipids. *J. Lipid Res.* **1996**, *37*, 999–1011. [CrossRef] [PubMed]
56. McIntosh, T.J.; Stewart, M.E.; Downing, D.T. X-ray diffraction analysis of isolated skin lipids: Reconstitution of intercellular lipid domains. *Biochemistry* **1996**, *35*, 3649–3653. [CrossRef]
57. McIntosh, T.J. Organization of skin stratum corneum extracellular lamellae: Diffraction evidence for asymmetric distribution of cholesterol. *Biophys. J.* **2003**, *85*, 1675–1681. [CrossRef]
58. de Jager, M.; Groenink, W.; van der Spek, J.; Janmaat, C.; Gooris, G.; Ponec, M.; Bouwstra, J. Preparation and characterization of a stratum corneum substitute for in vitro percutaneous penetration studies. *Biochim. Biophys. Acta* **2006**, *1758*, 636–644. [CrossRef] [PubMed]
59. Kessner, D.; Kiselev, M.; Dante, S.; Hauss, T.; Lersch, P.; Wartewig, S.; Neubert, R.H. Arrangement of ceramide [EOS] in a stratum corneum lipid model matrix: New aspects revealed by neutron diffraction studies. *Eur. Biophys. J.* **2008**, *37*, 989–999. [CrossRef] [PubMed]
60. Schroter, A.; Kessner, D.; Kiselev, M.A.; Hauss, T.; Dante, S.; Neubert, R.H. Basic nanostructure of stratum corneum lipid matrices based on ceramides [EOS] and [AP]: A neutron diffraction study. *Biophys. J.* **2009**, *97*, 1104–1114. [CrossRef] [PubMed]
61. Kiselev, M.A.; Ryabova, N.Y.; Balagurov, A.M.; Dante, S.; Hauss, T.; Zbytovska, J.; Wartewig, S.; Neubert, R.H. New insights into the structure and hydration of a stratum corneum lipid model membrane by neutron diffraction. *Eur. Biophys. J.* **2005**, *34*, 1030–1040. [CrossRef] [PubMed]
62. Groen, D.; Gooris, G.S.; Bouwstra, J.A. New insights into the stratum corneum lipid organization by X-ray diffraction analysis. *Biophys. J.* **2009**, *97*, 2242–2249. [CrossRef] [PubMed]
63. Schroeter, A.; Kiselev, M.A.; Hauss, T.; Dante, S.; Neubert, R.H. Evidence of free fatty acid interdigitation in stratum corneum model membranes based on ceramide [AP] by deuterium labelling. *Biochim. Biophys. Acta* **2009**, *1788*, 2194–2203. [CrossRef] [PubMed]
64. Kessner, D.; Kiselev, M.A.; Hauss, T.; Dante, S.; Wartewig, S.; Neubert, R.H. Localisation of partially deuterated cholesterol in quaternary SC lipid model membranes: A neutron diffraction study. *Eur. Biophys. J.* **2008**, *37*, 1051–1057. [CrossRef] [PubMed]
65. Zemlyanaya, E.V.; Kiselev, M.A.; Neubert, R.; Kohlbrecher, J.; Aksenov, V.L. Investigation of the structure and properties of model membranes of the stratum corneum by small-angle neutron scattering. *J. Surf. Investig. X-ray Synchrotron Neutron Tech.* **2008**, *2*, 884–889. [CrossRef]
66. Zbytovská, J.; Vávrová, K.; Kiselev, M.A.; Lessieur, P.; Wartewig, S.; Neubert, R.H.H. The effects of transdermal permeation enhancers on thermotropic phase behaviour of a stratum corneum lipid model. *Colloids Surf. A* **2009**, *351*, 30–37. [CrossRef]
67. Zbytovská, J.; Kiselev, M.A.; Funari, S.S.; Garamus, V.M.; Wartewig, S.; Palát, K.; Neubert, R. Influence of cholesterol on the structure of stratum corneum lipid model membrane. *Colloids Surf. A* **2008**, *328*, 90–99. [CrossRef]
68. Chen, H.; Mendelsohn, R.; Rerek, M.E.; Moore, D.J. Effect of cholesterol on miscibility and phase behavior in binary mixtures with synthetic ceramide 2 and octadecanoic acid. Infrared studies. *Biochim. Biophys. Acta* **2001**, *1512*, 345–356. [CrossRef] [PubMed]
69. Fenske, D.B.; Thewalt, J.L.; Bloom, M.; Kitson, N. Models of stratum corneum intercellular membranes: 2H NMR of macroscopically oriented multilayers. *Biophys. J.* **1994**, *67*, 1562–1573. [CrossRef] [PubMed]
70. Lafleur, M. Phase behaviour of model stratum corneum lipid mixtures: An infrared spectroscopy investigation. *Can. J. Chem.* **1998**, *76*, 1501–1511. [CrossRef]
71. Moore, D.J.; Rerek, M.E. Insights into the molecular organization of lipids in the skin barrier from infrared spectroscopy studies of stratum corneum lipid models. *Acta Derm.-Venereol. Suppl.* **2000**, *208*, 16–22. [CrossRef] [PubMed]
72. Kitson, N.; Thewalt, J.; Lafleur, M.; Bloom, M. A model membrane approach to the epidermal permeability barrier. *Biochemistry* **1994**, *33*, 6707–6715. [CrossRef] [PubMed]
73. Velkova, V.; Lafleur, M. Influence of the lipid composition on the organization of skin lipid model mixtures: An infrared spectroscopy investigation. *Chem. Phys. Lipids* **2002**, *117*, 63–74. [CrossRef] [PubMed]
74. Abraham, W.; Downing, D.T. Deuterium NMR investigation of polymorphism in stratum corneum lipids. *Biochim. Biophys. Acta* **1991**, *1068*, 189–194. [CrossRef] [PubMed]
75. Rerek, M.E.; Van Wyck, D.; Mendelsohn, R.; Moore, D.J. FTIR spectroscopic studies of lipid dynamics in phytosphingosine ceramide models of the stratum corneum lipid matrix. *Chem. Phys. Lipids* **2005**, *134*, 51–58. [CrossRef]
76. Engstrom, S.; Ekelund, K.; Engblom, J.; Eriksson, L.; Sparr, E.; Wennerstrom, H. The skin barrier from a lipid perspective. *Acta Derm.-Venereol. Suppl.* **2000**, *208*, 31–35. [CrossRef]
77. Brief, E.; Kwak, S.; Cheng, J.T.; Kitson, N.; Thewalt, J.; Lafleur, M. Phase behavior of an equimolar mixture of N-palmitoyl-D-erythro-sphingosine, cholesterol, and palmitic acid, a mixture with optimized hydrophobic matching. *Langmuir* **2009**, *25*, 7523–7532. [CrossRef]
78. ten Grotenhuis, E.; Demel, R.A.; Ponec, M.; Boer, D.R.; van Miltenburg, J.C.; Bouwstra, J.A. Phase behavior of stratum corneum lipids in mixed Langmuir-Blodgett monolayers. *Biophys. J.* **1996**, *71*, 1389–1399. [CrossRef] [PubMed]

79. Groen, D.; Gooris, G.S.; Ponec, M.; Bouwstra, J.A. Two new methods for preparing a unique stratum corneum substitute. *Biochim. Biophys. Acta* **2008**, *1778*, 2421–2429. [CrossRef] [PubMed]
80. Sinko, B.; Kokosi, J.; Avdeef, A.; Takacs-Novak, K. A PAMPA study of the permeability-enhancing effect of new ceramide analogues. *Chem. Biodivers.* **2009**, *6*, 1867–1874. [CrossRef] [PubMed]
81. Norlen, L.; Engblom, J. Structure-related aspects on water diffusivity in fatty acid-soap and skin lipid model systems. *J. Control. Release* **2000**, *63*, 213–226. [CrossRef] [PubMed]
82. Lopez-Castellano, A.; Cortell-Ivars, C.; Lopez-Carballo, G.; Herraez-Dominguez, M. The influence of Span20 on stratum corneum lipids in langmuir monolayers: Comparison with Azone. *Int. J. Pharm.* **2000**, *203*, 245–253. [CrossRef] [PubMed]
83. Laugel, C.; Yagoubi, N.; Baillet, A. ATR-FTIR spectroscopy: A chemometric approach for studying the lipid organisation of the stratum corneum. *Chem. Phys. Lipids* **2005**, *135*, 55–68. [CrossRef] [PubMed]
84. Lee, D.; Ashcraft, J.N.; Verploegen, E.; Pashkovski, E.; Weitz, D.A. Permeability of model stratum corneum lipid membrane measured using quartz crystal microbalance. *Langmuir* **2009**, *25*, 5762–5766. [CrossRef] [PubMed]
85. Moore, D.J.; Snyder, R.G.; Rerek, M.E.; Mendelsohn, R. Kinetics of membrane raft formation: Fatty acid domains in stratum corneum lipid models. *J. Phys. Chem. B* **2006**, *110*, 2378–2386. [CrossRef] [PubMed]
86. Sparr, E.; Eriksson, L.; Bouwstra, J.A.; Ekelund, K. AFM Study of Lipid Monolayers: III. Phase Behavior of Ceramides, Cholesterol and Fatty Acids. *Langmuir* **2000**, *17*, 164–172. [CrossRef]
87. Neubert, R.; Rettig, W.; Wartewig, S.; Wegener, M.; Wienhold, A. Structure of stratum corneum lipids characterized by FT-Raman spectroscopy and DSC. II. Mixtures of ceramides and saturated fatty acids. *Chem. Phys. Lipids* **1997**, *89*, 3–14. [CrossRef]
88. Wartewig, S.; Neubert, R.; Rettig, W.; Hesse, K. Structure of stratum corneum lipids characterized by FT-Raman spectroscopy and DSC. IV. Mixtures of ceramides and oleic acid. *Chem. Phys. Lipids* **1998**, *91*, 145–152. [CrossRef]
89. Trommer, H.; Wartewig, S.; Bottcher, R.; Poppl, A.; Hoentsch, J.; Ozegowski, J.H.; Neubert, R.H. The effects of hyaluronan and its fragments on lipid models exposed to UV irradiation. *Int. J. Pharm.* **2003**, *254*, 223–234. [CrossRef] [PubMed]
90. Trommer, H.; Bottcher, R.; Poppl, A.; Hoentsch, J.; Wartewig, S.; Neubert, R.H. Role of ascorbic acid in stratum corneum lipid models exposed to UV irradiation. *Pharm. Res.* **2002**, *19*, 982–990. [CrossRef] [PubMed]
91. Narayanan, S.; Hunerbein, A.; Getie, M.; Jackel, A.; Neubert, R.H. Scavenging properties of metronidazole on free oxygen radicals in a skin lipid model system. *J. Pharm. Pharmacol.* **2007**, *59*, 1125–1130. [CrossRef] [PubMed]
92. Rowat, A.C.; Kitson, N.; Thewalt, J.L. Interactions of oleic acid and model stratum corneum membranes as seen by 2H NMR. *Int. J. Pharm.* **2006**, *307*, 225–231. [CrossRef] [PubMed]
93. Pilgram, G.S.K.; Pelt, A.M.E.-v.; Oostergetel, G.T.; Koerten, H.K.; Bouwstra, J.A. Study on the lipid organization of stratum corneum lipid models by (cryo-) electron diffraction. *J. Lipid Res.* **1998**, *39*, 1669–1676. [CrossRef] [PubMed]
94. Plasencia, I.; Norlen, L.; Bagatolli, L.A. Direct visualization of lipid domains in human skin stratum corneum's lipid membranes: Effect of pH and temperature. *Biophys. J.* **2007**, *93*, 3142–3155. [CrossRef] [PubMed]
95. Glombitza, B.; Muller-Goymann, C.C. Influence of different ceramides on the structure of in vitro model lipid systems of the stratum corneum lipid matrix. *Chem. Phys. Lipids* **2002**, *117*, 29–44. [CrossRef] [PubMed]
96. Kim, C.-K.; Hong, M.-S.; Kim, Y.-B.; Han, S.-K. Effect of penetration enhancers (pyrrolidone derivatives) on multilamellar liposomes of stratum corneum lipid: A study by UV spectroscopy and differential scanning calorimetry. *Int. J. Pharm.* **1993**, *95*, 43–50. [CrossRef]
97. de la Maza, A.; Manich, A.M.; Coderch, L.; Bosch, P.; Parra, J.L. The formation of liposomes in vitro by mixtures of lipids modeling the composition of the stratum corneum. *Colloids Surf. A* **1995**, *101*, 9–19. [CrossRef]
98. Hatfield, R.M.; Fung, L.W. Molecular properties of a stratum corneum model lipid system: Large unilamellar vesicles. *Biophys. J.* **1995**, *68*, 196–207. [CrossRef] [PubMed]
99. Trommer, H.; Bottcher, R.; Huschka, C.; Wohlrab, W.; Neubert, R.H. Further investigations on the role of ascorbic acid in stratum corneum lipid models after UV exposure. *J. Pharm. Pharmacol.* **2005**, *57*, 963–972. [CrossRef] [PubMed]
100. de Jager, M.W.; Gooris, G.S.; Dolbnya, I.P.; Bras, W.; Ponec, M.; Bouwstra, J.A. Novel lipid mixtures based on synthetic ceramides reproduce the unique stratum corneum lipid organization. *J. Lipid Res.* **2004**, *45*, 923–932. [CrossRef] [PubMed]
101. Bouwstra, J.A.; Dubbelaar, F.E.; Gooris, G.S.; Weerheim, A.M.; Ponec, M. The role of ceramide composition in the lipid organisation of the skin barrier. *Biochim. Biophys. Acta* **1999**, *1419*, 127–136. [CrossRef]
102. Janssens, M.; Gooris, G.S.; Bouwstra, J.A. Infrared spectroscopy studies of mixtures prepared with synthetic ceramides varying in head group architecture: Coexistence of liquid and crystalline phases. *Biochim. Biophys. Acta* **2009**, *1788*, 732–742. [CrossRef] [PubMed]
103. Ohta, N.; Hatta, I. Interaction among molecules in mixtures of ceramide/stearic acid, ceramide/cholesterol and ceramide/stearic acid/cholesterol. *Chem. Phys. Lipids* **2002**, *115*, 93–105. [CrossRef] [PubMed]
104. Engelbrecht, T.N.; Schroeter, A.; Hauss, T.; Neubert, R.H. Lipophilic penetration enhancers and their impact to the bilayer structure of stratum corneum lipid model membranes: Neutron diffraction studies based on the example oleic acid. *Biochim. Biophys. Acta* **2011**, *1808*, 2798–2806. [CrossRef] [PubMed]
105. Gruzinov, A.Y.; Zabelin, A.V.; Kiselev, M.A. Short periodicity phase based on ceramide [AP] in the model lipid membranes of stratum corneum does not change during hydration. *Chem. Phys. Lipids* **2017**, *202*, 1–5. [CrossRef]

106. Engelbrecht, T.N.; Deme, B.; Dobner, B.; Neubert, R.H. Study of the influence of the penetration enhancer isopropyl myristate on the nanostructure of stratum corneum lipid model membranes using neutron diffraction and deuterium labelling. *Ski. Pharmacol. Physiol.* **2012**, *25*, 200–207. [CrossRef]
107. Sonnenberger, S.; Eichner, A.; Hauss, T.; Schroeter, A.; Neubert, R.H.; Dobner, B. Synthesis of specifically deuterated ceramide [AP]-C18 and its biophysical characterization using neutron diffraction. *Chem. Phys. Lipids* **2017**, *204*, 15–24. [CrossRef]
108. Ryabova, N.Y.; Kiselev, M.A.; Dante, S.; Hauss, T.; Balagurov, A.M. Investigation of stratum corneum lipid model membranes with free fatty acid composition by neutron diffraction. *Eur. Biophys. J.* **2010**, *39*, 1167–1176. [CrossRef] [PubMed]
109. Mueller, J.; Trapp, M.; Neubert, R.H.H. The effect of hydrophilic penetration/diffusion enhancer on stratum corneum lipid models: Part II*: DMSO. *Chem. Phys. Lipids* **2019**, *225*, 104816. [CrossRef] [PubMed]
110. Mueller, J.; Schroeter, A.; Steitz, R.; Trapp, M.; Neubert, R.H. Preparation of a New Oligolamellar Stratum Corneum Lipid Model. *Langmuir* **2016**, *32*, 4673–4680. [CrossRef] [PubMed]
111. Mueller, J.; Oliveira, J.S.L.; Barker, R.; Trapp, M.; Schroeter, A.; Brezesinski, G.; Neubert, R.H.H. The effect of urea and taurine as hydrophilic penetration enhancers on stratum corneum lipid models. *Biochim. Biophys. Acta* **2016**, *1858*, 2006–2018. [CrossRef] [PubMed]
112. Oliveira, J.S.; Lange, S.; Dobner, B.; Brezesinski, G. The effect of non-deuterated and deuterated isopropyl myristate on the thermodynamical and structural behavior of a 2D Stratum Corneum model with Ceramide [AP]. *Chem. Phys. Lipids* **2017**, *204*, 1–9. [CrossRef] [PubMed]
113. Ryabova, N.Y.; Kiselev, M.A.; Balagurov, A.M. Transition processes in stratum corneum model lipid membranes with a mixture of free fatty acids. *Biophysics* **2010**, *54*, 598–606. [CrossRef]
114. Watanabe, H.; Obata, Y.; Onuki, Y.; Ishida, K.; Takayama, K. Novel preparation of intercellular lipid models of the stratum corneum containing stereoactive ceramide. *Chem. Pharm. Bull.* **2010**, *58*, 312–317. [CrossRef] [PubMed]
115. Watanabe, H.; Obata, Y.; Onuki, Y.; Ishida, K.; Takayama, K. Different effects of l- and d-menthol on the microstructure of ceramide 5/cholesterol/palmitic acid bilayers. *Int. J. Pharm.* **2010**, *402*, 146–152. [CrossRef] [PubMed]
116. Schroeter, A.; Stahlberg, S.; Skolova, B.; Sonnenberger, S.; Eichner, A.; Huster, D.; Vavrova, K.; Hauss, T.; Dobner, B.; Neubert, R.H.; et al. Phase separation in ceramide[NP] containing lipid model membranes: Neutron diffraction and solid-state NMR. *Soft Matter* **2017**, *13*, 2107–2119. [CrossRef] [PubMed]
117. Kwak, S.; Brief, E.; Langlais, D.; Kitson, N.; Lafleur, M.; Thewalt, J. Ethanol perturbs lipid organization in models of stratum corneum membranes: An investigation combining differential scanning calorimetry, infrared and (2)H NMR spectroscopy. *Biochim. Biophys. Acta* **2012**, *1818*, 1410–1419. [CrossRef]
118. Sun, H.; Zielinska, K.; Resmini, M.; Zarbakhsh, A. Interactions of NIPAM nanogels with model lipid multi-bilayers: A neutron reflectivity study. *J. Colloid Interface Sci.* **2019**, *536*, 598–608. [CrossRef] [PubMed]
119. Chen, Y.; Liao, M.; Ma, K.; Wang, Z.; Deme, B.; Penfold, J.; Lu, J.R.; Webster, J.R.P.; Li, P. Implications of surfactant hydrophobic chain architecture on the Surfactant-Skin lipid model interaction. *J. Colloid Interface Sci.* **2022**, *608*, 405–415. [CrossRef] [PubMed]
120. Oguri, M.; Gooris, G.S.; Bito, K.; Bouwstra, J.A. The effect of the chain length distribution of free fatty acids on the mixing properties of stratum corneum model membranes. *Biochim. Biophys. Acta* **2014**, *1838*, 1851–1861. [CrossRef] [PubMed]
121. Ishida, K.; Obata, Y.; Akagi, C.; Onuki, Y.; Takayama, K. Practical syntheses of D-erythro and L-threo-ceramide [NDS] and difference in contribution of each isomer in microstructure of stratum corneum intercellular lipids. *J. Drug Deliv. Sci. Technol.* **2014**, *24*, 689–693. [CrossRef]
122. Mendelsohn, R.; Selevany, I.; Moore, D.J.; Mack Correa, M.C.; Mao, G.; Walters, R.M.; Flach, C.R. Kinetic evidence suggests spinodal phase separation in stratum corneum models by IR spectroscopy. *J. Phys. Chem. B* **2014**, *118*, 4378–4387. [CrossRef] [PubMed]
123. Garidel, P.; Folting, B.; Schaller, I.; Kerth, A. The microstructure of the stratum corneum lipid barrier: Mid-infrared spectroscopic studies of hydrated ceramide:palmitic acid:cholesterol model systems. *Biophys. Chem.* **2010**, *150*, 144–156. [CrossRef] [PubMed]
124. Groen, D.; Gooris, G.S.; Bouwstra, J.A. Model membranes prepared with ceramide EOS, cholesterol and free fatty acids form a unique lamellar phase. *Langmuir* **2010**, *26*, 4168–4175. [CrossRef]
125. Terayama, S.; Goto, Y.; Fukami, T. Amount of cholesterol in intercellular lipid models inversely correlates with hexagonal structure ratio in packing structures. *Chem. Eng. Sci.* **2022**, *259*, 117814. [CrossRef]
126. Skolova, B.; Hudska, K.; Pullmannova, P.; Kovacik, A.; Palat, K.; Roh, J.; Fleddermann, J.; Estrela-Lopis, I.; Vavrova, K. Different phase behavior and packing of ceramides with long (C16) and very long (C24) acyls in model membranes: Infrared spectroscopy using deuterated lipids. *J. Phys. Chem. B* **2014**, *118*, 10460–10470. [CrossRef]
127. Ramos, A.P.; Bouwstra, J.A.; Lafleur, M. Very Long Chain Lipids Favor the Formation of a Homogeneous Phase in Stratum Corneum Model Membranes. *Langmuir* **2020**, *36*, 13899–13907. [CrossRef]
128. Gorcea, M.; Hadgraft, J.; Moore, D.J.; Lane, M.E. Fourier transform infrared spectroscopy studies of lipid domain formation in normal and ceramide deficient stratum corneum lipid models. *Int. J. Pharm.* **2012**, *435*, 63–68. [CrossRef] [PubMed]
129. Lu, G.; Gillece, T.W.; Moore, D.J. Study of water vapor and surfactant absorption by lipid model systems using the quartz crystal microbalance. *Chem. Phys. Lipids* **2011**, *164*, 259–265. [CrossRef] [PubMed]

130. Jurek, I.; Goral, I.; Mierzynska, Z.; Moniuszko-Szajwaj, B.; Wojciechowski, K. Effect of synthetic surfactants and soapwort (*Saponaria officinalis* L.) extract on skin-mimetic model lipid monolayers. *Biochim. Biophys. Acta Biomembr.* **2019**, *1861*, 556–564. [CrossRef] [PubMed]
131. Eichner, A.; Stahlberg, S.; Sonnenberger, S.; Lange, S.; Dobner, B.; Ostermann, A.; Schrader, T.E.; Hauss, T.; Schroeter, A.; Huster, D.; et al. Influence of the penetration enhancer isopropyl myristate on stratum corneum lipid model membranes revealed by neutron diffraction and ^{2}H NMR experiments. *Biochim. Biophys. Acta Biomembr.* **2017**, *1859*, 745–755. [CrossRef] [PubMed]
132. Kiselev, M.A.; Ermakova, E.V.; Gruzinov, A.Y.; Zabelin, A.V. Formation of the long-periodicity phase in model membranes of the outermost layer of skin (*Stratum corneum*). *Crystallogr. Rep.* **2014**, *59*, 112–116. [CrossRef]
133. Curikova-Kindlova, B.A.; Diat, O.; Stepanek, F.; Vavrova, K.; Zbytovska, J. Probing the interactions among sphingosine and phytosphingosine ceramides with non- and alpha-hydroxylated acyl chains in skin lipid model membranes. *Int. J. Pharm.* **2019**, *563*, 384–394. [CrossRef] [PubMed]
134. Pullmannova, P.; Ermakova, E.; Kovacik, A.; Opalka, L.; Maixner, J.; Zbytovska, J.; Kucerka, N.; Vavrova, K. Long and very long lamellar phases in model stratum corneum lipid membranes. *J. Lipid Res.* **2019**, *60*, 963–971. [CrossRef] [PubMed]
135. Badhe, Y.; Schmitt, T.; Gupta, R.; Rai, B.; Neubert, R.H.H. Investigating the nanostructure of a CER[NP]/CER[AP]-based stratum corneum lipid matrix model: A combined neutron diffraction & molecular dynamics simulations approach. *Biochim. Biophys. Acta Biomembr.* **2022**, *1864*, 184007. [CrossRef] [PubMed]
136. Pham, Q.D.; Mojumdar, E.H.; Gooris, G.S.; Bouwstra, J.A.; Sparr, E.; Topgaard, D. Solid and fluid segments within the same molecule of stratum corneum ceramide lipid. *Q. Rev. Biophys.* **2018**, *51*, e7. [CrossRef] [PubMed]
137. Schmitt, T.; Lange, S.; Sonnenberger, S.; Dobner, B.; Deme, B.; Neubert, R.H.H.; Gooris, G.; Bouwstra, J.A. Determination of the influence of C24 D/(2R)- and L/(2S)-isomers of the CER[AP] on the lamellar structure of stratum corneum model systems using neutron diffraction. *Chem. Phys. Lipids* **2017**, *209*, 29–36. [CrossRef]
138. Yoshida, S.; Obata, Y.; Onuki, Y.; Utsumi, S.; Ohta, N.; Takahashi, H.; Takayama, K. Molecular Interaction between Intercellular Lipids in the Stratum Corneum and l-Menthol, as Analyzed by Synchrotron X-ray Diffraction. *Chem. Pharm. Bull.* **2017**, *65*, 134–142. [CrossRef] [PubMed]
139. Paz Ramos, A.; Gooris, G.; Bouwstra, J.; Lafleur, M. Evidence of hydrocarbon nanodrops in highly ordered stratum corneum model membranes. *J. Lipid Res.* **2018**, *59*, 137–143. [CrossRef] [PubMed]
140. Schmitt, T.; Gupta, R.; Lange, S.; Sonnenberger, S.; Dobner, B.; Hauss, T.; Rai, B.; Neubert, R.H.H. Impact of the ceramide subspecies on the nanostructure of stratum corneum lipids using neutron scattering and molecular dynamics simulations. Part I: Impact of CER[NS]. *Chem. Phys. Lipids* **2018**, *214*, 58–68. [CrossRef] [PubMed]
141. Sakdiset, P.; Todo, H.; Sugibayashi, K. Potential of Stratum Corneum Lipid Liposomes for Screening of Chemical Skin Penetration Enhancers. *Chem. Pharm. Bull.* **2017**, *65*, 776–783. [CrossRef] [PubMed]
142. Kovacik, A.; Vogel, A.; Adler, J.; Pullmannova, P.; Vavrova, K.; Huster, D. Probing the role of ceramide hydroxylation in skin barrier lipid models by ^{2}H solid-state NMR spectroscopy and X-ray powder diffraction. *Biochim. Biophys. Acta Biomembr.* **2018**, *1860*, 1162–1170. [CrossRef] [PubMed]
143. Eichner, A.; Sonnenberger, S.; Dobner, B.; Hauss, T.; Schroeter, A.; Neubert, R.H.H. Localization of methyl-branched ceramide [EOS] species within the long-periodicity phase in stratum corneum lipid model membranes: A neutron diffraction study. *Biochim. Biophys. Acta* **2016**, *1858*, 2911–2922. [CrossRef] [PubMed]
144. Engelbrecht, T.; Hauß, T.; Süß, K.; Vogel, A.; Roark, M.; Feller, S.E.; Neubert, R.H.H.; Dobner, B. Characterisation of a new ceramide EOS species: Synthesis and investigation of the thermotropic phase behaviour and influence on the bilayer architecture of stratum corneum lipid model membranes. *Soft Matter* **2011**, *7*, 8998–9011. [CrossRef]
145. Madsen, H.B.; Arboe-Andersen, H.M.; Rozlosnik, N.; Madsen, F.; Ifversen, P.; Kasimova, M.R.; Nielsen, H.M. Investigation of the interaction between modified ISCOMs and stratum corneum lipid model systems. *Biochim. Biophys. Acta* **2010**, *1798*, 1779–1789. [CrossRef] [PubMed]
146. Beddoes, C.M.; Gooris, G.S.; Bouwstra, J.A. Preferential arrangement of lipids in the long-periodicity phase of a stratum corneum matrix model. *J. Lipid Res.* **2018**, *59*, 2329–2338. [CrossRef] [PubMed]
147. Schmitt, T.; Lange, S.; Dobner, B.; Sonnenberger, S.; Hauss, T.; Neubert, R.H.H. Investigation of a CER[NP]- and [AP]-Based Stratum Corneum Modeling Membrane System: Using Specifically Deuterated CER Together with a Neutron Diffraction Approach. *Langmuir* **2018**, *34*, 1742–1749. [CrossRef]
148. Paz Ramos, A.; Gooris, G.; Bouwstra, J.; Molinari, M.; Lafleur, M. Raman and AFM-IR chemical imaging of stratum corneum model membranes. *Can. J. Chem.* **2020**, *98*, 495–501. [CrossRef]
149. Beddoes, C.M.; Gooris, G.S.; Barlow, D.J.; Lawrence, M.J.; Dalgliesh, R.M.; Malfois, M.; Deme, B.; Bouwstra, J.A. The importance of ceramide headgroup for lipid localisation in skin lipid models. *Biochim. Biophys. Acta Biomembr.* **2022**, *1864*, 183886. [CrossRef] [PubMed]
150. Schmitt, T.; Lange, S.; Sonnenberger, S.; Dobner, B.; Deme, B.; Langner, A.; Neubert, R.H.H. The long periodicity phase (LPP) controversy part I: The influence of a natural-like ratio of the CER[EOS] analogue [EOS]-br in a CER[NP]/[AP] based stratum corneum modelling system: A neutron diffraction study. *Biochim. Biophys. Acta Biomembr.* **2019**, *1861*, 306–315. [CrossRef]
151. Nadaban, A.; Gooris, G.S.; Beddoes, C.M.; Dalgliesh, R.M.; Bouwstra, J.A. Phytosphingosine ceramide mainly localizes in the central layer of the unique lamellar phase of skin lipid model systems. *J. Lipid Res.* **2022**, *63*, 100258. [CrossRef] [PubMed]

152. Groen, D.; Gooris, G.S.; Barlow, D.J.; Lawrence, M.J.; van Mechelen, J.B.; Deme, B.; Bouwstra, J.A. Disposition of ceramide in model lipid membranes determined by neutron diffraction. *Biophys. J.* **2011**, *100*, 1481–1489. [CrossRef]
153. Mojumdar, E.H.; Groen, D.; Gooris, G.S.; Barlow, D.J.; Lawrence, M.J.; Deme, B.; Bouwstra, J.A. Localization of cholesterol and fatty acid in a model lipid membrane: A neutron diffraction approach. *Biophys. J.* **2013**, *105*, 911–918. [CrossRef] [PubMed]
154. Beddoes, C.M.; Gooris, G.S.; Foglia, F.; Ahmadi, D.; Barlow, D.J.; Lawrence, M.J.; Deme, B.; Bouwstra, J.A. Arrangement of Ceramides in the Skin: Sphingosine Chains Localize at a Single Position in Stratum Corneum Lipid Matrix Models. *Langmuir* **2020**, *36*, 10270–10278. [CrossRef]
155. Mojumdar, E.H.; Gooris, G.S.; Bouwstra, J.A. Phase behavior of skin lipid mixtures: The effect of cholesterol on lipid organization. *Soft Matter* **2015**, *11*, 4326–4336. [CrossRef] [PubMed]
156. Beddoes, C.M.; Rensen, D.E.; Gooris, G.S.; Malfois, M.; Bouwstra, J.A. The Importance of Free Fatty Chain Length on the Lipid Organization in the Long Periodicity Phase. *Int. J. Mol. Sci.* **2021**, *22*, 3679. [CrossRef] [PubMed]
157. de Sousa Neto, D.; Gooris, G.; Bouwstra, J. Effect of the omega-acylceramides on the lipid organization of stratum corneum model membranes evaluated by X-ray diffraction and FTIR studies (Part I). *Chem. Phys. Lipids* **2011**, *164*, 184–195. [CrossRef]
158. Mojumdar, E.H.; Gooris, G.S.; Barlow, D.J.; Lawrence, M.J.; Deme, B.; Bouwstra, J.A. Skin lipids: Localization of ceramide and fatty acid in the unit cell of the long periodicity phase. *Biophys. J.* **2015**, *108*, 2670–2679. [CrossRef] [PubMed]
159. Mojumdar, E.H.; Gooris, G.S.; Groen, D.; Barlow, D.J.; Lawrence, M.J.; Deme, B.; Bouwstra, J.A. Stratum corneum lipid matrix: Location of acyl ceramide and cholesterol in the unit cell of the long periodicity phase. *Biochim. Biophys. Acta* **2016**, *1858*, 1926–1934. [CrossRef] [PubMed]
160. Stahlberg, S.; Lange, S.; Dobner, B.; Huster, D. Probing the Role of Ceramide Headgroup Polarity in Short-Chain Model Skin Barrier Lipid Mixtures by ^{2}H Solid-State NMR Spectroscopy. *Langmuir* **2016**, *32*, 2023–2031. [CrossRef] [PubMed]
161. Fandrei, F.; Havrisak, T.; Opalka, L.; Engberg, O.; Smith, A.A.; Pullmannova, P.; Kucerka, N.; Ondrejcekova, V.; Deme, B.; Novakova, L.; et al. The intriguing molecular dynamics of Cer[EOS] in rigid skin barrier lipid layers requires improvement of the model. *J. Lipid Res.* **2023**, *64*, 100356. [CrossRef] [PubMed]
162. Kiselev, M.A. Conformation of ceramide 6 molecules and chain-flip transitions in the lipid matrix of the outermost layer of mammalian skin, the stratum corneum. *Crystallogr. Rep.* **2007**, *52*, 525–528. [CrossRef]
163. Bouwstra, J.; Gooris, G.; Ponec, M. The lipid organisation of the skin barrier: Liquid and crystalline domains coexist in lamellar phases. *J. Biol. Phys.* **2002**, *28*, 211–223. [CrossRef] [PubMed]
164. Mojumdar, E.H.; Kariman, Z.; van Kerckhove, L.; Gooris, G.S.; Bouwstra, J.A. The role of ceramide chain length distribution on the barrier properties of the skin lipid membranes. *Biochim. Biophys. Acta* **2014**, *1838*, 2473–2483. [CrossRef] [PubMed]
165. Rerek, M.E.; Chen; Markovic, B.; Van Wyck, D.; Garidel, P.; Mendelsohn, R.; Moore, D.J. Phytosphingosine and Sphingosine Ceramide Headgroup Hydrogen Bonding: Structural Insights through Thermotropic Hydrogen/Deuterium Exchange. *J. Phys. Chem. B* **2001**, *105*, 9355–9362. [CrossRef]
166. Raudenkolb, S.; Wartewig, S.; Neubert, R.H. Polymorphism of ceramide 6: A vibrational spectroscopic and X-ray powder diffraction investigation of the diastereomers of N-(alpha-hydroxyoctadecanoyl)-phytosphingosine. *Chem. Phys. Lipids* **2005**, *133*, 89–102. [CrossRef]
167. Bouwstra, J.A.; Gooris, G.S.; Bras, W.; Downing, D.T. Lipid organization in pig stratum corneum. *J. Lipid Res.* **1995**, *36*, 685–695. [CrossRef]
168. Vávrová, K.; Kováčik, A.; Opálka, L. Ceramides in the skin barrier. *Eur. Pharm. J.* **2017**, *64*, 28–35. [CrossRef]
169. Uche, L.E.; Gooris, G.S.; Beddoes, C.M.; Bouwstra, J.A. New insight into phase behavior and permeability of skin lipid models based on sphingosine and phytosphingosine ceramides. *Biochim. Biophys. Acta Biomembr.* **2019**, *1861*, 1317–1328. [CrossRef] [PubMed]
170. Stahlberg, S.; Skolova, B.; Madhu, P.K.; Vogel, A.; Vavrova, K.; Huster, D. Probing the role of the ceramide acyl chain length and sphingosine unsaturation in model skin barrier lipid mixtures by ^{2}H solid-state NMR spectroscopy. *Langmuir* **2015**, *31*, 4906–4915. [CrossRef] [PubMed]
171. Groen, D.; Poole, D.S.; Gooris, G.S.; Bouwstra, J.A. Investigating the barrier function of skin lipid models with varying compositions. *Eur. J. Pharm. Biopharm.* **2011**, *79*, 334–342. [CrossRef]
172. Groen, D.; Poole, D.S.; Gooris, G.S.; Bouwstra, J.A. Is an orthorhombic lateral packing and a proper lamellar organization important for the skin barrier function? *Biochim. Biophys. Acta* **2011**, *1808*, 1529–1537. [CrossRef] [PubMed]
173. Ochalek, M.; Heissler, S.; Wohlrab, J.; Neubert, R.H. Characterization of lipid model membranes designed for studying impact of ceramide species on drug diffusion and penetration. *Eur. J. Pharm. Biopharm.* **2012**, *81*, 113–120. [CrossRef] [PubMed]
174. Groen, D.; Berthaud, F.; Bouwstra, J.A.; Chapuis, C.; Gooris, G.S.; Boncheva, M. In vitro model systems for studying the impact of organic chemicals on the skin barrier lipids. *Biochim. Biophys. Acta* **2014**, *1838*, 310–318. [CrossRef] [PubMed]
175. Ghonaim, H.; Periasamy, N.; Noro, M.G.; Anwar, J. Towards a simplified model membrane of skin lipids: Preparation and characterisation of a ternary lipid mixture. *Int. J. Pharm. Pharm. Sci.* **2014**, *6*, 148–152.
176. Mojumdar, E.H.; Helder, R.W.; Gooris, G.S.; Bouwstra, J.A. Monounsaturated fatty acids reduce the barrier of stratum corneum lipid membranes by enhancing the formation of a hexagonal lateral packing. *Langmuir* **2014**, *30*, 6534–6543. [CrossRef]
177. Skolova, B.; Janusova, B.; Vavrova, K. Ceramides with a pentadecasphingosine chain and short acyls have strong permeabilization effects on skin and model lipid membranes. *Biochim. Biophys. Acta* **2016**, *1858*, 220–232. [CrossRef]

178. Uchiyama, M.; Oguri, M.; Mojumdar, E.H.; Gooris, G.S.; Bouwstra, J.A. Free fatty acids chain length distribution affects the permeability of skin lipid model membranes. *Biochim. Biophys. Acta* **2016**, *1858*, 2050–2059. [CrossRef]
179. Curikova, B.A.; Prochazkova, K.; Filkova, B.; Diblikova, P.; Svoboda, J.; Kovacik, A.; Vavrova, K.; Zbytovska, J. Simplified stratum corneum model membranes for studying the effects of permeation enhancers. *Int. J. Pharm.* **2017**, *534*, 287–296. [CrossRef]
180. Uche, L.E.; Gooris, G.S.; Bouwstra, J.A.; Beddoes, C.M. Barrier Capability of Skin Lipid Models: Effect of Ceramides and Free Fatty Acid Composition. *Langmuir* **2019**, *35*, 15376–15388. [CrossRef] [PubMed]
181. Sochorova, M.; Audrlicka, P.; Cervena, M.; Kovacik, A.; Kopecna, M.; Opalka, L.; Pullmannova, P.; Vavrova, K. Permeability and microstructure of cholesterol-depleted skin lipid membranes and human stratum corneum. *J. Colloid Interface Sci.* **2019**, *535*, 227–238. [CrossRef] [PubMed]
182. Uche, L.E.; Gooris, G.S.; Bouwstra, J.A.; Beddoes, C.M. Increased Levels of Short-Chain Ceramides Modify the Lipid Organization and Reduce the Lipid Barrier of Skin Model Membranes. *Langmuir* **2021**, *37*, 9478–9489. [CrossRef]
183. Uche, L.E.; Gooris, G.S.; Bouwstra, J.A.; Beddoes, C.M. High concentration of the ester-linked omega-hydroxy ceramide increases the permeability in skin lipid model membranes. *Biochim. Biophys. Acta Biomembr.* **2021**, *1863*, 183487. [CrossRef] [PubMed]
184. Opalka, L.; Kovacik, A.; Pullmannova, P.; Maixner, J.; Vavrova, K. Effects of omega-O-acylceramide structures and concentrations in healthy and diseased skin barrier lipid membrane models. *J. Lipid Res.* **2020**, *61*, 219–228. [CrossRef]
185. Opalka, L.; Meyer, J.M.; Ondrejcekova, V.; Svatosova, L.; Radner, F.P.W.; Vavrova, K. omega-O-Acylceramides but not omega-hydroxy ceramides are required for healthy lamellar phase architecture of skin barrier lipids. *J. Lipid Res.* **2022**, *63*, 100226. [CrossRef]
186. Fandrei, F.; Engberg, O.; Opalka, L.; Jancalkova, P.; Pullmannova, P.; Steinhart, M.; Kovacik, A.; Vavrova, K.; Huster, D. Cholesterol sulfate fluidizes the sterol fraction of the stratum corneum lipid phase and increases its permeability. *J. Lipid Res.* **2022**, *63*, 100177. [CrossRef]
187. Nakov, A.; Magalhães, C.; Faria-Silva, C.; Ascensão, L.; Marto, J.; Ribeiro, H.M.; Simões, S. Lipid-coated membranes as skin surrogates for permeability assessment. *Materialia* **2022**, *21*, 101354. [CrossRef]
188. Sinko, B.; Garrigues, T.M.; Balogh, G.T.; Nagy, Z.K.; Tsinman, O.; Avdeef, A.; Takacs-Novak, K. Skin-PAMPA: A new method for fast prediction of skin penetration. *Eur. J. Pharm. Sci.* **2012**, *45*, 698–707. [CrossRef] [PubMed]
189. Engesland, A.; Skar, M.; Hansen, T.; Skalko-Basnet, N.; Flaten, G.E. New applications of phospholipid vesicle-based permeation assay: Permeation model mimicking skin barrier. *J. Pharm. Sci.* **2013**, *102*, 1588–1600. [CrossRef] [PubMed]
190. Palac, Z.; Engesland, A.; Flaten, G.E.; Skalko-Basnet, N.; Filipovic-Grcic, J.; Vanic, Z. Liposomes for (trans)dermal drug delivery: The skin-PVPA as a novel in vitro stratum corneum model in formulation development. *J. Liposome Res.* **2014**, *24*, 313–322. [CrossRef] [PubMed]
191. Engesland, A.; Skalko-Basnet, N.; Flaten, G.E. Phospholipid vesicle-based permeation assay and EpiSkin® in assessment of drug therapies destined for skin administration. *J. Pharm. Sci.* **2015**, *104*, 1119–1127. [CrossRef] [PubMed]
192. Engesland, A.; Skalko-Basnet, N.; Flaten, G.E. In vitro models to estimate drug penetration through the compromised stratum corneum barrier. *Drug Dev. Ind. Pharm.* **2016**, *42*, 1742–1751. [CrossRef] [PubMed]
193. Ma, M.; Di, H.J.; Zhang, H.; Yao, J.H.; Dong, J.; Yan, G.J.; Qiao, H.Z.; Chen, J. Development of phospholipid vesicle-based permeation assay models capable of evaluating percutaneous penetration enhancing effect. *Drug Dev. Ind. Pharm.* **2017**, *43*, 2055–2063. [CrossRef] [PubMed]
194. Zhang, H.; Zhu, X.; Shen, J.; Xu, H.; Ma, M.; Gu, W.; Jiang, Q.; Chen, J.; Duan, J. Characterization of a liposome-based artificial skin membrane for in vitro permeation studies using Franz diffusion cell device. *J. Liposome Res.* **2017**, *27*, 302–311. [CrossRef] [PubMed]
195. Shakel, Z.; Nunes, C.; Costa Lima, S.A.; Reis, S. Development of a novel human stratum corneum model, as a tool in the optimization of drug formulations. *Int. J. Pharm.* **2019**, *569*, 118571. [CrossRef] [PubMed]
196. Moniz, T.; Costa Lima, S.A.; Reis, S. Application of the human stratum corneum lipid-based mimetic model in assessment of drug-loaded nanoparticles for skin administration. *Int. J. Pharm.* **2020**, *591*, 119960. [CrossRef]
197. Karadzovska, D.; Riviere, J.E. Assessing vehicle effects on skin absorption using artificial membrane assays. *Eur. J. Pharm. Sci.* **2013**, *50*, 569–576. [CrossRef]
198. Flaten, G.E.; Dhanikula, A.B.; Luthman, K.; Brandl, M. Drug permeability across a phospholipid vesicle based barrier: A novel approach for studying passive diffusion. *Eur. J. Pharm. Sci.* **2006**, *27*, 80–90. [CrossRef] [PubMed]
199. Flaten, G.E.; Bunjes, H.; Luthman, K.; Brandl, M. Drug permeability across a phospholipid vesicle-based barrier 2. Characterization of barrier structure, storage stability and stability towards pH changes. *Eur. J. Pharm. Sci.* **2006**, *28*, 336–343. [CrossRef] [PubMed]
200. Franz, T.J.; Lehman, P.A.; Raney, S.G. Use of excised human skin to assess the bioequivalence of topical products. *Ski. Pharmacol. Physiol.* **2009**, *22*, 276–286. [CrossRef]
201. *76/768/EEC*; Council Directive of 27 July 1976 on the Approximation of the Laws of the Member States Relating to Cosmetic Products. European Parliament, The Council of the European Communities: Brussels, Belgium, 1976.
202. Wever, B.D.; Petersohn, D.; Mewes, K.R. Overview of human three-dimensional (3D) skin models used for dermal toxicity assessment Part 1. *HPC Today* **2013**, *8*, 18–22.
203. Di Cagno, M.P.; Bauer-Brandl, A. Assembly for Assessing Drug Permeability with Adjustable Biomimetic Properties. WO/2016/078667, 26 May 2016.

204. Liu, Y.; Thomas, A. Biomimetic Device. US20150361386A1, 17 December 2015.
205. Voelcker, N.; Tong, Z.; Tong, W.; Oddo, A.; Thissen, H.; Peng, B. Microfluidic Device for Investigating Interactions of Substances with Cells. WO/2021/168511A1, 2 September 2021.
206. Sun, S.X.; Choudhury, M.lK. A Novel, Organ-on-a-Chip Device for Assessing Trans-Epithelial Transport and Uses Thereof. US/2022/265176A1/WO/2021/007300A1, 25 August 2022.
207. Chen, J.; Wei, X.; Dong, J.; Gu, W. Keratin Liposome Artificial Skin Membrane and Application Thereof in External Preparations and Cosmetics. CN/114577678A, 3 June 2022.
208. Duan, Y. Biomimetic Cell Membrane of Phospholipid Polymer, Preparation Method and Use Thereof. WO/2008/011812A1, 31 January 2008.

antioxidants

MDPI

Article

Bioactive Potential of a Grape Stem Blend: A Sustainable Approach to Skin Regeneration

Mónica Serra [1,2], Claudia Botelho [1,3,4], Diana Sousa [1,2], Hugo Almeida [2,5,6], Ana Casas [2], José António Teixeira [1,3,4] and Ana Novo Barros [7,*]

1 CEB—Centre of Biological Engineering, University of Minho, 4710-057 Braga, Portugal; quality4@mesosystem.com (M.S.); claudiabotelho@deb.uminho.pt (C.B.); id5@mesosystem.com (D.S.); jateixeira@deb.uminho.pt (J.A.T.)

2 Mesosystem Investigação & Investimentos by Spinpark, Barco, 4805-017 Guimarães, Portugal; cto@mesosystem.com (H.A.); ana@mesosystem.com (A.C.)

3 LABBELS—Associate Laboratory, 4710-057 Braga, Portugal

4 Centre of Molecular and Environmental Biology (CBMA), Aquatic Research Network (ARNET), Associate Institute of Science and Innovation for Sustainability (IB-S), University of Minho, Campus de Gualtar, 4710-057 Braga, Portugal

5 UCIBIO, Laboratory of Pharmaceutical Technology, Faculty of Pharmacy, University of Porto, 4051-401 Porto, Portugal

6 Associate Laboratory i4HB Institute for Health and Bioeconomy, Faculty of Pharmacy, University of Porto, 4051-401 Porto, Portugal

7 Centre for the Research and Technology of Argo-Environmental and Biological Sciences (CITAB), Institute for Innovation, Capacity Building and Sustainability of Agri-Food Production (Inov4Agro), University of Trás-os-Montes and Alto Douro (UTAD), Quinta de Prados, 5000-801 Vila Real, Portugal

* Correspondence: abarros@utad.pt

Abstract: The European wine industry is embracing sustainability through circular economy principles, particularly by valorizing by-products, such as grape stems. Grape stems are rich in phenolic compounds with recognized health benefits. This study investigates the bioactive potential of molecules extracted from a blend of grape stems (GS blend extract). The GS blend extract was chemically characterized in terms of total phenolic content (TPC), *ortho*-diphenol content (ODC), and flavonoid content (FC), with key compounds identified via HPLC-MS. The extract's antioxidant capacity was assessed using ABTS, FRAP, and DPPH assays, while its anti-aging and depigmenting properties were evaluated through elastase and tyrosinase inhibition assays. Additionally, in vitro assays were conducted to assess its effects on skin cells, including morphology, metabolic activity, cell cycle, and cell migration. The GS blend extract was found to be rich in proanthocyanidins and exhibited notable antioxidant and depigmenting properties. In vitro assays demonstrated that the extract had no significant impact on cellular metabolic activity or cell morphology, although a reorganization of the cell monolayer was observed. Furthermore, deviations in cell migration and cell cycle regulation suggest that the GS blend extract may aid in scar formation management. Notably, the extract arrested fibroblasts in the Sub G0-G1 phase and inhibited HaCaT cell migration, supporting its potential application in cosmetic and pharmaceutical formulations aimed at scar modulation and skin health.

Keywords: circular economy; sustainability; by-product valorization; grape stems; phenolic compounds; chemical characterization; in vitro assays; biological activities; skin regeneration; antioxidant properties

1. Introduction

The European wine sector holds a prominent position in the global economy, with countries such as France, Italy, Spain, Germany, and Portugal identified by the International Organization of Vine and Wine (OIV) as key contributors to worldwide wine production [1,2].

The global wine industry, deeply rooted in tradition and economic significance, has increasingly gained attention for its potential to support sustainable practices, particularly within the framework of circular economy principles [3,4]. Grapes, essential for wine production, not only serve as the foundation of the winemaking process but also generate valuable by-products, such as grape stems (or stalks), seeds, and skins, which hold significant potential as bioactive resources [5–8].

In the context of the circular economy [9], which emphasizes minimizing waste and maximizing resource efficiency, the valorisation of by-products becomes crucial. Grape stems, one of the most abundant by-products of winemaking (accounting for up to 14% of total by-product volume), are often overlooked despite their high content of bioactive molecules, particularly phenolics [10].

Phenolic compounds are secondary metabolites synthesized by plants in response to various environmental challenges, acting as defence mechanisms against pathogens, parasites, UV radiation, and other biotic and abiotic stressors [10,11]. Grape stems are a rich source of these compounds, including flavanols (epicatechin, catechin, quercetin-3-*O*-rutinoside), hydroxycinnamic acids (caftaric acid, coumaric acid), hydroxybenzoic acids (gallic acid), and stilbenes (viniferin), among others [12–17]. While the phenolic profile of grape stem extracts is similar across different varieties, the proportion of each compound can vary significantly, depending on factors such as soil type, maturation stage, grape size, harvest time [18–21], and extraction method [22]. The distinct proportions of phenolics in each extract significantly impact their biological potential, particularly their antioxidant, anti-inflammatory, anti-aging, anticancer, and antimicrobial properties [11,22,23].

In terms of chemical antioxidant efficacy, phenolic compounds extracted from grape stems, particularly from Touriga Nacional [13], Sousa [14], Cabernet Moravia [17], and Mazuelo [12] varieties, exhibit notable free radical scavenging capacity. Similarly, extracts from grape stems of the Mandilaria variety have demonstrated cellular antioxidant properties, especially in endothelial and muscle cells [24,25].

Beyond grape stems antioxidant activity, *Vitis vinifera* L. grape stem extracts have shown anti-aging properties by inhibiting elastase and tyrosinase enzymes, contributing to the prevention of age spots and the loss of skin elasticity and strength [26]. The inhibition of these enzymes is particularly relevant due to the role of tyrosinase in melanin synthesis, which is associated with hyperpigmentation and age spots, as well as other pigmentation-related skin disorders. Similarly, elastase is responsible for the degradation of elastin, a key structural protein in the skin. Elastin breakdown leads to reduced mechanical strength in connective tissues, wrinkle formation, and other skin deformations [26–29]. Therefore, the inhibition of these enzymes is of paramount importance in anti-aging skincare applications.

These findings highlight the diverse and promising bioactive properties of phenolic compounds extracted from grape stems, emphasizing their potential applications in health and wellness.

While previous studies have primarily focused on evaluating isolated grape varieties, this study takes a different approach by analysing the extraction of phenolic compounds from a blend of grape stems, specifically from Tinta Roriz, Touriga Nacional, Castelão, Syrah, Arinto, and Fernão Pires. Blending extracts from multiple grape varieties may have increased benefits in cosmetic skincare formulations, as different grape varieties contain distinct bioactive compounds. Their combination can result in synergistic effects that amplify skin regeneration potential [30]. For instance, resveratrol, abundant in red grape varieties, exhibits potent antioxidant and anti-inflammatory properties [31], while flavonoids, such as quercetin and catechins, found in both red and white grapes, play crucial roles in angiogenesis and extracellular matrix pathways [32–35]. Proanthocyanidins, commonly present in grape seeds and stems, contribute to collagen crosslinking [36]. and improved skin elasticity, which can further aid in minimizing scar formation.

In addition to enhancing bioactivity, blended grape extracts may offer improved stability and bioavailability. Certain polyphenols are prone to oxidation and degradation, limiting their efficacy in topical formulations. A multi-variety blend can counteract these limitations by providing a more chemically balanced and stable composition, ensuring prolonged bioactivity when applied to the skin. Moreover, a combination of bioactive molecules targets multiple wound healing pathways, creating a comprehensive approach to skin regeneration that fosters smooth, resilient, and scar-free tissue repair.

Therefore, this study explores the potential of blended grape stem extracts as an advanced cosmetic ingredient for skin regeneration. By investigating their bioactive composition, stability, and effects on skin cell function, we aim to provide insights into how a multi-variety extract can contribute to the development of next-generation skincare products.

2. Materials and Methods

2.1. Chemicals and Reagents

Reagents

Gallic acid (3,4,5-trihydroxybenzoic acid), Folin–Ciocalteu reagent, and acetic acid (all extra pure, >99%) were obtained from Panreac (Panreac Química S.L.U., Barcelona, Spain). Sodium hydroxide (98%), sodium nitrite, sodium carbonate (purity > 99%), aluminum chloride (purity > 99%), and ethanol were acquired from Merck (Merck, Darmstadt, Germany). Sodium molybdate (99.5%) was purchased from Chem-Lab (Chem-Lab N.V., Zedelgem, Belgium).

Additionally, catechin (98%), TROLOX (6-hydroxy-2,5,7,8-tetramethylchroman-2-carboxylic acid, purity $\geq$ 98.0%), ABTS ((2,2′-azino-bis-(3-ethylbenzothiazoline-6-sulfonic) acid) diammonium salt, purity $\geq$ 98.0%), DPPH (2,2-diphenyl-1-picrylhydrazyl radical, 100.0%), potassium persulfate (purity $\geq$ 99.0%), TPTZ (2,4,6-tripyridyl-s-triazine, purity $\geq$ 98.0%), iron (III) chloride (purity $\geq$ 99.9%), tyrosinase and elastase enzymes, and all reagents used in enzymatic activity assays were sourced from Sigma-Aldrich (Steinheim, Germany), while hydrochloric acid ($\approx$37%) was obtained from Honeywell Fluka (Berlin, Germany).

All reagents used in the cell assays were obtained from PAN-Biotech. Distilled water was generated using a Millipore water purification system (Millipore, Bedford, MA, USA).

2.2. Preparation of GS Blend Extract

Equal amounts of grape stems from various varieties (Tinta Roriz, Touriga Nacional, Castelão, Syrah, Arinto, and Fernão Pires) were dried in an oven (Memmert, Schwabach, Germany) at 40 °C for 72 h. After drying, the stems were ground, and the resulting samples from different varieties were combined in equal proportions to create the GS blend extract.

The extraction of phenolic compounds followed the methodology defined by Dias Costa et al. [24], with certain modifications. A 1 g sample was homogenized and stirred in 50 mL of extraction solvent (ethanol/water, 70:30) for 30 min in an orbital shaker (GFL 3005, GEMINI, Apeldoorn, The Netherlands). Following homogenization, the mixture was subjected to centrifugation (Sigma 2-16KL Refrigerated Centrifuge, Sigma Laborzentrifugen,

Berlin, Germany) at 10,000× *g* for 15 min at 4 °C, a process that was repeated three times. The resulting supernatants from the repeated centrifugations were pooled and adjusted to a final volume of 50 mL in a volumetric flask using the extraction solvent.

2.3. Determination of Phenolic Content

The phenolic content of the GS blend extract was determined using adapted spectrophotometric methodologies designed for 96-well microplates (PrimeSurface MS-9096MZ, Frilabo, Maia, Portugal), following the protocol of Gouvinhas et al. [27]. Absorbance measurements were performed using a microplate reader (Multiskan GO Microplate Photometer, Thermo Fisher Scientific, Vantaa, Finland).

2.4. Determination of Total Phenolic Content

The total phenolic content (TPC) of the GS blend extract was determined by mixing 1 mL of the sample with 0.5 mL of Folin–Ciocalteu reagent. Next, 2.0 mL of Na_2CO_3 (7.5% *w*/*v*) and 6.5 mL of water were added to the mixture. The reaction was incubated at 70 °C for 30 min, protected from light, and then cooled under running water. Absorbance was measured at 750 nm. An analytical curve was constructed using various concentrations of gallic acid, and the total phenolic content was quantified and expressed as milligrams of gallic acid equivalent per gram of sample (mg GAE/g sample).

2.5. Determination of Ortho-Diphenol Content

The *ortho*-diphenol content (ODC) of the GS blend extract was assessed by adding 1 mL of Na_2MoO_4 (50 g/L) to 4 mL of the appropriately diluted samples. After vortexing, the mixtures were left to stand at room temperature for 15 min, protected from light. Absorbance was then measured at 370 nm. Quantification was performed using gallic acid as the standard, and the results were expressed as milligrams of gallic acid equivalent per gram of dry weight (mg GAE/g dw).

2.6. Determination of Flavonoid Content

The method used to determine flavonoid content (FC) involved the formation of a flavonoid–aluminum complex. In this process, 0.5 mL of the diluted sample was added to a test tube, followed by 150 µL of a 5% sodium nitrite solution. After a 5 min reaction, 150 µL of a 10% aluminum chloride solution was introduced. Following a 6 min reaction period, 1 mL of a 1 M sodium hydroxide solution was added to the well. The microplate was then stirred for 30 s, and absorbance was measured at 510 nm. To construct the calibration curve, various concentrations of catechin were used. The results were expressed as milligrams of catechin per gram of dry weight of the sample (mg CAT/g dw).

2.7. Identification of Phenolic Compounds by HPLC-MS

The quantitative (poly)phenolic profile of the grape stem extract was determined using the methodology described by Costa-Pérez et al. [24]. Chromatographic separation of (poly)phenols was performed on a Luna C18 column (150.0 × 4.6 mm, 5.0 µm particle size, Phenomenex, Macclesfield, UK) using an Agilent HPLC 1100 series system (Agilent Technologies, Waldbronn, Germany) consisting of a binary pump (model G1312A), an autosampler (model G1313A), a degasser (model G1322A), a photodiode array (PDA) detector (model G1315B), and an ion trap spectrometer (model G2445A) with an electrospray ionization interface. The system was controlled using LCMSD software, v. 4.1 (Agilent Technologies), following the specifications detailed by Barros et al. [15].

The mobile phases used were H_2O/formic acid (99.0:1.0, *v*/*v*) (solvent A) and acetonitrile/formic acid (9.0:1.0, *v*/*v*) (solvent B). Spectral data from all peaks were detected in the 200–600 nm range, with chromatograms recorded at 280 nm for proanthocyanidins, 330 nm for phenolic acids and stilbenes, and 360 nm for flavonols. Mass spectrometry data were acquired in negative mode. Phenolic compound identification was carried out by examining retention time (min), parent ions, and MS^2 fragmentation patterns.

Phenolic compounds were quantified using PDA chromatograms recorded at 280 nm for proanthocyanidins, 330 nm for phenolic acids, and 360 nm for flavonols. Daily prepared calibration curves were applied using catechin (for proanthocyanidins and catechin derivatives), chlorogenic acid (for phenolic acids), and quercetin glucoside (for flavonols). The results were expressed as μg/g dw.

2.8. *Determination of Antioxidant Capacity*

The antioxidant capacity of the GS blend extract was assessed using spectrophotometric methods, specifically the ABTS ((2,2′-azino-bis-(3-ethylbenzothiazoline-6-sulfonic) acid) diammonium salt), DPPH (2,2-diphenyl-1-picrylhydrazyl), and FRAP (Ferric-Reducing Antioxidant Power) assays, according to Santos et al. [28]. All methods are based on the principle of antioxidants in the sample donating electrons to reduce the oxidized forms of the reagents.

In the ABTS and DPPH assays, the reaction involves a radical formed by the oxidation of ABTS and DPPH, respectively. In contrast, the FRAP assay is based on the reduction of a ferric complex.

2.8.1. ABTS Assay

The first step of this assay involved oxidizing the ABTS salt by adding 88 μL of a 148 mM potassium persulfate solution to 5 mL of a 7 mM ABTS solution. The mixture was left to stand for 12 to 16 h at room temperature and was protected from light to achieve its most stable oxidative state.

Subsequently, the ABTS working solution was prepared by diluting the cationic ABTS solution in a 20 mM sodium acetate buffer (pH 4.5) until an absorbance of 0.700 ± 0.020 at 734 nm was obtained.

The antioxidant capacity of the samples was evaluated by adding 188 μL of the ABTS working solution to each microplate well, followed by the addition of 12 μL of the diluted sample and blank. After a 30 min of incubation at room temperature, protected from light, absorbance was measured at 734 nm [Abs (734 nm)].

For this assay, various concentrations of TROLOX were used to construct the analytical curve, with water serving as the blank. The ABTS assay measured the ability of the samples to scavenge ABTS radicals, and the inhibition percentage was calculated using the following formula:

$$\%\ \text{inhibition}_{\text{Sample}} = \frac{\text{Abs (734 nm)}_{\text{Blank}} - \text{Abs (734 nm)}_{\text{Sample}}}{\text{Abs (734 nm)}_{\text{Blank}}} \times 100$$

The results were presented as TROLOX Equivalent Antioxidant Capacity (TEAC) and expressed in millimoles of TROLOX per gram of dry weight of the sample (mmol TROLOX/g dw).

2.8.2. FRAP Assay

The evaluation of antioxidant activity using the FRAP method began with the preparation of the FRAP working solution. This solution was prepared by combining 10 volumes of 220 mM acetate buffer (pH 3.6), 1 volume of 40 mM TPTZ (2,4,6-Tripyridyl-s-Triazine) solution dissolved in 40 mM HCl, and 1 volume of 20 mM ferric chloride. The FRAP working solution, prepared daily, was preheated to 37 °C for 10 min before use.

Subsequently, 20 µL of the sample was introduced into the microplate well, followed by the addition of 280 µL of the FRAP working solution. The microplate was then shaken and incubated at 37 °C in darkness for 30 min, after which the absorbance was measured at 593 nm. To establish the calibration curve, varying concentrations of TROLOX were used. The results were expressed in millimoles of TROLOX per gram of dry weight of the sample (mmol TROLOX/g dw).

2.8.3. DPPH Assay

In the initial phase of this assay, the DPPH stock solution (8.87 mM) was diluted in ethanol/water (70:30, *v*/*v*) until an absorbance close to 1.000 at 520 nm was achieved, creating the DPPH working solution.

To evaluate the antioxidant capacity of the samples, 190 µL of the DPPH working solution was added to each microplate well, followed by 10 µL of the diluted sample and blank (ethanol/water, 70:30, *v*/*v*). The mixture was incubated at room temperature for 30 min and was protected from light, and the absorbance was measured at 520 nm [Abs (520 nm)].

TROLOX, at various concentrations, was used as the standard to construct the analytical curve. The inhibition percentage was calculated using the following formula:

$$\%\ \text{inhibition}_{\text{Sample}} = \frac{\text{Abs (520 nm)}_{\text{Blank}} - \text{Abs (520 nm)}_{\text{Sample}}}{\text{Abs (520 nm)}_{\text{Blank}}} \times 100$$

The results were presented as TROLOX Equivalent Antioxidant Capacity (TEAC) and expressed in millimoles of TROLOX per gram of dry weight of the sample (mmol TROLOX/g dw).

2.9. Evaluation of Anti-Aging and Depigmenting Activities

The evaluation of anti-aging activity was conducted through assays measuring the inhibition percentage of elastase and tyrosinase enzymes, following the methodology described by Taghouti et al. [29].

2.9.1. Tyrosinase Inhibition Assay

The tyrosinase inhibition assay was performed to determine the percentage of tyrosinase inhibition. In this procedure, 20 µL of 1000 U/mL tyrosinase and 170 µL of a mixture containing 1 mM L-tyrosine solution, 50 mM phosphate buffer (pH 6.5), and water (in a 10:10:9 proportion) were added to the sample.

Subsequently, the microplate was incubated at 37 °C for 10 min, and the absorbance was measured at 490 nm. Positive and negative controls—1 mg/mL Kojic acid and 10% DMSO, respectively—were used.

The percentage of inhibition was calculated using the following equation:

$$\%\ \text{inhibition}_{\text{Sample}} = \frac{\text{Abs (490 nm)}_{\text{Negative Control}} - \text{Abs (490 nm)}_{\text{Sample}}}{\text{Abs (490 nm)}_{\text{Negative Control}}} \times 100$$

2.9.2. Elastase Inhibition Assay

The elastase inhibition assay was conducted to determine the percentage of elastase inhibition.

A 50 µL aliquot of the sample was combined with 160 µL of 0.20 mM Tris-HCl buffer (pH 8) and 20 µL of 0.80 mM N-Succinyl-Ala-Ala-Ala-p-nitroanilide substrate (prepared in Tris-HCl buffer). The mixture was incubated at room temperature for 10 min.

Subsequently, 20 µL of 1 U/L elastase enzyme (in Tris-HCl buffer) was added, and the microplate was incubated for a further 20 min at room temperature. The absorbance was then measured at 410 nm. Tris-HCl buffer was used as the negative control.

The percentage of inhibition was calculated using the following equation:

$$\%\ \text{inhibition}_{\text{Sample}} = \frac{\text{Abs (410 nm)}_{\text{Negative Control}} - \text{Abs (410 nm)}_{\text{Sample}}}{\text{Abs (410 nm)}_{\text{Negative Control}}} \times 100$$

2.10. Cell Maintenance and Assays

2.10.1. Cell Morphology Assay

HaCaT and BJ-5ta cells were first plated in a 48-well plate at an optimized density of 3.5×10^4 cells/mL and incubated for 24 h.

Following this incubation period, the extracts were added to the wells, and incubation continued for another 24 h. The medium was then removed, and the cells were washed once with PBS. To fix the cells, 100 µL of 4% paraformaldehyde was added, followed by a 40 min incubation at room temperature (RT). The cells were then washed again with PBS.

For permeabilization, 100 µL of 0.1% Triton X-100 in PBS was added, and the cells were incubated for 30 min at RT. The permeabilization solution was prepared by mixing 5 µL of Triton X-100 with 4995 µL of PBS. After permeabilization, the cells were washed again with PBS.

Phalloidin staining was performed by diluting 0.25 µL of Alexa Fluor 568 Phalloidin (Invitrogen, A12380, Waltham, MA, USA) from a 400x stock solution into 100 µL of PBS. The staining solution was prepared by combining 3.8 µL of the Phalloidin stock solution, 1496.2 µL of PBS, and 15 mg of BSA. The cells were then incubated with this Phalloidin solution for 60 min at RT and were protected from light.

After staining, the cells were washed twice with PBS. Subsequently, 50 µL of DAPI solution was added, and the cells were incubated for 15 min. Finally, the cells were washed twice with PBS and examined using a Leica Inverted Microscope (DMI 3000B, Wetzlar, Germany, 2012).

2.10.2. Cell Metabolism Assay

Cells were monitored until they reached 80% confluency. Once confluency was achieved, the DMEM medium was removed, and the cells were washed with filtered PBS (1 mL for T25 flasks). To detach the cells, 500 µL of trypsin was added to the T25 flasks, followed by incubation for 5–10 min. Detachment was confirmed by microscopy or gentle shaking. After detachment, double the volume of the medium was added to neutralize the trypsin.

For further culture, 1000 µL of the cell suspension was added to 5 mL of medium in a T25 flask. The flasks were then incubated under standard conditions.

Resazurin Reduction Assay as a Measurement of Cellular Metabolic Activity/Viability

In this work, the resazurin reduction assay, which is an established and widely used alternative for evaluating cell viability through metabolic activity was used. Resazurin is a non-toxic, cell-permeable compound that is reduced to the highly fluorescent resorufin by metabolically active cells. This reduction process is dependent on cellular respiration and mitochondrial function, making it a reliable indicator of cell viability.

After subculturing, the cell suspension was transferred to 15 mL Falcon tubes. A 10 μL sample was used for cell counting with a Neubauer Chamber. The average cell count from four quadrants was used to adjust the cell concentration to 4.0×10^4 cells/mL.

For the assay, 200 μL of the cell suspension was plated in triplicate in a 96-well microplate and incubated for 24 h. Following incubation, the medium was removed, and 200 μL of extract at various concentrations (20.84; 41.49; 83.38; 166.75; 333.50; 667.00 μg/mL) was added, alongside controls (medium alone and medium with 0.5% DMSO). Cells were incubated for another 24 h.

The next day, the cells were washed with 1× PBS and incubated with a resazurin solution (10 mL resazurin in 1× PBS + 40 mL medium) for 2 h in the dark. Resazurin reduction was assessed by measuring the fluorescence of its reduced product, resorufin, with an excitation wavelength of 560 nm and an emission wavelength of 590 nm. The obtained results were then used to evaluate cellular viability.

2.10.3. In Vitro Cell Migration Assay

The cell lines were seeded into a 24-well plate at a density of 8.0×10^4 cells/mL and incubated at 37 °C in a 5% CO_2 atmosphere until confluence was reached. A scratch was introduced into the confluent cell layer using a 10 μL pipette tip. The culture medium was then replaced with the different extract dissolved in the culture medium at a concentration of 0.5 mg/mL.

Cell migration was monitored microscopically, and images were acquired at 0 h, 6 h, and 24 h for HaCaT cells and at 0 h, 24 h, and 48 h for BJ-5ta cells. The scratch area was measured at different time points using ImageJ's MRI Wound Healing Tool plugin (updated version 2020) [36,37].

2.10.4. Evaluation of Cell Cycle by Flow Cytometry

For cell cycle analysis, cells were cultured until they reached over 90% confluency. The medium was then removed, and cells were washed with PBS. Cells were detached with 1 mL trypsin, incubated for 10 min at 37 °C in a 5% CO_2 atmosphere, and neutralized with 2 mL of DMEM medium.

After cell counting, cells were plated in a 6-well plate in triplicate at the following densities: HaCaT: 2.64×10^5 cells/well and BJ-5ta: 1.40×10^5 cells/well.

Cells were incubated for 24 h to allow adherence, followed by treatment with GS blend extract for an additional 24 h. The concentrations tested were 20.84, 83.38, and 166.75 μg/mL.

Post-treatment, the medium was removed, and cells were washed with PBS. Cells were detached with 500 μL trypsin, incubated for 10 min, and neutralized with 1 mL of medium. Cells from three wells per condition were transferred into 15 mL Falcon tubes.

After cell counting, cells were centrifuged at 400× *g* for 7 min. The cell pellet was resuspended in 2 mL of an ice-cold Binding Buffer and centrifuged again at 400× *g* for 7 min. Cells were then fixed by adding 2 mL of ice-cold 70% ethanol dropwise while vortexing, followed by incubation on ice for 40 min.

Following fixation, cells were centrifuged at 400× *g* for 7 min and resuspended in a Binding Buffer, followed by another centrifugation step. The final cell pellet was resuspended in 500 μL of a staining solution (100 μL Enzyme A, 400 μL Nuclear Dye, and 10 mL Binding Buffer) and incubated in the dark for 30 min. Stained cells were analyzed for cell cycle distribution using flow cytometry.

2.11. Statistical Analysis

The data were analyzed using IBM SPSS 29.0 statistical software. Analysis of variance (ANOVA) was performed, followed by Tukey's multiple range test, with significance set at $p < 0.05$.

Pearson correlation analysis was conducted using Microsoft Excel (version 16.87).

3. Results and Discussion

3.1. Phenolic Content

The phenolic (TPC), *ortho*-diphenol (ODC), and flavonoid (FC) contents of the extract obtained from the blend of grape stem varieties (GS blend extract) are presented in Table 1.

Table 1. Total phenolic, *ortho*-diphenol, and flavonoid content of the GS blend extract determined by colorimetric chemical assays.

TPC (mg GAE/g)	ODC (mg GAE/g)	FC (mg CAT/g)
78.55 ± 1.67	80.83 ± 2.69	66.77 ± 1.96

The results are presented as mean ± standard deviation (n = 3).

Consistent with the existing literature [24,26,38], the GS blend extract exhibits a high concentration of phenolic compounds.

It is important to take into consideration that the proportions of each grape variety present in the blend are equal, making it possible to infer the contribution of each variety to the final effect.

Leal et al. [13] characterized methanolic extracts from distinct grape stem varieties, namely, Tinta Roriz, Touriga Nacional, Castelão, Syrah, Arinto, and Fernão Pires based on their TPC, ODC, and FC contents. The authors reported that each variety exhibited distinct concentrations, ranging from 30.91 ± 0.73 to 96.12 ± 8.14 mg GAE/g dw for TPC, 32.17 ± 1.04 to 77.26 ± 5.31 mg GAE/g dw for ODC, and 25.76 ± 1.14 to 65.14 ± 0.65 mg CAT/g dw for FC.

Similarly, Silva et al. [16] determined TPC in ethanolic extracts of grape stems from Touriga Nacional and Preto Martinho, reporting values between 45.9 ± 10.7 µg/mg and 226.8 ± 6.9 µg/mg. As shown in Table 1, the chemical characterization of the GS blend extract falls within the range of methanolic extracts. However, compared to ethanolic extracts, the GS blend extract contains approximately 1.7 times higher TPC than Touriga Nacional, but it is 2.8 times lower than Preto Martinho.

The combination of grape stems can induce chemical reactions that lead to the formation of new compounds, potentially resulting in synergistic or antagonistic effects [39]. El-Belaga et al. [40] described an increase in total phenolic and flavonoid contents when grape seed and green tea extracts were mixed in equal proportions. TPC increased by approximately 4% (from 22% to 26%), while FC increased by 7% (from 53% to 60%). Conversely, Giovanelli-Vicuña et al. [34] demonstrated that mixing methanolic extracts from different fruits (grape, lemon, and blueberry) led to a significant decrease of approximately 87% in TPC and 63% in FC, highlighting the unpredictability of compound composition in a blend.

3.2. Identification and Quantification of Phenolic Compounds by HPLC-MS Given the Biological Relevance of Phenolic Compounds, the Detailed Phenolic Composition of the GS Blend Extract Was Analyzed Using HPLC-MS (Table 2)

As shown in Table 2, the GS blend extract is rich in flavonoids, containing 1270.47 ± 2.88 µg/g dw of flavan-3-ols, including proanthocyanidins and catechin derivatives, and 108.36 ± 0.49 μg/g dw of flavonols. These results align with the flavonoid content assay, which yielded 66.77 ± 1.96 mg CAT/g dw, demonstrating consistency between individual flavonoid quantifications and the total flavonoid content.

The *ortho*-diphenol content was confirmed through the identification of catechin and proanthocyanidins. The predominant compounds in the GS blend extract include proanthocyanidin trimers (B-type) (Isomers 1, 2, and 3), proanthocyanidin trimer monogallates (Isomers 1 and 2), and various catechin derivatives.

Leal et al. [13] identified flavan-3-ols (catechin, epicatechin), flavonols (rutin), and stilbenes (resveratrol, viniferin) in methanolic extracts of grape stems from Tinta Roriz, Touriga Nacional, Castelão, Syrah, Arinto, and Fernão Pires. Similarly, Silva et al. [16] identified flavan-3-ols (catechin, epicatechin, gallo-catechin gallate, catechin gallate), flavonols (rutin, quercetin 3-*O*-galactoside, quercetin 3-*O*-glucoside, quercetin 3-*O*-rhamnoside), and stilbenes (trans-resveratrol) in ethanolic extracts of grape stems from Touriga Nacional and Preto Martinho.

Table 2. Compounds identified in the GS blend extract and their retention times (Rt) by HPLC-MS.

Identified Compounds	Rt (min)	m/z [M-H]$^-$	Quantification (μg/g dw)
Proanthocyanidins and Catechin Derivatives			
Proanthocyanidin trimer (B-type) (Isomer 1)	3.75	865	605.13 ± 1.48 ^{a}
Proanthocyanidin trimer (B-type) (Isomer 2)	4.11	865	131.55 ± 10.68 ^{c}
Proanthocyanidin trimer (B-type) (Isomer 3)	6.16	865	229.67 ± 1.7 ^{b}
Proanthocyanidin trimer (B-type) (Isomer 4)	6.33	865	48.48 ± 1.27 ^{e}
Proanthocyanidin trimer monogallate (Isomer 1)	7.11	1017	145.83 ± 3.56 ^{c}
Proanthocyanidin trimer monogallate (Isomer 2)	9.95	1017	67.43 ± 0.27 ^{d}
Epicatechin-3-*O*-glucoside	11.35	449	42.38 ± 1.19 ^{e}
Total			1270.47 ± 2.88
Flavonols			
Rutin	10.10	609	9.82 ± 0.08 ^{c}
Quercetin-3-*O*-glucoside (Isomer 1)	10.42	463	20.16 ± 0.29 ^{b}
Isorhamnetin-3-*O*-glucoside	10.64	301	66.38 ± 0.22 ^{a}
Kampferol-3-*O*-glucoside	11.42	447	12.01 ± 1.36 ^{c}
Total			108.36 ± 0.49
Stilbenes			
Resveratrol tetramer (Isomer 1)	14.75	905	14.41 ± 0.39
Total			14.41 ± 0.39

Rt, retention time; dw, dry weight, LOQ, limit of quantification Distinct letters in the same row indicate significant differences between varieties ($p < 0.05$), according to the Welch test followed by a post-hoc Games-Howell test.

In the GS blend extract, the formation of derivatives is particularly notable, with the identification of proanthocyanidin trimers and resveratrol tetramer. This suggests that chemical interactions during the blending process resulted in the formation of complex compounds, which may explain the absence of catechin, epicatechin, and resveratrol while their derivatives are present. This reinforces the assumption that these compounds underwent complexation, forming trimers and tetramers.

Additionally, Leal et al. [13] identified phenolic acids in different grape stem extracts, including gallic acid (Syrah, Arinto, Fernão Pires), protocatechuic acid (Touriga Nacional, Castelão, Arinto, Fernão Pires), and cinnamic and caftaric acids (all varieties). Similarly, Silva et al. [16] identified gallic acid, coumaric acid, vanillic acid, chlorogenic acid, protocatechuic acid, and ferulic acid in grape stem extracts from Touriga Nacional and Preto Martinho. However, no phenolic acids were detected in the GS blend extract. This discrepancy may be due to chemical interactions or transformations occurring during the blending process, potentially leading to the alteration or degradation of these compounds, impairing their detection. El-Beltagi et al. [40] suggested that proanthocyanidins can combine with gallic acid, forming gallate esters and glycosides, as previously observed in grape seed and green tea extract mixtures. The presence of proanthocyanidin derivatives, such as proanthocyanidin trimer monogallate, supports this assumption.

3.3. *Radical Scavenging Capacity and Reducing Power of GS Blend Ectract*

The antioxidant activity of the GS blend extract is a key indicator of its biological potential. To evaluate this, three distinct methods were employed: ABTS, DPPH, and FRAP assays. ABTS and DPPH assays assess a sample's ability to scavenge free radicals. However, DPPH is more responsive to lipophilic compounds, while ABTS is a more versatile method, measuring both hydrophilic and lipophilic antioxidants. The FRAP assay evaluates antioxidant potential by measuring the extract's ability to reduce ferric (Fe^{3+}) ions to ferrous (Fe^{2+}) ions. The results of these analyses are presented in Table 3.

Table 3. Evaluation of antioxidant activities of the GS blend extract by distinct colorimetric chemical assays.

ABTS (mmol TE/g)	DPPH (mmol TE/g)	FRAP (mmol TE/g)
1.035 ± 0.04	0.574 ± 0.03	0.689 ± 0.02

DPPH, DPPH• scavenging capacity; ABTS, ABTS•+ scavenging capacity; FRAP, Ferric-reducing antioxidant power; TE, Trolox equivalents. Data are presented as mean ± SD (n = 3).

The GS blend extract demonstrated strong antioxidant activity, particularly in the ABTS assay, indicating a rich presence of hydrophilic phenolic compounds capable of scavenging free radicals, including glucosides. Several hydrophilic compounds were identified in the GS blend extract, such as rutin, quercetin-3-*O*-glucoside (Isomer 1), isorhamnetin-3-*O*-glucoside, kaempferol-3-*O*-glucoside, and epicatechin-3-*O*-glucoside.

The antioxidant potential of various grape stem varieties, Tinta Roriz, Touriga Nacional, Castelão, Syrah, Arinto, and Fernão Pires, has been reported to vary across different assays: 0.35 ± 0.00 to 0.84 ± 0.06 mmol T/g in the ABTS assay, 0.15 ± 0.01 to 0.64 ± 0.05 mmol T/g in the DPPH assay, and 0.35 ± 0.02 to 1.03 ± 0.06 mmol T/g in the FRAP assay. Costa et al. [24] conducted a similar study on another set of grape stem varieties, including Arinto, Códega do Larinho, Esgana Cão, Folgasão, Fernão Pires, and Moscatel, and found comparable results using the ABTS, DPPH, and FRAP assays.

The difference in antioxidant activity between the GS blend extract and isolated grape varieties can be attributed to the complex interactions among the phenolic compounds present in each grape variety. Phenolic compounds, such as proanthocyanidins and resveratrol, can form complex structures that either enhance or diminish antioxidant activity. Kotov et al. [41] observed a synergistic effect when combining plant extracts, such as bur-marigold and hawthorn, leading to a 110.6% increase in antioxidant activity. Similarly, another three-component mixture, bur-marigold, calendula, and hawthorn, showed a 112% increase. Aslam et al. [42] also reported enhanced antioxidant potential when combining plant extracts, particularly mixtures of *Terminalia arjuna*, *Rauvolfia serpentina*, *Elettaria cardamomum*, and *Crataegus oxyacantha*.

However, not all extract combinations result in increased antioxidant activity. Giovagnoli-Vicuña et al. [43] found an antagonistic interaction in a fruit mixture extract (grape, lemon, and blueberry). Similarly, Peixoto et al. [44] noted that grape seeds exhibited a higher concentration of phenolic compounds and greater antioxidant potential compared to pomace (a mixture of grape skins and seeds), indicating that the combination of these components can result in diminished antioxidant capability.

Plumb et al. [45] explored how polymerization, galloylation, and glycosylation affect the antioxidant activity of catechins and proanthocyanidins. Using the TBARS assay for lipid-phase antioxidant activity and the ABTS assay for aqueous-phase activity, they discovered that catechins polymerized from monomers to tetramers, decreasing the ability to prevent free radical damage in lipidic systems. A monomeric catechin presented an IC50 of 3.4 μM in the lipid peroxidation assay, while trimers had a higher IC50 of 6.5 μM, indicating reduced effectiveness.

In contrast, antioxidant activity in the aqueous phase increased with polymerization. Monomers exhibited a TROLOX Equivalent Antioxidant Capacity (TEAC) value of 2.47, whereas trimers displayed a significantly higher TEAC value of 4.87. This trend clearly demonstrates that polymerization enhances hydrophilic antioxidant activity while diminishing lipophilic protection, suggesting a structural influence on antioxidant partitioning between lipid and aqueous environments.

These findings have important implications for the bioavailability and functional application of catechin-based antioxidants. In lipid-rich environments, such as cell membranes and lipoproteins, monomeric catechins may provide more effective protection against oxidative stress. However, in aqueous systems, like blood plasma and intracellular compartments, polymerized catechins may exhibit superior antioxidant potential. This suggests that the degree of polymerization plays a crucial role in determining the efficacy of catechin-derived antioxidants in different biological contexts.

The results obtained in this study align with the findings of Plumb et al. [45]. The elevated antioxidant activity measured by ABTS, uncommon in isolated varieties, is notably higher in the GS blend extract, likely due to the presence of proanthocyanidin trimers.

These findings suggest that phenolic compounds exert their antioxidant effects through different mechanisms, depending on the environment. Therefore, antioxidant effectiveness cannot be solely explained by the ability of phenolic hydroxyl groups to donate hydrogen. Antagonistic effects in mixtures may occur due to interactions between phenolic compounds, such as the formation of hydrogen bonds. As the mixture oxidizes, it is possible to regenerate into free radicals, which may then accept electrons or hydrogen atoms from other antioxidants within the mixture, further influencing antioxidant activity.

3.4. Assessment of Anti-Ageing and Skin Depigmentation Effects

The anti-aging and depigmentation properties of the GS blend extract were evaluated using enzyme inhibition assays for elastase and tyrosinase.

The evaluation of anti-aging and depigmentation properties using elastase and tyrosinase inhibition assays is scientifically justified due to the critical roles these enzymes play in skin aging and pigmentation.

Elastase is an enzyme responsible for breaking down elastin, a key structural protein that maintains skin elasticity and firmness. Overactive elastase leads to the degradation of elastin fibers, contributing to visible signs of aging such as wrinkles, sagging, and loss of skin resilience [46,47]. Similarly, tyrosinase is a crucial enzyme involved in melanin synthesis, catalyzing the oxidation of tyrosine to dopaquinone, which eventually leads to pigment formation in the skin. Excessive tyrosinase activity results in hyperpigmentation conditions such as dark spots, melasma, and uneven skin tone. Compounds that inhibit tyrosinase can effectively reduce melanin production, leading to skin brightening and depigmentation [48].

The results of elastase and tyrosinase inhibition by the GS blend extract are presented in Table 4.

Table 4. Evaluation of anti-aging and depigmenting activities determined by colorimetric enzymatic assays.

% Inhibition Elastase	% Inhibition Tyrosinase
37.78 ± 2.26	45.94 ± 0.63

The results are presented as mean ± standard deviation (n = 3) and are expressed in % inhibition.

As seen in Table 4, the GS blend extract exhibits low elastase inhibition ability but demonstrates an average tyrosinase inhibition capability (approximately 46%). When comparing these results with the literature, the limited ability of the GS blend extract to inhibit elastase becomes evident. Leal et al. [13] reported that isolated grape varieties exhibit elastase inhibition percentages ranging from 73% to 98%. Similarly, Jiratchayamaethasakul et al. (2020) showed that *S. europaea* extract had the highest anti-elastase activity (74.88 ± 4.84%) among 22 halophyte plant extracts, further highlighting the weak elastase inhibition capability of the GS blend extract.

On the other hand, the studies conducted by Leal et al. [13] and Jiratchayamaethasakul et al. [47] support the good performance of the GS blend extract in inhibiting tyrosinase, with inhibition values ranging from 44% to 54% for isolated grape stem extracts and 58.62 ± 6.08% for *S. anglica* [47].

This difference can be attributed to the complex interactions among phenolic compounds present in the extract. Yu et al. [49] demonstrated the inhibitory capacity of isolated phenolic compounds, such as cinnamic acid, ferulic acid, quercetin, isorhamnetin, and gallic acid, both individually and in combination. The distinct combination of these phenolic compounds led to a variety of results due to chemical interactions, which may or may not exert a synergistic effect on tyrosinase inhibition. Yu et al. [49] also demonstrated that quercetin, cinnamic acid, and ferulic acid have a synergistic effect on tyrosinase inhibition.

These findings suggest that phenolic compounds can interact in ways that either enhance or diminish the function of tyrosinase. The low elastase inhibition observed in the GS blend extract may be due to less optimal combinations or interactions of phenolic compounds compared to those found in isolated grape varieties.

3.5. Pearson Correlation

The Pearson correlation coefficient, also known as the product moment correlation coefficient, is dimensionless and ranges from −1 to +1. A positive value indicates a positive relationship, while a negative value indicates a negative relationship. A perfect correlation, where all data points on a scatter plot lie on a straight line, is represented by a coefficient of +1 (perfect positive) or −1 (perfect negative). A coefficient of zero indicates no linear relationship between the variables, meaning they are uncorrelated [50].

To understand the correlation between the different properties assessed in this study, a Pearson correlation analysis was performed (Table 5).

Table 5. Pearson correlation. (TPC: total phenolic content; ODC: *ortho*-diphenol content; FC: flavonoid content; FRAP assay; DPPH assay; ABTS assay; TYR: anti-tyrosinase assay; ELAS: anti-elastase assay).

	TPC	ODC	FC	FRAP	DPPH	ABTS	TYR	ELAS
TPC	1.00	0.89	0.86	0.54	−0.37	0.96	0.24	−0.96
ODC	0.89	1.00	1.00	0.09	−0.76	0.98	0.66	−0.72
FC	0.86	1.00	1.00	0.02	−0.80	0.96	0.71	−0.68

As observed in Table 5, TPC is positively correlated with ODC and FC, showing a near-perfect positive correlation. This result was expected, as *ortho*-diphenols and flavonoids belong to the phenolic compound family, which directly contributes to the increase in TPC. Consequently, as TPC increases, there is a simultaneous rise in ODC and FC, demonstrating a co-dependent relationship. This interdependence arises from the structural similarities and shared biosynthetic pathways of phenolic compounds, which collectively influence antioxidant properties and enzyme inhibition potential.

The total phenolic content (TPC, ODC, and FC) also appears to be positively correlated with antioxidant activity as measured by the ABTS assay, though it is only moderately correlated with tyrosinase enzyme inhibition. This indicates that phenolic compounds play a significant role in scavenging free radicals in aqueous-phase systems, which is effectively captured by the ABTS method. The moderate correlation with tyrosinase inhibition suggests that while phenolics contribute to the extract's depigmenting properties, other bioactive compounds or synergistic interactions may further enhance tyrosinase inhibition.

Conversely, a negative correlation is observed between phenolic content and both elastase enzyme inhibition and antioxidant activity, as measured by the DPPH assay. This suggests that while the GS blend extract effectively neutralizes radicals in hydrophilic environments (as shown by the ABTS assay), its efficacy in lipophilic antioxidant systems (as measured by DPPH) may be limited. This could be due to the hydrophilic nature of many phenolic compounds, which may have a lower affinity for lipid-phase radical scavenging. Similarly, the negative correlation with elastase inhibition suggests that phenolics alone may not be the primary contributors to elastase inhibition, implying that other classes of compounds may be responsible for anti-aging effects related to elastin preservation.

Overall, these findings demonstrate that the phenolic compounds present in the GS blend extract are primarily responsible for its antioxidant activity against free radicals in aqueous environments and its inhibitory effect on tyrosinase, supporting its potential application in skin-brightening formulations. However, the weaker correlation between elastase inhibition and lipid-phase antioxidant activity suggests that additional bioactive compounds may be necessary to enhance its anti-aging properties. Further studies focusing on the identification and characterization of these compounds, as well as their interactions with phenolics, would be valuable for optimizing the extract's efficacy in both antioxidant and anti-aging applications.

The work of Ali et al. [51] corroborates these findings, as they demonstrated a Pearson coefficient close to 1 when correlating total phenolic content with flavonoid content in a *Caragana brachyantha* extract. Additionally, phenolic content showed a moderate correlation with tyrosinase inhibition. However, contrary to the results of this study, phenolic content appeared to be strongly correlated with antioxidant activity, as measured by the FRAP assay. Similarly, Teixeira et al. [52] reported a significant correlation between TPC and ABTS (0.912) as well as FRAP (0.944), along with a notable inverse correlation in DPPH assays (−0.848).

3.6. In Vitro Assays

The skin consists of multiple layers, each with distinct characteristics and specific cell populations. Among these, keratinocytes and fibroblasts were selected for this study due to their fundamental roles in skin integrity and repair. Keratinocytes, located in the epidermis, are essential for maintaining epithelial cohesion and providing protection against external factors. In contrast, fibroblasts, found in the dermis, play a key role in synthesizing structural proteins, such as collagen and elastin, which contribute to skin firmness and elasticity [53].

With aging, the epidermis becomes 10–50% thinner, primarily due to keratinocyte atrophy. Another hallmark of aging is the accumulation of senescent cells and an increase in cellular apoptosis [54–56]. In the dermis, aging is also characterized by a reduction in fibroblast number, size, and cytoplasmic extensions. These changes are associated with alterations in the endoplasmic reticulum, diminished proliferative and metabolic activity, and a decreased expression of extracellular matrix (ECM) components, particularly collagen and elastin [57–59].

In this context, several cellular assays were performed using a keratinocyte cell line (HaCaT) and a fibroblast cell line (BJ-5ta) to assess the effects of the GS blend extract on distinct aspects of cellular behaviour.

3.6.1. Influence of the GS Blend Extract on Cell Morphology

Keratinocytes have a polygonal shape, appearing cuboidal in the basal layer and becoming increasingly flattened as they migrate towards the surface. These cells contain a prominent nucleus that gradually diminishes in size during maturation, reflecting their transition to a keratinized state [60,61]. Given the critical role of keratinocyte morphology in maintaining skin structure, the impact of the GS blend extract on this feature was examined. The results are illustrated in Figure 1.

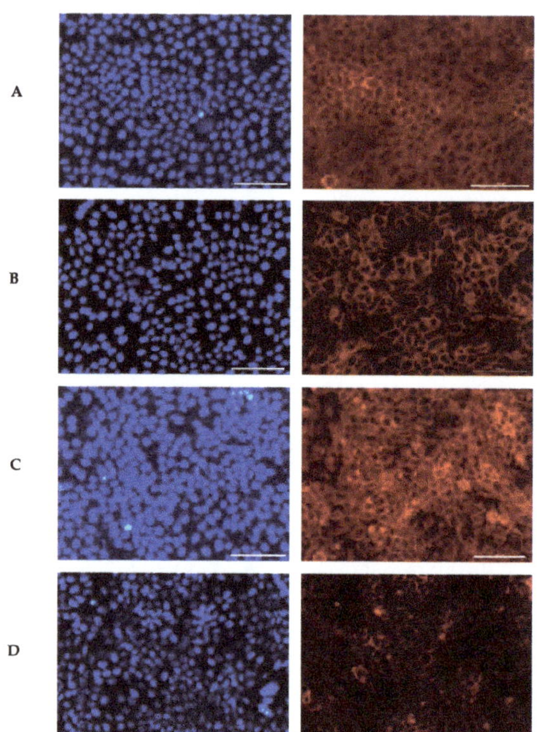

Figure 1. Influence of the GS blend extract on keratinocyte (HaCaT cells) morphology after 24 h of exposure. (**A**) DMSO control, showing a uniform distribution of well-spread, flattened cells with intact cytoskeletal organization, (**B**) 20.84 μg/mL GS blend extract, where slight alterations in cell shape and distribution begin to appear, (**C**) 83.38 μg/mL GS blend extract, showing increased cytoskeletal disorganization and reduced cell spreading, (**D**) 166.75 μg/mL GS blend extract, exhibiting pronounced loss of cell–cell interaction and altered spatial orientation. The actin cytoskeleton is stained in red (Alexa Fluor 568 Phalloidin), and nuclei are stained in blue (DAPI). Scale Bar 100 μm.

In the control group, the cells appeared flattened, with a uniform distribution of their prominent nuclei (blue) and cytoplasmic proteins (red). The presence of the GS blend extract appeared to induce disorganization in cell spatial orientation in a dose-dependent manner.

Similarly, George et al. [62] examined the impact of luteolin (5–80 μM) on the same keratinocyte line used in this study, HaCaT. Their study revealed that luteolin treatment led to a significant loss of cell–cell adhesion, causing the cells to lose their normal connections with each other. Additionally, they observed the development of membrane protrusions and distended areas on the cell surface, as well as the formation of fluid-filled spaces within the cells, indicative of cellular distress. Likewise, Hsu et al. [63] reported that keratinocytes treated with epigallocatechin-3-gallate (50 μM) underwent morphological changes, including an increase in enlarged, flattened, squame-like cells, which was attributed to keratinocyte differentiation. D'Angelo et al. [64] also documented morphological alterations in HaCaT cells treated with 250 μM and 500 μM of Annurca Apple polyphenol extract, including dissociation from the growth substrate, nuclear disintegration, and a decrease in cellular size. These changes were linked to the extract's strong inhibitory effect on keratinocyte growth, highlighting its potential for treating psoriasis.

Fibroblasts are characterized by their plump, spindle-shaped, or stellate forms and centrally located oval or round nuclei [65]. A similar approach was used to evaluate the impact of the GS extract blend on a fibroblast cell line, as shown in Figure 2. The GS extract blend appeared to have a similar effect on fibroblast morphology as observed in keratinocytes.

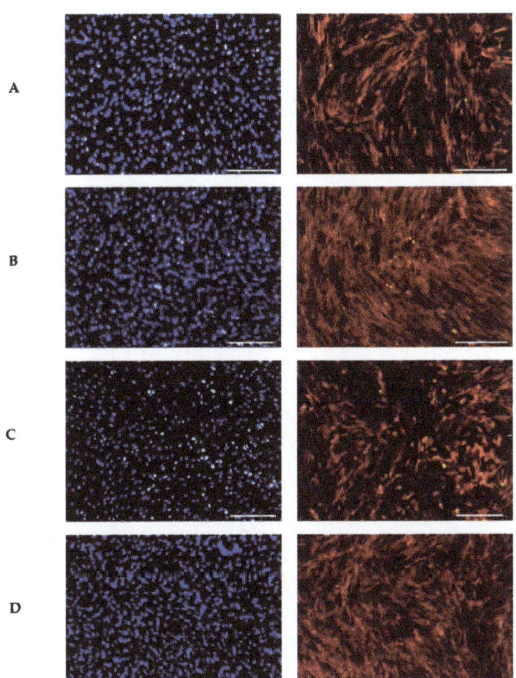

Figure 2. Influence of the GS blend extract on fibroblast (BJ-5ta cells) morphology after 24 h of exposure. (**A**) DMSO control, showing a uniform distribution of spindle-shaped cells with intact cytoskeletal organization, (**B**) 20.84 μg/mL GS blend extract, (**C**) 83.38 μg/mL GS blend extract, and (**D**) 166.75 μg/mL GS blend extract, exhibiting decreased cytoskeletal organization and altered spatial orientation in a dose-dependent manner. The actin cytoskeleton is stained in red (Alexa Fluor 568 Phalloidin), and nuclei are stained in blue (DAPI). Scale Bar 100 μm.

Czemplik et al. [66] reported comparable changes in normal human dermal fibroblast (NHDF) morphology when treated with *Linum usitatissimum* extract (140.4 µg TPC). The treated cells exhibited a rounded shape instead of their typical elongated form. However, the nuclei remained similar in size and shape to those in the control group, consistent with our observations. The authors attributed these morphological alterations to the toxicity of high concentrations of phenolic compounds.

Similarly, Kikowska et al. [67] observed that human skin fibroblasts treated with a callus extract at lower concentrations (12.5 µg/mL) maintained a proper elongated shape with large, round nuclei. In contrast, fibroblasts exposed to higher concentrations (100 µg/mL) exhibited abnormal morphology, lacking the typical elongations. Both studies corroborate the results observed in this study.

In conclusion, although the GS blend extract does not appear to significantly alter cell morphology, it does induce noticeable changes in the organization of the cell monolayer.

3.6.2. Influence of the GS Blend Extract on Cellular Metabolic Activity

To assess the influence of the GS blend extract on cells, the metabolic activity of keratinocytes (HaCaT) and fibroblasts (BJ-5ta) was evaluated following treatment with the extract.

The findings depicted in Figure 3 demonstrate that the metabolic activity of HaCaT cells increases by approximately 15% in the presence of the GS blend extract, with a threshold concentration of 166.75 µg/mL. Above this concentration, metabolic activity decreases, reaching a minimum of 30% at the highest concentration tested (667 µg/mL).

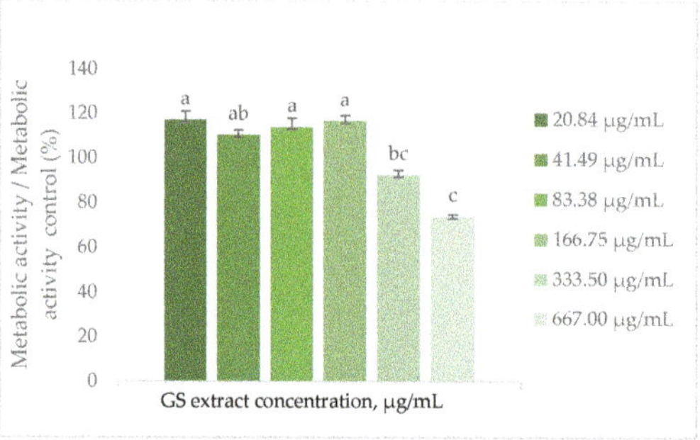

Figure 3. Effect of the GS blend extract on the metabolic activity of keratinocytes (HaCaT cell line) measured by the resazurin reduction assay after 24 h of exposure to different concentrations. The GS blend results are presented as mean ± standard error (n = 3). Significant differences between samples are denoted by different lowercase letters ($p < 0.05$).

As shown in Figure 4, the metabolic activity of BJ-5ta cells remains statistically unchanged when treated with the GS blend extract up to 667.00 µg/mL.

This observation is supported by Domínguez-Perles et al. [63], who investigated the cytotoxic effects on HaCaT keratinocytes treated with extracts from various grape stem varieties (Tinto Cão, Tinta Barroca, Moscatel Branco, Malvasia Fina) across a range of concentrations (0.04–30.60 ng GAE/mL). Their study demonstrated that the cells' metabolic activity exceeded 97.5% compared to the control (100%).

Similarly, Sangiovanni et al. [68] analysed cell metabolic activity upon contact with leaves of the *Teinturiers* variety in HaCaT cells. Their study revealed no signs of cytotoxicity in cells exposed to the extract within a concentration range of 5 to 500 µg/mL.

The metabolic activity results suggest that the GS blend extract is compatible with the evaluated cell lines. While it does not significantly affect fibroblast metabolism, it appears to promote metabolic activity in keratinocytes at moderate concentrations but reduces it at higher concentrations.

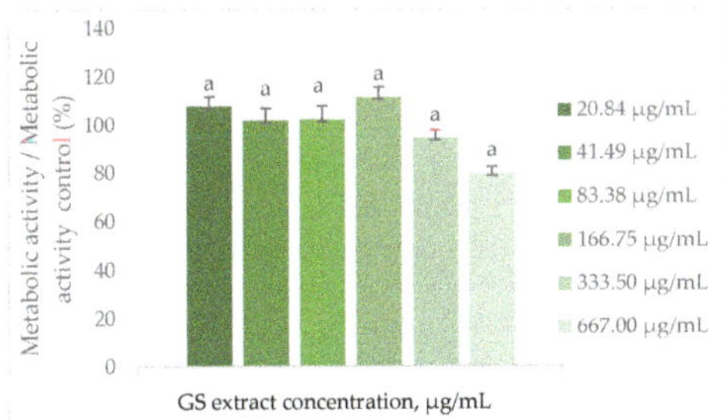

Figure 4. Effect of the GS blend extract on the metabolic activity of fibroblasts (BJ-5ta cell line) measured by the resazurin reduction assay after 24 h of exposure to different concentrations. The GS blend results are presented as mean ± standard error (n = 3). Significant differences between samples are denoted by a lowercase letter ($p < 0.05$).

3.6.3. Influence of the GS Blend Extract on the Keratinocyte and Fibroblast (HaCaT and BJ-5ta, Respectively) Cell Cycle

The cell cycle is a fundamental process by which cells grow, replicate their DNA, and divide into two daughter cells. This cycle is essential for the growth, development, and repair of multicellular organisms. It consists of several distinct phases that ensure precise replication and segregation of genetic material, as well as proper cell growth [69].

The Sub G0-G1 phase typically indicates the presence of apoptotic or dead cells, characterized by reduced DNA content due to fragmentation. In contrast, the G0-G1 phase includes cells that are either resting or in the initial stage of the cell cycle. During the S phase, cells undergo active DNA replication, reflecting a heightened level of proliferative activity as they prepare for division. Finally, in the G2-M phase, cells complete DNA synthesis and prepare for mitosis, the final step in cell division [70,71].

Understanding these phases is crucial for evaluating cell proliferation dynamics and the cellular response to treatments. Thus, the effect of the GS blend extract on the HaCaT cell cycle was assessed using flow cytometry. Fluorescence intensity, which is proportional to DNA content, allowed for the distinction of cells in different phases of the cycle. Apoptotic cells exhibited reduced DNA content compared to viable cells in the G0/G1 phase.

Flow cytometry analysis demonstrated that the GS blend extract does not significantly influence the HaCaT cell cycle at concentrations ranging from 20.84 to 83.38 μg/mL (Figure 5). However, at a higher concentration (166.75 μg/mL), an arrest in the G0-G1 phase was observed, leading to a reduction in the number of cells in the S phase. This suggests that at higher concentrations, the GS extract can inhibit cell proliferation by preventing cells from progressing to the S phase, thereby blocking DNA replication and subsequent cell division. Furthermore, when cells were treated with the GS extract at concentrations above 166.75 μg/mL, a 70–100% decrease in metabolic activity was observed, potentially indicating a reduction in cell number due to the inhibition of proliferation or induction of apoptosis.

The cell cycle of BJ-5ta fibroblast cells was also evaluated using flow cytometry, with the fibroblast gate selected in the same manner. The results are shown below (Figure 6).

Regarding the BJ-5ta cell cycle, the GS blend extract does not appear to significantly alter cell cycle progression. However, at higher concentrations (83.38 and 166.75 μg/mL), cell cycle arrest in the Sub G0-G1 phase was observed, suggesting the potential induction of apoptosis (Figure 6).

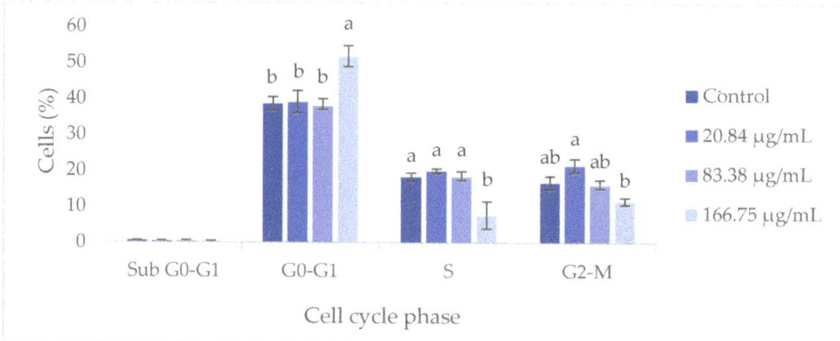

Figure 5. Flow cytometry analysis of the cell cycle in keratinocytes (HaCaT cell line) following exposure to different concentrations of the GS blend extract. The GS blend results are presented as mean ± standard error (n = 3). Significant differences between samples for each cell cycle phase are denoted by different lowercase letters ($p < 0.05$).

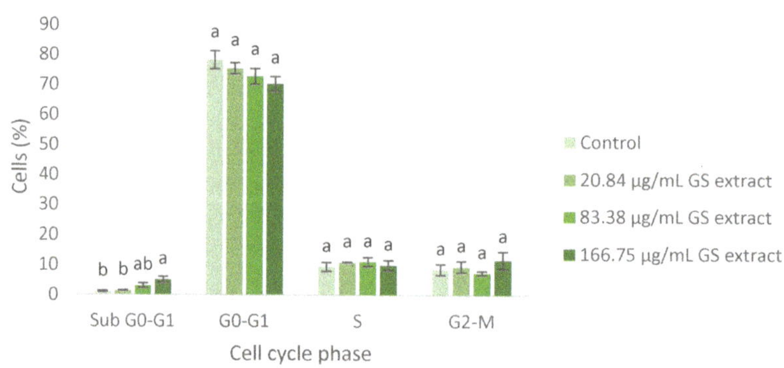

Figure 6. Flow cytometry analysis of the cell cycle in fibroblasts (BJ-5ta cell line) following exposure to different concentrations of the GS blend extract. The GS blend results are presented as mean ± standard error (n = 3). Significant differences between samples for each cell cycle phase are denoted by different lowercase letters ($p < 0.05$).

These findings support the potential of the GS extract in managing scar formation. The GS extract arrests keratinocytes and fibroblasts in the G0-G1 and Sub G0-G1 phases, impeding proliferation and promoting apoptosis. This is crucial for managing keloids by reducing excessive fibrotic responses [71–73]. Similarly, treating hypertrophic scars involves inhibiting fibroblast proliferation and inducing apoptosis [70].

Overall, these findings highlight the therapeutic potential of the GS extract in managing scar formation by regulating the cell cycle and promoting apoptosis.

3.6.4. Effect of the GS Blend Extract on Cell Migration of HaCaT Keratinocytes and BJ-5ta Fibroblasts

Cell migration is a crucial process in wound healing, involving the movement of keratinocytes to cover the wound site and fibroblasts to deposit collagen and remodel the extracellular matrix [74,75]. To evaluate the influence of the GS extract on wound healing, the migratory ability of HaCaT keratinocytes and BJ-5ta fibroblasts was assessed using the scratch method. Figure 7 demonstrates the migration performance of HaCaT cells upon contact with the GS blend extract.

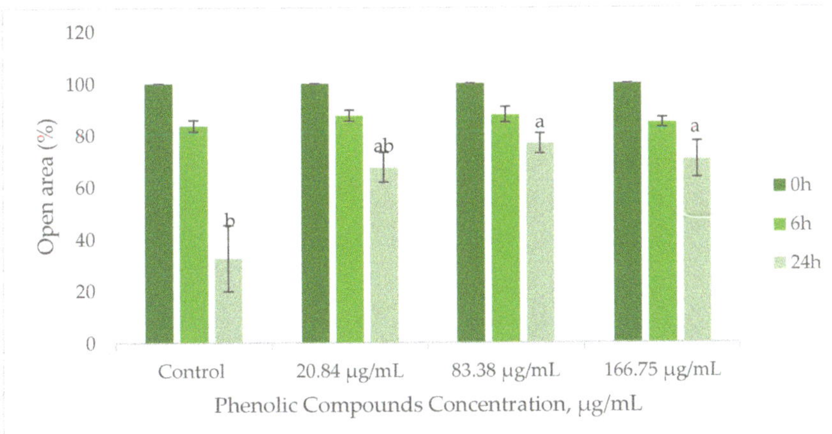

Figure 7. Influence of the GS blend extract on keratinocyte (HaCaT cell line) migration measured using the scratch assay at 0, 6, and 24 h. The GS blend results are presented as mean ± standard error (n = 3). Significant differences between samples at 24 h are denoted by different lowercase letters ($p < 0.05$).

The results suggest that the GS blend extract does not affect BJ-5ta fibroblast migration but reduces HaCaT cell migration in a concentration-dependent manner (Figure 7).

At a lower concentration (20.84 µg/mL), HaCaT cell migration remained comparable to that of the control group, indicating no adverse effects. However, at higher concentrations (83.38–166.75 µg/mL), the extract significantly reduced migration, suggesting potential applications in the controlled inhibition of cell movement.

As shown in Figure 8, the GS blend extract does not significantly impact BJ-5ta fibroblast migration, as migration rates remained similar across treated and untreated cells.

Research by Loggenberg et al. [72] demonstrated that plant-based extracts can suppress keratinocyte migration by reducing platelet-derived growth factor (PDGF-AA) expression, which plays a key role in keratinocyte proliferation and angiogenesis. Since angiogenesis is a major factor in hypertrophic scars and keloids, the inhibition of keratinocyte migration by the GS extract suggests potential anti-angiogenic properties, which could help control excessive scar formation [73,74].

Studies indicate that keloids and hypertrophic scars exhibit increased angiogenesis, suggesting that targeting angiogenic pathways may contribute to reducing excessive scar formation [75–78]. Thus, the observed reduction in HaCaT cell migration may be linked to anti-angiogenic mechanisms, supporting GS extract's potential application in scar therapy.

The GS blend extract does not promote wound healing through increased cell migration; its ability to inhibit migration and proliferation aligns with findings related to scar tissue regulation, suggesting potential therapeutic applications in controlling keloids and hypertrophic scars.

While our study provides preliminary evidence that the GS blend extract influences key cellular behaviours involved in scar modulation, particularly through the inhibition of keratinocyte migration and proliferation, it is important to mention that scar formation is largely influenced by extracellular matrix (ECM) remodelling. Abnormal ECM deposition, including altered collagen type I and III expression, matrix metalloproteinase (MMP) activity, and tissue inhibitors of metalloproteases (TIMPs), plays a pivotal role in fibrosis and scar formation. Taking into consideration the promising results obtained in this study, further research will be conducted to investigate GS extract's direct effects on ECM remodelling, namely, by assessing the extract's impact on collagen synthesis, MMP/TIMP balance, and collagenase.

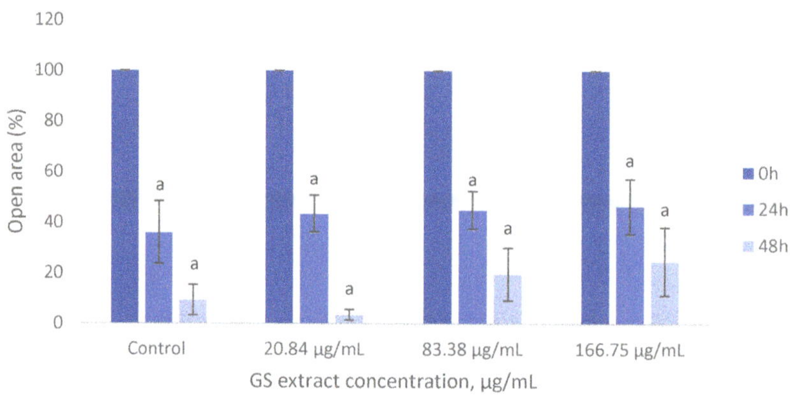

Figure 8. Influence of the GS blend extract on keratinocyte (BJ-5ta cell line) migration measured using the scratch assay at 0, 6, and 24 h GS blend. The results are presented as mean ± standard error (n = 3). Significant differences between samples at 24 h and samples at 48 h are denoted by a lowercase letter ($p < 0.05$).

4. Conclusions

This study comprehensively analysed the GS blend extract, focusing on its phenolic content, antioxidant properties, and biological effects. The extract was found to be rich in bioactive phenolic compounds, particularly proanthocyanidin trimers, catechins, rutin, quercetin, and resveratrol tetramers. The GS extract exhibited strong antioxidant activity, particularly in the ABTS assay, likely due to its high content of proanthocyanidin trimers, which are known for their potent radical scavenging properties.

In in vitro assays, the GS extract was non-cytotoxic at low concentrations. However, at higher concentrations, it inhibited cell proliferation and migration, particularly in HaCaT keratinocytes, induced cell cycle arrest in the G0-G1 and Sub G0-G1 phases, and increased fibroblast apoptosis.

These findings suggest that the GS extract may play a role in regulating cell proliferation. However, the data do not support its direct application in wound healing or scar formation, particularly in keloids, as no evidence of extracellular matrix remodelling was provided. Further studies are needed to clarify its mechanisms of action.

Author Contributions: M.S.: conceptualization, data curation, formal analysis, investigation, methodology, and writing—original draft. C.B.: methodology, supervision, writing—original draft, and writing—review and editing. D.S.: data curation and methodology. H.A.: data curation and methodology. A.C.: data curation and writing—review and editing. J.A.T.: supervision, and writing—review and editing. A.N.B.: funding acquisition, project administration, supervision, writing—original draft, and writing—review and editing. All authors have read and agreed to the published version of the manuscript.

Funding: This work was supported by the National Funds by the FCT, the Portuguese Foundation for Science and Technology, under project 2023.03608.BD and was funded by the Centre for the Research and Technology of Agro-Environmental and Biological Sciences (CITAB) research unit supported by the FCT under projects UID/04033 and LA/P/0126/2020 (https://doi.org/10.54499/LA/P/0126/2020) and WASTELESS funding from HORIZONRIA (HORIZON-CL6-2022-FARM2FORK-01) under Grant Agreement No. 101084222.

Data Availability Statement: Data is contained within the article.

Conflicts of Interest: Authors Monica Serra, Hugo Almeida, Diana Sousa and Ana Casas are employed by the company Mesosystem. The remaining authors declare that the research was conducted in the absence of any commercial and financial relationships that could be constructed as a potential conflict of interest.

References

1. OIV—International Organisation of Vine and Wine. *Annual Assessment of the World Vine and Wine Sector*; OIV: Dijon, France, 2021.
2. OIV—International Organisation of Vine and Wine. *Annual Assessment of the World Vine and Wine Sector*; OIV: Dijon, France, 2022.
3. Ferrer-Gallego, R.; Silva, P. The Wine Industry By-Products: Applications for Food Industry and Health Benefits. *Antioxidants* **2022**, *11*, 2025. [CrossRef] [PubMed]
4. Christ, K.L.; Burritt, R.L. Critical Environmental Concerns in Wine Production: An Integrative Review. *J. Clean. Prod.* **2013**, *53*, 232–242. [CrossRef]
5. Silva, A.; Silva, V.; Igrejas, G.; Aires, A.; Falco, V.; Valentão, P.; Poeta, P. Phenolic Compounds Classification and Their Distribution in Winemaking By-Products. *Eur. Food Res. Technol.* **2023**, *249*, 207–239. [CrossRef]
6. Maicas, S.; Mateo, J.J. Sustainability of Wine Production. *Sustainability* **2020**, *12*, 559. [CrossRef]
7. Graça, A.; Corbet-Milward, J.; Schultz, H.R.; Ozer, C.; de la Fuente, M. *Managing By-Products of Vitivinicultural Origin*; OIV COLLECTIVE EXPERTISE; OIV–International Organization of Vine and Wine: Paris, France, 2017.
8. Teixeira, A.; Baenas, N.; Dominguez-Perles, R.; Barros, A.; Rosa, E.; Moreno, D.; Garcia-Viguera, C. Natural Bioactive Compounds from Winery By-Products as Health Promoters: A Review. *Int. J. Mol. Sci.* **2014**, *15*, 15638–15678. [CrossRef]
9. European Parliament Circular Economy: Definition, Importance and Benefits. Available online: https://www.europarl.europa.eu/topics/en/article/20151201STO05603/circular-economy-definition-importance-and-benefits (accessed on 29 January 2025).
10. Spigno, G.; Marinoni, L.; Garrido, G.D. State of the Art in Grape Processing By-Products. In *Handbook of Grape Processing By-Products*; Galanakis, C.M., Ed.; Academic Press: Cambridge, MA, USA, 2017; pp. 1–27.
11. Gouvinhas, I.; Queiroz, M.; Rodrigues, M.; Barros, A.I.R.N.A. Evaluation of the Phytochemistry and Biological Activity of Grape (*Vitis vinifera* L.) Stems: Toward a Sustainable Winery Industry. In *Polyphenols in Plants*; Watson, R.R., Ed.; Academic Press: Cambridge, MA, USA, 2019; pp. 381–394.
12. Esparza, I.; Moler, J.A.; Arteta, M.; Jiménez-Moreno, N.; Ancín-Azpilicueta, C. Phenolic Composition of Grape Stems from Different Spanish Varieties and Vintages. *Biomolecules* **2021**, *11*, 1221. [CrossRef]
13. Leal, C.; Gouvinhas, I.; Santos, R.A.; Rosa, E.; Silva, A.M.; Saavedra, M.J.; Barros, A.I.R.N.A. Potential Application of Grape (*Vitis vinifera* L.) Stem Extracts in the Cosmetic and Pharmaceutical Industries: Valorization of a by-Product. *Ind. Crops Prod.* **2020**, *154*, 112675. [CrossRef]
14. Dias, C.; Domínguez-Perles, R.; Aires, A.; Teixeira, A.; Rosa, E.; Barros, A.; Saavedra, M.J. Phytochemistry and Activity against Digestive Pathogens of Grape (*Vitis vinifera* L.) Stem's (Poly)Phenolic Extracts. *LWT—Food Sci. Technol.* **2015**, *61*, 25–32. [CrossRef]
15. Barros, A.; Gironés-Vilaplana, A.; Teixeira, A.; Collado-González, J.; Moreno, D.A.; Gil-Izquierdo, A.; Rosa, E.; Domínguez-Perles, R. Evaluation of Grape (*Vitis vinifera* L.) Stems from Portuguese Varieties as a Resource of (Poly)Phenolic Compounds: A Comparative Study. *Food Res. Int.* **2014**, *65*, 375–384. [CrossRef]
16. Silva, V.; Igrejas, G.; Falco, V.; Santos, T.P.; Torres, C.; Oliveira, A.M.P.; Pereira, J.E.; Amaral, J.S.; Poeta, P. Chemical Composition, Antioxidant and Antimicrobial Activity of Phenolic Compounds Extracted from Wine Industry by Products. *Food Control* **2018**, *92*, 516–522. [CrossRef]
17. Prusova, B.; Licek, J.; Kumista, M.; Baron, M.; Sochor, J. Polyphenolic Composition of Grape Stems. *Not. Bot. Horti Agrobot. Cluj. Napoca* **2020**, *48*, 1543–1559. [CrossRef]
18. Mangione, R.; Simões, R.; Pereira, H.; Catarino, S.; Ricardo-da-Silva, J.; Miranda, I.; Ferreira-Dias, S. Potential Use of Grape Stems and Pomaces from Two Red Grapevine Cultivars as Source of Oligosaccharides. *Processes* **2022**, *10*, 1896. [CrossRef]
19. Blackford, M.; Comby, M.; Zeng, L.; Dienes-Nagy, Á.; Bourdin, G.; Lorenzini, F.; Bach, B. A Review on Stems Composition and Their Impact on Wine Quality. *Molecules* **2021**, *26*, 1240. [CrossRef] [PubMed]
20. González-Centeno, M.R.; Rosselló, C.; Simal, S.; Garau, M.C.; López, F.; Femenia, A. Physico-Chemical Properties of Cell Wall Materials Obtained from Ten Grape Varieties and Their Byproducts: Grape Pomaces and Stems. *LWT—Food Sci. Technol.* **2010**, *43*, 1580–1586. [CrossRef]

21. Baroi, A.M.; Popitiu, M.; Fierascu, I.; Sărdărescu, I.-D.; Fierascu, R.C. Grapevine Wastes: A Rich Source of Antioxidants and Other Biologically Active Compounds. *Antioxidants* **2022**, *11*, 393. [CrossRef]
22. Ferreira-Santos, P.; Nobre, C.; Rodrigues, R.M.; Genisheva, Z.; Botelho, C.; Teixeira, J.A. Extraction of Phenolic Compounds from Grape Pomace Using Ohmic Heating: Chemical Composition, Bioactivity and Bioaccessibility. *Food Chem.* **2024**, *436*, 137780. [CrossRef]
23. Charalampia, D.; Koutelidakis, A. Grape Pomace: A Challenging Renewable Resource of Bioactive Phenolic Compounds with Diversified Health Benefits. *MOJ Food Process. Technol.* **2016**, *3*, 00065. [CrossRef]
24. Costa-Pérez, A.; Medina, S.; Sánchez-Bravo, P.; Domínguez-Perles, R.; García-Viguera, C. The (Poly)Phenolic Profile of Separate Winery By-Products Reveals Potential Antioxidant Synergies. *Molecules* **2023**, *28*, 2081. [CrossRef]
25. Simonetti, G.; Brasili, E.; Pasqua, G. Antifungal Activity of Phenolic and Polyphenolic Compounds from Different Matrices of Vitis Vinifera L. against Human Pathogens. *Molecules* **2020**, *25*, 3748. [CrossRef]
26. Costa, R.D.; Domínguez-Perles, R.; Abraão, A.; Gomes, V.; Gouvinhas, I.; Barros, A.N. Exploring the Antioxidant Potential of Phenolic Compounds from Winery By-Products by Hydroethanolic Extraction. *Molecules* **2023**, *28*, 6660. [CrossRef]
27. Gouvinhas, I.; Santos, R.A.; Queiroz, M.; Leal, C.; Saavedra, M.J.; Domínguez-Perles, R.; Rodrigues, M.; Barros, A.I.R.N.A. Monitoring the Antioxidant and Antimicrobial Power of Grape (Vitis Vinifera L.) Stems Phenolics over Long-Term Storage. *Ind. Crops Prod.* **2018**, *126*, 83–91. [CrossRef]
28. Santos, S.; Maia, M.; Barros, A.; Gouvinhas, I. Assessment of Phenolic Content, Antioxidant and Anti-Aging Activities of Honey from Pittosporum Undulatum Vent. Naturalized in the Azores Archipelago (Portugal). *Appl. Sci.* **2023**, *13*, 1788. [CrossRef]
29. Taghouti, M.; Martins-Gomes, C.; Schäfer, J.; Félix, L.M.; Santos, J.A.; Bunzel, M.; Nunes, F.M.; Silva, A.M. Thymus Pulegioides L. as a Rich Source of Antioxidant, Anti-Proliferative and Neuroprotective Phenolic Compounds. *Food Funct.* **2018**, *9*, 3617–3629. [CrossRef]
30. Castro, M.L.; Azevedo-Silva, J.; Valente, D.; Machado, A.; Ribeiro, T.; Ferreira, J.P.; Pintado, M.; Ramos, O.L.; Borges, S.; Baptista-Silva, S. Elevating Skincare Science: Grape Seed Extract Encapsulation for Dermatological Care. *Molecules* **2024**, *29*, 3717. [CrossRef]
31. Sahiner, M.; Yilmaz, A.S.; Gungor, B.; Ayoubi, Y.; Sahiner, N. Therapeutic and Nutraceutical Effects of Polyphenolics from Natural Sources. *Molecules* **2022**, *27*, 6225. [CrossRef]
32. Constantin, O.E.; Stoica, F.; Rațu, R.N.; Stănciuc, N.; Bahrim, G.E.; Râpeanu, G. Bioactive Components, Applications, Extractions, and Health Benefits of Winery By-Products from a Circular Bioeconomy Perspective: A Review. *Antioxidants* **2024**, *13*, 100. [CrossRef]
33. Zhou, Y.; Wu, Y.; Jiang, X.; Zhang, X.; Xia, L.; Lin, K.; Xu, Y. The Effect of Quercetin on the Osteogenesic Differentiation and Angiogenic Factor Expression of Bone Marrow-Derived Mesenchymal Stem Cells. *PLoS ONE* **2015**, *10*, e0129605. [CrossRef]
34. Lupo, G.; Cambria, M.T.; Olivieri, M.; Rocco, C.; Caporarello, N.; Longo, A.; Zanghì, G.; Salmeri, M.; Foti, M.C.; Anfuso, C.D. Anti-angiogenic Effect of Quercetin and Its 8-methyl Pentamethyl Ether Derivative in Human Microvascular Endothelial Cells. *J. Cell Mol. Med.* **2019**, *23*, 6565. [CrossRef]
35. Zulkefli, N.; Che Zahari, C.N.M.; Sayuti, N.H.; Kamarudin, A.A.; Saad, N.; Hamezah, H.S.; Bunawan, H.; Baharum, S.N.; Mediani, A.; Ahmed, Q.U.; et al. Flavonoids as Potential Wound-Healing Molecules: Emphasis on Pathways Perspective. *Int. J. Mol. Sci.* **2023**, *24*, 4607. [CrossRef]
36. MRI-MRI_Wound_Healing_Tool Wound Healing Tool. Available online: https://github.com/AlejandraArnedo/Wound-healing-size-tool/wiki (accessed on 18 February 2025).
37. Suarez-Arnedo, A.; Torres Figueroa, F.; Clavijo, C.; Arbeláez, P.; Cruz, J.C.; Muñoz-Camargo, C. An Image J Plugin for the High Throughput Image Analysis of in Vitro Scratch Wound Healing Assays. *PLoS One* **2020**, *15*, e0232565. [CrossRef]
38. Han, B.; Jaurequi, J.; Tang, B.W.; Nimni, M.E. Proanthocyanidin: A Natural Crosslinking Reagent for Stabilizing Collagen Matrices. *J. Biomed. Mater. Res. A* **2003**, *65*, 118–124. [CrossRef] [PubMed]
39. Madesh, M.; Balasubramanian, K. Microtiter Plate Assay for Superoxide Dismutase Using MTT Reduction by Superoxide. *Indian. J. Biochem. Biophys.* **1998**, *35*, 184–188. [PubMed]
40. El-Beltagi, H.S.; El-Desouky, W.; Yousef, R.S. Synergistic Antioxidant Scavenging Activities of Grape Seed and Green Tea Extracts against Oxidative Stress. *Not. Bot. Horti Agrobot. Cluj Napoca* **2016**, *44*, 367–374. [CrossRef]
41. Kotov, S.A.; Gontova, T.M. Study of the Correlation between the Content of Polyphenols in Different Combinations of Plant Extracts and Their Antioxidant Activity of a Multicomponent Herbal Medicine with Antiallergic Effect. *Farmatsevtychnyi Zhurnal* **2022**, 65–73. [CrossRef]
42. Aslam, S.; Jahan, N.; Rahman, K.; Zafar, F.; Ashraf, M. Synergistic Interactions of Polyphenols and Their Effect on Antiradical Potential. *Pak. J. Pharm. Sci.* **2017**, *30*, 1297–1304.

43. Giovagnoli-Vicunã, C.; Pizarro, S.; Briones-Labarca, V.; Delgadillo, Á. A Square Wave Voltammetry Study on the Antioxidant Interaction and Effect of Extraction Method for Binary Fruit Mixture Extracts. *J. Chem.* **2019**, *2019*, 8706061. [CrossRef]
44. Peixoto, C.M.; Dias, M.I.; Alves, M.J.; Calhelha, R.C.; Barros, L.; Pinho, S.P.; Ferreira, I.C.F.R. Grape Pomace as a Source of Phenolic Compounds and Diverse Bioactive Properties. *Food Chem.* **2018**, *253*, 132–138. [CrossRef]
45. Plumb, G.W.; De Pascual-Teresa, S.; Santos-Buelga, C.; Cheynier, V.; Williamson, G. Antioxidant Properties of Catechins and Proanthocyanidins: Effect of Polymerisation, Galloylation and Glycosylation. *Free Radic. Res.* **1998**, *29*, 351–358. [CrossRef]
46. Imokawa, G.; Ishida, K. Biological Mechanisms Underlying the Ultraviolet Radiation-Induced Formation of Skin Wrinkling and Sagging I: Reduced Skin Elasticity, Highly Associated with Enhanced Dermal Elastase Activity, Triggers Wrinkling and Sagging. *Int. J. Mol. Sci.* **2015**, *16*, 7753–7775. [CrossRef]
47. Jiratchayamaethasakul, C.; Ding, Y.; Hwang, O.; Im, S.T.; Jang, Y.; Myung, S.W.; Lee, J.M.; Kim, H.S.; Ko, S.C.; Lee, S.H. In Vitro Screening of Elastase, Collagenase, Hyaluronidase, and Tyrosinase Inhibitory and Antioxidant Activities of 22 Halophyte Plant Extracts for Novel Cosmeceuticals. *Fish. Aquat. Sci.* **2020**, *23*, 6. [CrossRef]
48. Putri, S.; Maharani, R.; Maksum, I.; Siahaan, T. Peptide Design for Enhanced Anti-Melanogenesis: Optimizing Molecular Weight, Polarity, and Cyclization. *Drug Des. Devel Ther.* **2025**, *19*, 645–670. [CrossRef] [PubMed]
49. Yu, Q.; Fan, L.; Duan, Z. Five Individual Polyphenols as Tyrosinase Inhibitors: Inhibitory Activity, Synergistic Effect, Action Mechanism, and Molecular Docking. *Food Chem.* **2019**, *297*, 124910. [CrossRef] [PubMed]
50. Sedgwick, P. Pearson's Correlation Coefficient. *BMJ* **2012**, *345*, e4483. [CrossRef]
51. Ali, L.; Khan, S.; Nazir, M.; Raiz, N.; Naz, S.; Zengin, G.; Mukhtar, M.; Parveen, S.; Shazmeen, N.; Saleem, M.; et al. Chemical Profiling, in Vitro Biological Activities and Pearson Correlation between Phenolic Contents and Antioxidant Activities of Caragana Brachyantha Rech.f. *S. Afr. J. Bot.* **2021**, *140*, 189–193. [CrossRef]
52. Teixeira, T.S.; Vale, R.C.; Almeida, R.R.; Ferreira, T.P.S.; Guimarães, L.G.L. Antioxidant Potential and Its Correlation with the Contents of Phenolic Compounds and Flavonoids of Methanolic Extracts from Different Medicinal Plants. *Rev. Virtual Química* **2017**, *9*, 1546–1559. [CrossRef]
53. Barel, A.O.; Paye, M.; Maibach, H.I. (Eds.) *Handbook of Cosmetic Science and Technology*; CRC Press: Boca Raton, FL, USA, 2014. [CrossRef]
54. Bocheva, G.; Slominski, R.M.; Slominski, A.T. Neuroendocrine Aspects of Skin Aging. *Int. J. Mol. Sci.* **2019**, *20*, 2798. [CrossRef]
55. Chambers, E.S.; Vukmanovic-Stejic, M. Skin Barrier Immunity and Ageing. *Immunology* **2020**, *160*, 116–125. [CrossRef]
56. Csekes, E.; Račková, L. Skin Aging, Cellular Senescence and Natural Polyphenols. *Int. J. Mol. Sci.* **2021**, *22*, 12641. [CrossRef]
57. Eckhart, L.; Tschachler, E.; Gruber, F. Autophagic Control of Skin Aging. *Front. Cell Dev. Biol.* **2019**, *7*, 469277. [CrossRef]
58. Shin, J.W.; Kwon, S.H.; Choi, J.Y.; Na, J.I.; Huh, C.H.; Choi, H.R.; Park, K.C. Molecular Mechanisms of Dermal Aging and Antiaging Approaches. *Int. J. Mol. Sci.* **2019**, *20*, 2126. [CrossRef]
59. Russell-Goldman, E.; Murphy, G.F. The Pathobiology of Skin Aging: New Insights into an Old Dilemma. *Am. J. Pathol.* **2020**, *190*, 1356–1369. [CrossRef] [PubMed]
60. Roig-Rosello, E.; Rousselle, P. The Human Epidermal Basement Membrane: A Shaped and Cell Instructive Platform That Aging Slowly Alters. *Biomolecules* **2020**, *10*, 1607. [CrossRef] [PubMed]
61. Findlay, M.W.; Gurtner, G.C. Engineering Niches for Skin and Wound Healing. In *Biology and Engineering of Stem Cell Niches*; Academic Press: Cambridge, MA, USA, 2017; pp. 559–579. [CrossRef]
62. Koehler, M.J.; Zimmermann, S.; Springer, S.; Elsner, P.; König, K.; Kaatz, M. Keratinocyte Morphology of Human Skin Evaluated by in Vivo Multiphoton Laser Tomography. *Ski. Res. Technol.* **2011**, *17*, 479–486. [CrossRef] [PubMed]
63. George, V.C.; Kumar, D.R.N.; Suresh, P.K.; Kumar, S.; Kumar, R.A. Comparative Studies to Evaluate Relative in Vitro Potency of Luteolin in Inducing Cell Cycle Arrest and Apoptosis in HaCaT and A375 Cells. *Asian Pac. J. Cancer Prev.* **2013**, *14*, 631–637. [CrossRef]
64. Hsu, S.; Bollag, W.B.; Lewis, J.; Huang, Q.; Singh, B.; Sharawy, M.; Yamamoto, T.; Schuster, G. Green Tea Polyphenols Induce Differentiation and Proliferation in Epidermal Keratinocytes. *J. Pharmacol. Exp. Ther.* **2003**, *306*, 29–34. [CrossRef]
65. D'Angelo, S.; La Porta, R.; Napolitano, M.; Galletti, P.; Quagliuolo, L.; Boccellino, M. Effect of Annurca Apple Polyphenols on Human HaCaT Keratinocytes Proliferation. *J. Med. Food* **2012**, *15*, 1024–1031. [CrossRef]
66. Ravikanth, M.; Soujanya, P.; Manjunath, K.; Saraswathi, T.R.; Ramachandran, C.R. Heterogenecity of Fibroblasts. *J. Oral. Maxillofac. Pathol.* **2011**, *15*, 247–250. [CrossRef]
67. Czemplik, M.; Korzun-Chłopicka, U.; Szatkowski, M.; Działo, M.; Szopa, J.; Kulma, A. Optimization of Phenolic Compounds Extraction from Flax Shives and Their Effect on Human Fibroblasts. *Evid. Based Complement. Altern. Med.* **2017**, *2017*, 3526392. [CrossRef]

68. Kikowska, M.A.; Chmielewska, M.; Włodarczyk, A.; Studzinska-Sroka, E.; Zuchowski, J.; Stochmal, A.; Kotwicka, M.; Thiem, B. Effect of Pentacyclic Triterpenoids-Rich Callus Extract of Chaenomeles Japonica (Thunb.) Lindl. Ex Spach on Viability, Morphology, and Proliferation of Normal Human Skin Fibroblasts. *Molecules* **2018**, *23*, 3009. [CrossRef]
69. Sangiovanni, E.; Di Lorenzo, C.; Piazza, S.; Manzoni, Y.; Brunelli, C.; Fumagalli, M.; Magnavacca, A.; Martinelli, G.; Colombo, F.; Casiraghi, A.; et al. *Vitis vinifera* L. Leaf Extract Inhibits In Vitro Mediators of Inflammation and Oxidative Stress Involved in Inflammatory-Based Skin Diseases. *Antioxidants* **2019**, *8*, 134. [CrossRef]
70. Marassatti, D.A.; Ramiro, G.C.; Hassunuma, R.M.; Garcia, P.C.; Messias, S.H.N. *A Vida Da Célula: Uma Visão Morfológica Do Ciclo e Das Divisões Celulares*; Catanzaro, D.P., Sampaio, V.R.E., Eds.; FALTA: Bauru, Brazil, 2021.
71. Wolfram, D.; Tzankov, A.; Pülzl, P.; Piza-Katzer, H. Hypertrophic Scars and Keloids–A Review of Their Pathophysiology, Risk Factors, and Therapeutic Management. *Dermatol. Surg.* **2009**, *35*, 171–181. [CrossRef] [PubMed]
72. Loggenberg, S.R.; Twilley, D.; De Canha, M.N.; Meyer, D.; Mabena, E.C.; Lall, N. Evaluation of Wound Healing and Antibacterial Potential of Greyia Radlkoferi Szyszyl. Ethanolic Leaf Extract. *Front. Pharmacol.* **2022**, *13*, 806285. [CrossRef] [PubMed]
73. Korntner, S.; Lehner, C.; Gehwolf, R.; Wagner, A.; Grütz, M.; Kunkel, N.; Tempfer, H.; Traweger, A. Limiting Angiogenesis to Modulate Scar Formation. *Adv. Drug Deliv. Rev.* **2019**, *146*, 170–189. [CrossRef] [PubMed]
74. Lali, F.V.; Metcalfe, A.D. The Role of Angiogenesis in Wound Healing, Scarring and Tissue Regeneration. *Plast. Maxillofac. Surg. Aesthet. (PMFA) J.* **2014**, *2*, 1–4.
75. Han, C.; Barakat, M.; Dipietro, L.A. Angiogenesis in Wound Repair: Too Much of a Good Thing? *Cold Spring Harb. Perspect. Biol.* **2022**, *14*, a041225. [CrossRef]
76. Diao, J.S.; Xia, W.S.; Guo, S.Z. Bevacizumab: A Potential Agent for Prevention and Treatment of Hypertrophic Scar. *Burns* **2010**, *36*, 1136–1137. [CrossRef]
77. Gira, A.K.; Brown, L.F.; Washington, C.V.; Cohen, C.; Arbiser, J.L. Keloids Demonstrate High-Level Epidermal Expression of Vascular Endothelial Growth Factor. *J. Am. Acad. Dermatol.* **2004**, *50*, 850–853. [CrossRef]
78. Ren, H.T.; Hu, H.; Li, Y.; Jiang, H.F.; Hu, X.L.; Han, C.M. Endostatin Inhibits Hypertrophic Scarring in a Rabbit Ear Model. *J. Zhejiang Univ. Sci. B* **2013**, *14*, 224–230. [CrossRef]

pharmaceuticals

Article

Combined Bipolar Radiofrequency and Non-Crosslinked Hyaluronic Acid Mesotherapy Protocol to Improve Skin Appearance and Epidermal Barrier Function: A Pilot Study

Anna Płatkowska [1,*], Szymon Korzekwa [1], Bartłomiej Łukasik [2] and Nicola Zerbinati [3]

[1] Anclara Health & Aesthetics, 02624 Warsaw, Poland; skorzekwa@anclara.pl
[2] Medical Department, Matex Lab Switzerland SA, 1228 Geneva, Switzerland; bartlomiej.lukasik@neauvia.com
[3] Dermatologic Unit, University of Insubria, 21100 Varese, Italy; nicola.zerbinati@uninsubria.it
* Correspondence: dr.platkowska@gmail.com

Abstract: Background: Age-associated changes in epidermal hydration, pigmentation, thickness and cell renewal influence skin appearance and can lead to laxity, dryness and poor skin tone. The aim of this pilot study was to assess the synergistic effects of a new bipolar radiofrequency plus non-crosslinked hyaluronic acid (HA) mesotherapy protocol compared with radiofrequency alone on skin appearance and markers of epidermal function. Methods: This prospective, single-center, split-face pilot study recruited women aged 25–65 years with dryness and laxity of the facial skin defined by a trans-epidermal water loss (TEWL) value of $\geq$26 g/m^2/h. Subjects were treated with a bipolar radiofrequency device on both sides of the face. This was immediately followed by needle hyaluronic acid (HA) treatment on one side of the face with 2.5 mL of a non-crosslinked HA. Photographic documentation, analysis of epidermal barrier function parameters, and high frequency (HF) ultrasound analysis were performed prior to treatment and at 28 days. Results: Twenty female subjects with a mean age of 46 (range 29 to 54) years and dry and lax facial skin were included. TEWL was reduced and skin hydration improved to a greater extent with the combined radiofrequency plus mesotherapy protocol compared with radiofrequency alone (−5.8% vs. +3.9% and +23.1% vs. +1.0%, respectively). The combined protocol was also associated with greater improvements in melanin (−7.5% vs. −1.5%) and erythema values (−7.2% vs. +3.0%), respectively. Ultrasound measures of epidermal thickness and epidermal density were greater after the combined protocol compared with radiofrequency alone (12.0% vs. 5.6% and 57.7% vs. 7.1%, respectively). Both treatments were well-tolerated. Conclusions: The combined bipolar radiofrequency and HA mesotherapy protocol provided greater improvements in skin hydration, firmness and tone compared with radiofrequency alone. The combination treatment was also associated with greater epidermal thickness and density and increased keratinocyte differentiation suggesting a synergistic effect of both treatments on epidermal homeostasis and barrier function. Both treatments were well-tolerated and led to improvements in facial appearance.

Keywords: skin aging; bipolar radiofrequency; non-crosslinked hyaluronic acid (HA) mesotherapy

1. Introduction

Skin aging is one of the most visible indicators of advancing age with the first changes to the face generally becoming apparent in the third decade of life [1]. As a result, there is high consumer demand for anti-ageing treatments in all age groups. Individuals are increasingly requesting aesthetic procedures that improve skin quality as part of a global facial approach that addresses skin laxity, hydration and overall skin tone, as well as reducing lines and wrinkles.

The aging of human skin is a complex process that involves genetically-determined intrinsic factors and age-related hormonal changes, the effects of which are accelerated by cumulative exposure to external environmental factors, mainly ultraviolet (UV) light exposure, but also factors such as air pollution, smoking, and poor nutrition [2].

The skin comprises the dermis and epidermis overlying an inner layer of subcutaneous tissue. The dermis consists mostly of extracellular matrix (ECM) with collagen, elastin and glycosaminoglycans as the principal components, all of which are produced by fibroblasts. The dermal ECM is arranged in a dense network of interlinked collagen (types I and III) and elastin fibers containing water-retaining proteoglycans, and provides the skin with both strong mechanical resistance and elasticity [2]. The outer layer of skin, the epidermis, is attached to the dermis by specialized proteins of the basement membrane. It is a stratified epithelium consisting mainly of keratinocytes, which undergo terminal differentiation and form the stratum corneum, which serves as a barrier against environmental damage.

Intrinsic skin aging is characterized by a progressive loss of tissue with a thinning of all skin layers [2,3]. In the dermis, the collagen network undergoes re-organization, resulting in a reduction in collagen, proteoglycan and hyaluronic acid (HA) content, accompanied by an increase in denatured and degraded collagen. These changes lead to a loss of ECM volume, and reduced mechanical tension on the fibroblasts, which is required for optimal regulation of collagen and other ECM components. The dermis is particularly affected by ageing as it has a low turn-over compared to other skin layers [4]. The most noticeable microstructural modifications of the dermis with age include decreased collagen and elastin content, a reduction in type I collagen leading to an increased ratio of type III to type I, and a deterioration of ECM proteoglycans.

Within the epidermis, the turnover of keratinocytes is reduced resulting in decreased epidermal thickness, flattening of the dermal-epidermal junction, and a reduced capacity to rapidly restore the skin barrier [2,3].

Extrinsic aging further impairs the structural integrity of the skin creating disorganized and fragmented collagen networks [5], and increased amounts of abnormal and disordered elastin [6]. Skin also requires adequate levels of hydration and transepidermal water loss (TEWL) must be carefully regulated. The prevention of TEWL is dependent on two major components: the presence of natural hygroscopic agents, such as HA, and an effective stratum corneum epidermal barrier. The relative permeability of the different epithelia reflect the extent of their exposure to the external environment with skin epithelia being the least permeable and intestinal the most [7]. A greater susceptibility to dry, rough and fragile skin in older individuals is associated with degradation and slower recovery of skin epithelial barrier function.

Hyaluronic acid is abundant in the dermis, where it contributes to the turgor and flexibility of healthy skin. It is also found in the epidermis where in addition to acting as a humectant, it is involved in maintaining normal stratum corneum structure and epidermal barrier function and serves a number of bioregulatory functions [8,9]. Both intrinsic and extrinsic factors appear to downregulate HA synthases, resulting in a reduction in HA levels, particularly in the epidermis [10–12].

The epidermis is important from an aesthetic standpoint, because it is the layer that provides the skin with its texture and moisture, and contributes to skin color. If the surface of the epidermis is dry or rough, the skin appears aged. A healthy epidermis continually regenerates itself to maintain a functional barrier and to respond to cutaneous insults. It is primarily composed of keratinocytes, which migrate upwards from basal to cornified layers and undergo a program of differentiation that ultimately leads to their transformation into corneocytes before they desquamate from the skin surface [13]. With age, epidermal cell turnover decreases and there is reduced nutrient transfer between the dermis and epidermis, which contributes to less effective desquamation and increased skin fragility [14].

Each stage of keratinocyte differentiation is associated with the expression of specific proteins: Ki67 (marker of cell proliferation), K10 (required for keratin filaments), and K17 (regulates many biological processes, including cell proliferation and growth, skin inflammation and hair follicle cycling) [15]. Epidermal homeostasis is dependent on the regenerative capacity of these cells and the complex signaling pathways that orchestrate it. Aged epidermis displays multiple alterations in keratinocyte differentiation and function which have been linked to a decrease in the integrity of the stratum corneum and a delayed ability to recover in older compared with younger age groups [16]. Reduced levels of hyaluronic acid are also observed in aged skin [17], and studies have shown that topical application of hyaluronic acid can stimulate keratinocyte differentiation as well as lipid production, leading to enhanced epidermal permeability barrier function [18].

To counteract these aging-related skin changes, a number of strategies have been developed that aim to rejuvenate the dermal and epidermal layers to reflect those of young and healthy skin [19]. Topical agents can help improve skin barrier function, but to stop the degradation and rejuvenate the skin primary structural constituents, such as collagen and elastin, technologies that deliver treatments deeper into the skin are required. These include chemical peelings, visible light devices, intense pulsed light, ablative and nonablative laser photo–rejuvenation, radiofrequency, and injectable skin biostimulation and rejuvenation in the form of mesotherapy.

Radiofrequency and mesotherapy protocols, in particular, provide skin rejuvenating effects with minimal to no downtime. Radiofrequency produces heat when the electrical resistance of tissues converts the electric current to thermal energy. In vitro studies have shown that radiofrequency currents can trigger the proliferation and migration of keratinocytes and fibroblasts, the main cell types involved in skin regeneration [20]. Both monopolar and bipolar radiofrequency treatments have demonstrated direct collagen contraction and immediate skin tightening. Subsequent remodeling and reorientation of collagen bundles, and the formation of new collagen, is achieved over months of treatment, which leads to an improvement in skin laxity [21–24].

The introduction of HA-based formulations as multiple micro-injections promotes skin hydration as well as keratinocyte differentiation to provide an optimal environment for healthy skin homeostasis and protect against extrinsic aging-related risk factors [25–28].

Combining treatments that act on multiple aspects of skin aging allows physicians to address the visible signs based on an understanding of their underlying cause, and is becoming increasingly popular to maximize outcomes and improve patient satisfaction. The main objective of this pilot study was to assess the synergistic effects of a new bipolar radiofrequency plus non-crosslinked hyaluronic acid mesotherapy protocol compared with radiofrequency alone on skin appearance and markers of epidermal function.

2. Results

The study recruited 20 female subjects, mean age 46 years (range 29 to 54 years), with Fitzpatrick skin phototype I–IV. All women presented with dry and lax facial skin with a TEWL value of at least 26 $g/m^2/h$.

All subjects completed the study, and showed clear clinical improvement of skin tightening and rhytides in the cheek, chin, crow's feet and forehead regions (Figure 1).

TEWL was reduced and skin hydration improved to a greater extent with the combined radiofrequency plus mesotherapy protocol compared with radiofrequency alone (−5.8% vs. +3.9% and +23.1% vs. +1.0%, respectively) (Table 1).

2.1. Skin Pigmentation

Subjects' melanin and erythema values, reflecting the subjects' skin tone, were improved with the combined protocol compared with radiofrequency alone (Table 1). Further analysis of skin pigmentation assessed with the Smart Mirror-Skin Analysis System confirmed that the addition of mesotherapy to the radiofrequency protocol resulted in greater improvements in all parameters analyzed compared with radiofrequency alone (Table 1).

2.2. Epidermal Thickness and Density

Mean changes in skin thickness and epidermal thickness measured by ultrasound were approximately twofold greater after the combined protocol compared with radiofrequency alone (Table 1).

The change in epidermal density was particularly pronounced with the combined protocol with an improvement of 57.7% compared with baseline, versus 7.1% compared with baseline for radiofrequency alone.

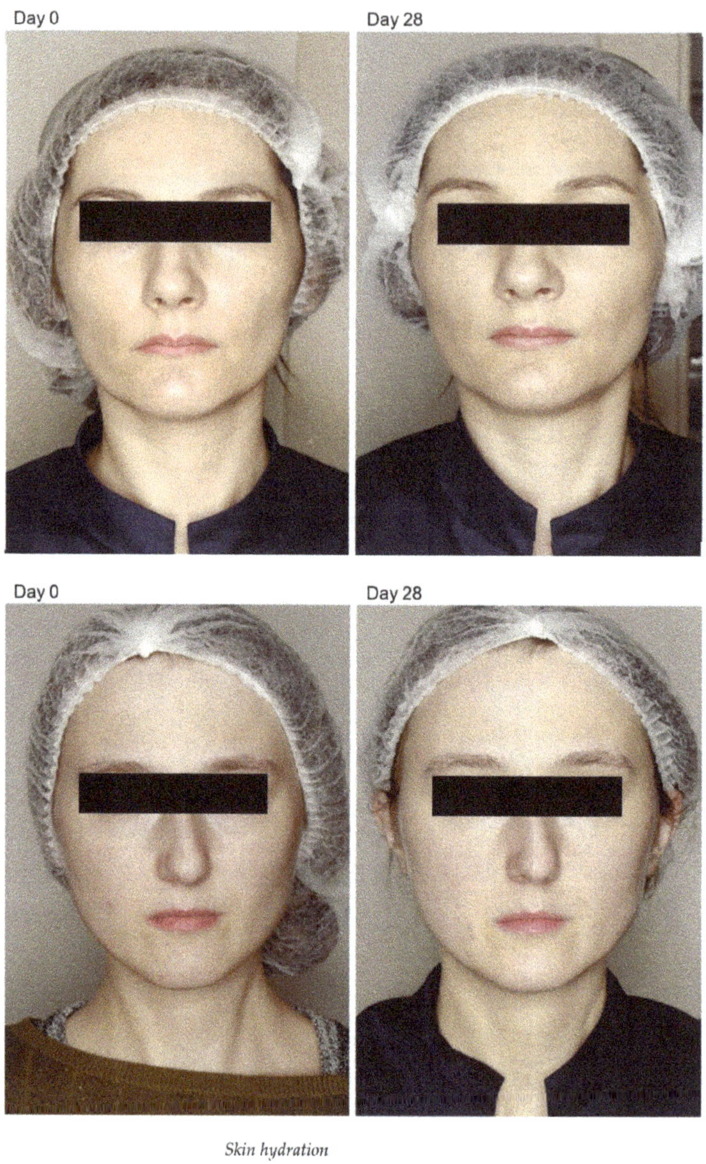

Figure 1. Bipolar radiofrequency and non-crosslinked hyaluronic acid mesotherapy clinical study results. Top image: condition before treatment; condition 28 days after treatment, bottom image: condition before treatment; condition 28 days after treatment.

Table 1. Changes in parameters of skin barrier function and homeostasis compared with pretreatment (baseline) values.

Parameter	Radiofrequency (% Change)	Radiofrequency + HA Mesotherapy (% Change)
	Cutometer analysis	
Stratum corneum hydration	+1.0	+23.1
TEWL	+3.9	−5.8
Melanin	−1.5	−7.5
Erythema	+3.0	−7.2
	Smart Mirror-Skin analysis	
RGB pore	-	−15.6
RGB spots	-	−11.9
UV pigmentation	-	−22.0
UV damage	-	−10.8
Pigmentation changes	-	−27.4
	Ultrasound analysis	
Change in skin thickness	+1.4	+2.7
Change in epidermal thickness	+5.6	+12.0
Change in skin density	−18.1	+12.4
Change in epidermal density	+7.1	+57.7

2.3. Adverse Effects

All patients tolerated the procedures well with no adverse events related to treatment. No palpable nodules, erythema, or post-inflammatory hyperpigmentation were observed.

3. Materials and Methods

3.1. Participant Selection

This prospective, single-center, split-face pilot study recruited women aged 25–65 years in good general health with dryness and laxity of the facial skin defined by a TEWL value of at least 26 $g/m^2/h$. Skin firmness was assessed by facial imaging and palpation, with the presence of gravitational wrinkles in the area of the tear trough, nasolabial folds, and marionette lines taken as evidence of a decrease in firmness.

Participants with the following conditions or treatments in their medical history were excluded: current skin diseases (autoimmune disorders involving the skin, infections and inflammation); severe allergies; use of treatments that may influence skin healing in the 6 months prior to study entry (tetracyclines, immunosuppressants, systemic steroids, anticoagulants, cosmetics containing thyme extract, herbs containing St. John's wort); facial aesthetic treatments involving fillers and/or energy-based devices in the 6 months prior to study entry; hypersensitivity to hyaluronic acid; and pregnancy or breastfeeding. Participants with a history of permanent metal implant devices or pacemakers were also excluded due to the use of radiofrequency.

3.2. Treatment Protocol

The study consisted of two visits, 4 weeks apart. Subjects were asked not to use moisturizing cosmetics on the day before each visit. Prior to performing facial imaging and skin measurements, subjects were asked to clean their face and remove all makeup. Subsequently, subjects were asked to sit quietly for 30 min to allow equilibration of their skin to the ambient conditions. At the first visit (V1), photographic documentation and analysis of the surface and color of the skin was performed. Objective assessment of epidermal barrier function was obtained using the Cutometer® Dual MPA 580 (Courage + Khazaka electronic GmbH, Cologne, Germany) based on analysis of the following biophysical indicators: TEWL; skin hydration; melanin concentration; and intensity of erythema and skin elasticity. High frequency ultrasound analysis was also performed with images obtained at six points on both sides of the face covering the infraorbital, zygomatic, and corner of the mouth areas.

Participants were then treated with radiofrequency, according to manufacturer's instructions, and adjusted to each patient. This was followed by needle mesotherapy treatment on one (randomly selected) side of the face with 2.5 mL of a non-cross-linked hyaluronic acid (Neauvia Hydro Deluxe, Matex Lab, Geneva, Switzerland).

At the second visit (V2), repeat photographic documentation, photographic analysis, analysis of skin biophysical indicators, and ultrasound analysis was performed, followed by needle mesotherapy on the side of the face not treated at V1.

3.3. *Radiofrequency Treatment*

Radiofrequency was performed once, prior to treatment with the mesotherapy product, using a Sectum device (Berger & Kraft Medical Sp. z o.o., Warsaw, Poland) and a FACE hand piece (Berger & Kraft Medical Sp. z o.o., Warsaw, Poland). Output frequency was 480 kHz. During the treatment, a current is released in a continuous mode and the applicator is set in motion mode using slow circular movement over the facial skin. Treatment times for the different areas of the face were as follows: left face cheek and chin (10–15 min), right face cheek and chin (10–15 min), forehead and crow's feet (10 min). The energy was set to between 10 and 40 W and temperature to 38–41 °C and was adjusted based on patient feedback (comfortable or uncomfortable). The temperature for treatment of the forehead and crow's feet was generally 1 °C lower than that used to treat the rest of the face.

3.4. *Hyaluronic Acid Treatment*

The non-crosslinked HA used in the study contained sodium hyaluronate 18 mg/mL, calcium hydroxyapatite microspheres (0.01%), and glycine and L-proline in buffered non-pyrogenic water. The product was supplied in a pre-filled 2.5 mL disposable syringe with a luer-lock connector. Mesotherapy treatment consisted of multiple intradermal or subcutaneous injections of HA using a 30G needle and the pre-filled syringe. To minimize pain, local anesthetic cream (EMLA®, Astra Zeneca, Cambridge, UK) was applied under occlusion for 30 min to the treated area before starting the procedure.

3.5. *Post Procedure Recommendations*

Subjects were advised not to apply pressure or use cosmetic products on the treated areas in the first 24 h after treatment. They were also advised to avoid surgery and dental procedures for 2 weeks after injection, and cosmetic procedures such as laser therapy or exfoliation of the treated area for 3 weeks after injection. They were recommended to avoid exposure to direct sunlight and UVA-B radiation (e.g., in a solarium) for 2 weeks after treatment.

3.6. *Facial Skin Parameters*

The facial skin of all patients was assessed before treatment and at 28 days using non-invasive cutometric, high-frequency ultrasound, and photographic analyses. Measurements of TEWL, skin hydration, melanin and erythema were obtained with Tewameter®, Corneometer®, and Mexameter® probes, respectively (Courage and Khazaka Multi Probe Adaptor [MPA] System, Cologne, Germany). Changes in parameters compared with pretreatment values (baseline) are presented.

The Smart Mirror-Skin Analysis System (Langdai International Beauty Co., Ltd., Hong Kong, China), which uses the RGB, ultraviolet and pulsed light imaging spectrum, was used to assess a range of skin parameters including: RGB pore (larger than normal pore size indicating blocked pores), RGB spot (spot of darker than normal skin color), RGB wrinkle (current wrinkle condition), PL texture (current skin texture), UV porphyrin (sebum secretion and blackhead distribution), UV pigmentation (current dermal pigmentation), UV moisture (deep skin moisture state), red area, brown spot (skin metabolism condition), and UV damage (sun damage to deep skin layers).

Epidermal thickness (measured from the outermost epidermis [stratum corneum] to the basal cells) and epidermal density (resistance of medium to ultrasound) were assessed with a high frequency and high-resolution diagnostic ultrasound system (DUB® SkinScanner, Taberna Pro Medicum, Lueneburg, Germany).

3.7. Ethical Considerations

The study was conducted in accordance with the Declaration of Helsinki and was approved by the Ethics Committee of Medical Chamber in Warsaw, Poland (protocol code: 05/22, 10 March 2022). Informed, written consent was obtained from all patients.

4. Discussion

Aging skin is subject to a number of cellular and functional alterations that influence quality and appearance, and which are amenable to treatment. In this split-face pilot study, a treatment protocol combining bipolar radiofrequency with a non-crosslinked hyaluronic acid mesotherapy procedure improved epidermal barrier function parameters including stratum corneum hydration, TEWL, and pigmentation to a greater extent than treatment with radiofrequency alone. Ultrasound analysis confirmed that the combined protocol was also associated with significantly improved epidermal thickness and cell density compared with radiofrequency alone.

TEWL is a key component when considering the skin barrier as it assesses stratum corneum function [29,30]. High TEWL values indicate that the permeability of the skin is increased, which means that its integrity is disrupted and skin hydration is reduced. Gradual dehydration of the epidermal layers starting with the stratum corneum can lead to the release of inflammatory mediators and disruption of epidermal differentiation, leading to further impairments in barrier function [31]. The combined treatment protocol in this study led to improved epidermal hydration and reduced TEWL. The hyaluronic acid mesotherapy component of the treatment replenished endogenous HA, levels of which are reduced with age and whose humectant properties are required to maintain epidermal water content. In animal studies, radiofrequency treatment has also been shown to increase epidermal HA, and to reverse age-related epidermal dysfunction [32]. It is hypothesized that when combined, the two treatments act synergistically to increase levels of epidermal HA and skin hydration.

The mechanical stress generated by the hyaluronic acid may also enhance the biochemical response of dermal cells. Mechanical tension is required by fibroblasts for optimal collagen and elastin homeostasis, but is reduced with age as HA content falls and skin collagen production becomes disordered and diminished. Skin biopsies obtained after injection of HA filler into human forearm indicate that mechanical stress generated by injection of the hyaluronic acid may enhance mechanical stretching and collagen production in dermal ECM [33].

With aging, there is a gradual decrease in the number of melanocytes, estimated at 10–20% per decade, which also become unevenly distributed in the epidermis [34,35]. The loss of functional melanocytes and alterations in the interactions between melanocytes and keratinocytes leads to protective barrier dysfunction against UV radiation [36]. Uneven skin tone as a result of melanin deposits and erythema was improved when mesotherapy was added to the radiofrequency protocol, but erythema was slightly increased with radiofrequency alone. The benefits of the combined treatment on skin appearance were confirmed by the Smart Mirror Skin-Analysis System, which showed that 28 days after initial treatment, a range of epidermal skin problems including large pore size, pigmentation spots and UV damage were improved compared with baseline.

Aged human skin is characterized by a thinned epidermis, caused in part by decreased proliferation of basal keratinocytes [37]. Both epidermal thickness and overall skin thickness were improved with radiofrequency and HA mesotherapy. With the combined protocol, epidermal thickness was increased by 12% compared with baseline, twice that achieved with radiofrequency alone. In addition to improving skin laxity, epidermal thickening can promote protection against UV damage through enhanced keratinocyte proliferation, independent of an increase in pigmentation [38]. Epidermal density was particularly increased with the combined protocol by 57.7% compared with 7.1% for radiofrequency alone.

Together, these findings suggest that radiofrequency and HA mesotherapy were able to increase epidermal thickness by enhancing keratinocyte proliferation, with the greatest effects observed when the two procedures were combined. Importantly, no evidence of pathologic hyperkeratosis or an immune response was observed. The exact mechanisms by which these treatments may enhance keratinocyte proliferation have not been confirmed, but research with human skin substitutes subjected to radiofrequency treatment has demonstrated upregulation of keratin genes involved in the epidermal differentiation process with downregulation of matrix metalloprotein genes [39].

Several studies, all of which used crosslinked HA fillers, have reported that radiofrequency treatment prior to HA filler injection may provide synergistic and long-lasting effects for the reduction of nasolabial fold wrinkles [40], marionette lines [41], and forehead lines [42]. One study used a radiofrequency hydro-injector device to simultaneously deliver microneedle intradermal radiofrequency and a non-crosslinked HA to the periorbital area [43]. Improvements in wrinkles, roughness, and pore volume were noted after two to three treatments.

Radiofrequency devices create electrical currents that convert to thermal energy when they encounter tissue resistance and are widely used in the aesthetic field to induce dermal remodeling. As their effects do not depend on the chromophore, but on the electrical properties of the target tissue, they can be safely used in all Fitzpatrick's skin types. The current study used bipolar radiofrequency, which is designed to treat more superficial levels of the skin than monopolar treatment, and combined it with HA mesotherapy. Mesotherapy is a treatment approach for skin rejuvenation that introduces a range of compounds for optimal biorevitalization. In the current study, the mesotherapy product used combined a highly biocompatible non-crosslinked HA (18 mg/mL) enriched with 0.01% calcium hydroxyapatite (CaHA) and the amino acids glycine and L-proline. Hyperdiluted CaHA is an established biostimulatory agent to improve skin quality and firmness via its ability to induce neocollagenesis [44]. Glycine and L-proline comprise a major portion of the primary amino acid sequence of collagen, and had the greatest effect on collagen production in a comparison of 20 amino acids added to human dermal fibroblasts [45]. The addition of glycine and l-proline to hyaluronic acid has also been shown to improve hydrogel elasticity and osmoregulation in a series of in vitro viscoelastic studies [46]. The synergistic actions of the radiofrequency treatment combined with those of the mesotherapy contents improve epidermal homeostasis [47] and have both visible and preventative skin benefits. These include improved skin appearance (increased firmness, even skin tone, reduced dryness) and improved epidermal barrier function and skin homeostasis to support optimal skin renewal.

The treatment protocol employed radiofrequency immediately prior to HA mesotherapy and not vice versa. This sequence is important as previous research has suggested that radiofrequency treatment after HA filler injection may lead to degradation of the HA, especially if performed on the same day [48]. A systematic review of monopolar and bipolar radiofrequency devices has shown that they produce measurable improvements in skin laxity with an acceptable adverse event profile [49]. The majority of reported complications were minor and transient in nature; major complication rates were higher with the use of monopolar devices than with the use of bipolar devices. In the current study, no adverse events related to treatment were reported.

A major advantage of this study was the split-face design, which allowed objective assessment of the synergistic effect of combination treatment compared with radiofrequency alone. The study was limited by its relatively small sample size. In addition, the duration of follow-up was not long enough to evaluate the long-term effect of the combination treatment. Further studies, with a larger sample size and a long duration of follow-up in subjects with a range of skin types, are now warranted to confirm the synergistic effect and safety of treatment with this novel radiofrequency and HA mesotherapy combination.

5. Conclusions

Radiofrequency devices and HA mesotherapy products represent two separate modalities for skin rejuvenation with synergistic effects on epidermal rejuvenation. In this pilot study, in subjects with dry and lax facial skin, a treatment protocol combining radiofrequency with HA mesotherapy was associated with greater improvements in skin hydration, firmness and tone compared with radiofrequency alone. At a cellular level this was linked to increased epidermal thickness and density. Both treatments were well-tolerated and led to improvements in facial appearance.

Author Contributions: Conceptualization, A.P.; methodology, A.P. and B.Ł.; validation; formal analysis, A.P. and S.K.; investigation, A.P. and S.K.; data curation, S.K.; writing—original draft preparation, B.Ł.; writing—review and editing, A.P. and B.Ł.; visualization, A.P. and S.K.; supervision, N.Z.; project administration, N.Z. All authors have read and agreed to the published version of the manuscript.

Funding: This research received no external funding.

Institutional Review Board Statement: The study was conducted in accordance with the Declaration of Helsinki, and approved by the Ethics Committee of Medical Chamber in Warsaw, Poland (protocol code: 05/22, 10 March 2022).

Informed Consent Statement: Informed consent was obtained from all subjects involved in the study. Written informed consent was obtained from the patients to publish this paper.

Data Availability Statement: The data presented in this study are available on request from the corresponding author. The data are not publicly available due to privacy restrictions.

Conflicts of Interest: Nicola Zerbinati is the Scientific Director of Matex Lab. Bartłomiej Łukasik is the Medical Manager of Matex Lab. The authors declare no other conflict of interest.

References

1. Baroni Edo, R.; Biondo-Simões, M.d.L.; Auersvald, A.; Auersvald, L.A.; Montemor Netto, M.R.; Ortolan, M.C.; Kohler, J.N. Influence of aging on the quality of the skin of white women: The role of collagen. *Acta Cir. Bras.* **2012**, *27*, 736–740. [CrossRef]
2. Zhang, S.; Duan, E. Fighting against Skin Aging: The Way from Bench to Bedside. *Cell Transplant.* **2018**, *27*, 729–738. [CrossRef]
3. Biniek, K.; Kaczvinsky, J.; Matts, P.; Dauskardt, R.H. Understanding age-induced alterations to the biomechanical barrier function of human stratum corneum. *J. Dermatol. Sci.* **2015**, *80*, 94–101. [CrossRef] [PubMed]
4. Naylor, E.C.; Watson, R.E.; Sherratt, M.J. Molecular aspects of skin ageing. *Maturitas* **2011**, *69*, 249–256. [CrossRef] [PubMed]
5. El-Domyati, M.; Attia, S.; Saleh, F.; Brown, D.; Birk, D.E.; Gasparro, F.; Ahmad, H.; Uitto, J. Intrinsic aging vs. photoaging: A comparative histopathological, immunohistochemical, and ultrastructural study of skin. *Exp. Dermatol.* **2002**, *11*, 398–405. [CrossRef]
6. Farkas, J.P.; Pessa, J.E.; Hubbard, B.; Rohrich, R.J. The Science and Theory behind Facial Aging. *Plast. Reconstr. Surg. Glob. Open* **2013**, *1*, e8–e15. [CrossRef] [PubMed]
7. Mohammed, Y.; Holmes, A.; Kwok, P.C.L.; Kumeria, T.; Namjoshi, S.; Imran, M.; Matteucci, L.; Ali, M.; Tai, W.; Benson, H.A.; et al. Advances and future perspectives in epithelial drug delivery. *Adv. Drug Deliv. Rev.* **2022**, *186*, 114293. [CrossRef]
8. Lee, J.Y.; Spicer, A.P. Hyaluronan: A multifunctional, megaDalton, stealth molecule. *Curr. Opin. Cell Biol.* **2000**, *12*, 581–586. [CrossRef]
9. Maytin, E.V.; Chung, H.H.; Seetharaman, V.M. Hyaluronan Participates in the Epidermal Response to Disruption of the Permeability Barrier In Vivo. *Am. J. Pathol.* **2004**, *165*, 1331–1341. [CrossRef]
10. Meyer, L.J.; Stern, R. Age-Dependent Changes of Hyaluronan in Human Skin. *J. Investig. Dermatol.* **1994**, *102*, 385–389. [CrossRef]
11. Bernstein, E.F.; Underhill, C.B.; Hahn, P.J.; Brown, D.B.; Uitto, J. Chronic sun exposure alters both the content and distribution of dermal glycosaminoglycans. *Br. J. Dermatol.* **1996**, *135*, 255–262. [CrossRef] [PubMed]

12. Tzellos, T.G.; Klagas, I.; Vahtsevanos, K.; Triaridis, S.; Printza, A.; Kyrgidis, A.; Karakiulakis, G.; Zouboulis, C.C.; Papakonstantinou, E. Extrinsic ageing in the human skin is associated with alterations in the expression of hyaluronic acid and its metabolizing enzymes. *Exp. Dermatol.* **2009**, *18*, 1028–1035. [CrossRef]
13. Fuchs, E.; Raghavan, S. Getting under the skin of epidermal morphogenesis. *Nat. Rev. Genet.* **2002**, *3*, 199–209. [CrossRef]
14. Khavkin, J.; Ellis, D.A. Aging skin: Histology, physiology, and pathology. *Facial Plast. Surg. Clin.* **2011**, *19*, 229–234. [CrossRef] [PubMed]
15. Yang, L.; Zhang, S.; Wang, G. Keratin 17 in disease pathogenesis: From cancer to dermatoses. *J. Pathol.* **2019**, *247*, 158–165. [CrossRef]
16. Ghadially, R.; Brown, B.E.; Sequeira-Martin, S.M.; Feingold, K.R.; Elias, P.M. The aged epidermal permeability barrier. Structural, functional, and lipid biochemical abnormalities in humans and a senescent murine model. *J. Clin. Investig.* **1995**, *95*, 2281–2290. [CrossRef] [PubMed]
17. Oh, J.-H.; Kim, Y.K.; Jung, J.-Y.; Shin, J.-E.; Chung, J.H. Changes in glycosaminoglycans and related proteoglycans in intrinsically aged human skin in vivo. *Exp. Dermatol.* **2011**, *20*, 454–456. [CrossRef]
18. Bourguignon, L.Y.; Wong, G.; Xia, W.; Man, M.-Q.; Holleran, W.M.; Elias, P.M. Selective matrix (hyaluronan) interaction with CD44 and RhoGTPase signaling promotes keratinocyte functions and overcomes age-related epidermal dysfunction. *J. Dermatol. Sci.* **2013**, *72*, 32–44. [CrossRef]
19. Tapfumaneyi, P.; Imran, M.; Mohammed, Y.; Roberts, M.S. Recent advances and future prospective of typical and transdermal delivery systems. *Front. Drug Deliv.* **2022**, *2*, 957732. [CrossRef]
20. Hernández-Bule, M.L.; Toledano-Macías, E.; Naranjo, A.; de Andrés-Zamora, M.; Úbeda, A. In vitro stimulation with radiofrequency currents promotes proliferation and migration in human keratinocytes and fibroblasts. *Electromagn. Biol. Med.* **2021**, *40*, 338–352. [CrossRef]
21. El-Domyati, M.; El-Ammawi, T.S.; Medhat, W.; Moawad, O.; Brennan, D.; Mahoney, M.G.; Uitto, J. Radiofrequency facial rejuvenation: Evidence-based effect. *J. Am. Acad. Dermatol.* **2011**, *64*, 524–535. [CrossRef] [PubMed]
22. Taub, A.F.; Tucker, R.; Palange, A. Facial tightening with an advanced 4-MHz monopolar radiofrequency device. *J. Drugs Dermatol.* **2012**, *11*, 1288–1294.
23. Bloom, B.S.; Emer, J.; Goldberg, D.J. Assessment of safety and efficacy of a bipolar fractionated radiofrequency device in the treatment of photodamaged skin. *J. Cosmet. Laser Ther.* **2012**, *14*, 208–211. [CrossRef]
24. Nelson, A.A.; Beynet, D.; Lask, G.P. A novel non-invasive radiofrequency dermal heating device for skin tightening of the face and neck. *J. Cosmet. Laser Ther.* **2015**, *17*, 307–312. [CrossRef]
25. Zerbinati, N.; Sommatis, S.; Maccario, C.; Capillo, M.C.; Di Francesco, S.; Rauso, R.; Protasoni, M.; D'este, E.; Gasperina, D.D.; Mocchi, R. In Vitro Evaluation of the Effect of a Not Cross-Linked Hyaluronic Acid Hydrogel on Human Keratinocytes for Mesotherapy. *Gels* **2021**, *7*, 15. [CrossRef] [PubMed]
26. Narurkar, V.A.; Fabi, S.G.; Bucay, V.W.; Tedaldi, R.; Downie, J.B.; Zeichner, J.A.; Butterwick, K.; Taub, A.; Kadoya, K.; Makino, E.T.; et al. Rejuvenating Hydrator: Restoring Epidermal Hyaluronic Acid Homeostasis with Instant Benefits. *J. Drugs Dermatol.* **2016**, *15* (Suppl. S2), s24–s37. [PubMed]
27. Makino, E.T.; Huang, P.C.; Emmerich, T.; Jiang, L.I.; Mehta, R.C. Efficacy and Tolerability of Cosmetic Serums Enriched with Five Forms of Hyaluronic Acid as Part of Biweekly Diamond Tip Microdermabrasion Treatments for Facial Skin Dryness and Age-Associated Features. *Clin. Cosmet. Investig. Dermatol.* **2023**, *16*, 1123–1134. [CrossRef]
28. Hsieh, D.M.-Y.; Bi, X.; Zhong, S.; Wu, Y. In Vivo Investigation of the Biostimulatory and Rejuvenating Effects of Small-Particle Cross-linked Hyaluronic Acid Injections in Photoaged Human Skin. *Aesthetic Surg. J.* **2023**, *43*, 595–608. [CrossRef]
29. Zhang, Q.; Murawsky, M.; LaCount, T.; Kasting, G.B.; Li, S.K. Transepidermal water loss and skin conductance as barrier integrity tests. *Toxicol. Vitr.* **2018**, *51*, 129–135. [CrossRef]
30. van Rensburg, S.J.; Franken, A.; Du Plessis, J.L. Measurement of transepidermal water loss, stratum corneum hydration and skin surface pH in occupational settings: A review. *Skin Res. Technol.* **2019**, *25*, 595–605. [CrossRef]
31. Rawlings, A.V.; Matts, P.J. Stratum Corneum Moisturization at the Molecular Level: An Update in Relation to the Dry Skin Cycle. *J. Investig. Dermatol.* **2005**, *124*, 1099–1110. [CrossRef] [PubMed]
32. Lee, H.J.; Seo, S.R.; Yoon, M.S.; Song, J.Y.; Lee, E.Y.; Lee, S.E. Microneedle fractional radiofrequency increases epidermal hyaluronan and reverses age-related epidermal dysfunction. *Lasers Surg. Med.* **2016**, *48*, 140–149. [CrossRef]
33. Turlier, V.; Delalleau, A.; Casas, C.; Rouquier, A.; Bianchi, P.; Alvarez, S.; Josse, G.; Briant, A.; Dahan, S.; Saint-Martory, C.; et al. Association between collagen production and mechanical stretching in dermal extracellular matrix: In vivo effect of cross-linked hyaluronic acid filler. A randomised, placebo-controlled study. *J. Dermatol. Sci.* **2013**, *69*, 187–194. [CrossRef]
34. Ortonne, J.-P. Pigmentary changes of the ageing skin. *Br. J. Dermatol.* **1990**, *122* (Suppl. S35), 21–28. [CrossRef] [PubMed]
35. Sun, K.-L.; Liu, W.; Gao, X.-M.; Yang, M.; Chang, J.-M. A Study of Normal Epidermal Melanocyte Distribution. *Int. J. Dermatol. Venereol.* **2021**, *4*, 32–35. [CrossRef]
36. Brenner, M.; Hearing, V.J. The Protective Role of Melanin Against UV Damage in Human Skin. *Photochem. Photobiol.* **2008**, *84*, 539–549. [CrossRef]
37. Zouboulis, C.C.; Makrantonaki, E. Clinical aspects and molecular diagnostics of skin aging. *Clin. Dermatol.* **2011**, *29*, 3–14. [CrossRef]

38. Scott, T.L.; Christian, P.A.; Kesler, M.V.; Donohue, K.M.; Shelton, B.; Wakamatsu, K.; Ito, S.; D'Orazio, J. Pigment-independent cAMP-mediated epidermal thickening protects against cutaneous UV injury by keratinocyte proliferation. *Exp. Dermatol.* **2012**, *21*, 771–777. [CrossRef]
39. Louis, F.; Fujii, N.; Katsuyama, M.; Okumoto, S.; Matsusaki, M. Effects of radiofrequency and ultrasound on the turnover rate of skin aging components (skin extracellular matrix and epidermis) via HSP47-induced stimulation. *Biochem. Biophys. Res. Commun.* **2020**, *525*, 73–79. [CrossRef]
40. Choi, S.Y.; Lee, Y.H.; Kim, H.; Koh, H.-J.; Park, S.-Y.; Park, W.-S.; Bae, I.-H.; Park, K.Y.; Kim, B.J. A combination trial of intradermal radiofrequency and hyaluronic acid filler for the treatment of nasolabial fold wrinkles: A pilot study. *J. Cosmet. Laser Ther.* **2014**, *16*, 37–42. [CrossRef] [PubMed]
41. Ko, E.J.; Choi, H.G.; Kim, H.; Park, W.-S.; Kim, B.J.; Kim, M.N. Novel Treatment Using Intradermal Radiofrequency and Hyaluronic Acid Filler to Correct Marionette Lines. *Ann. Dermatol.* **2015**, *27*, 351–352. [CrossRef]
42. Hyun, M.Y.; Mun, S.K.; Kim, B.J.; Kim, H.; Park, W.-S. Efficacy and Safety of Hyaluronic Acid with and without Radiofrequency for Forehead Augmentation: A Pilot Study Using Three-Dimensional Imaging Analysis. *Ann. Dermatol.* **2016**, *28*, 107–109. [CrossRef] [PubMed]
43. Lim, Y.-K.; Jung, C.-J.; Lee, M.-Y.; Moon, I.-J.; Won, C.-H. The Evaluation of Efficacy and Safety of A Radiofrequency Hydro-Injector Device for the Skin around the Eye Area. *J. Clin. Med.* **2021**, *10*, 2582. [CrossRef] [PubMed]
44. de Almeida, A.T.; Figueredo, V.; da Cunha, A.L.G.; Casabona, G.; Costa de Faria, J.R.; Alves, E.V.; Sato, M.; Branco, A.; Guarnieri, C.; Palermo, E. Consensus Recommendations for the Use of Hyperdiluted Calcium Hydroxyapatite (Radiesse) as a Face and Body Biostimulatory Agent. *Plast. Reconstr. Surg. Glob. Open* **2019**, *7*, e2160. [CrossRef] [PubMed]
45. Lee, J.E.; Boo, Y.C. Combination of Glycinamide and Ascorbic Acid Synergistically Promotes Collagen Production and Wound Healing in Human Dermal Fibroblasts. *Biomedicines* **2022**, *10*, 1029. [CrossRef]
46. Martina, V.; Gallo, A.; Tarantino, E.; Esposito, C.; Zerbinati, U.; Mocchi, R.; Monticelli, D.; Lotti, T.; Tirant, M.; Van Thuong, N.; et al. Viscoelastic properties and thermodynamic balance improvement of a hyaluronic acid hydrogel enriched with proline and glycyne. *J. Biol. Regul. Homeost. Agents* **2019**, *33*, 1955–1959. [PubMed]
47. Kubik, P.; Jankau, J.; Rauso, R.; Galadari, H.; Protasoni, M.; Gruszczyński, W.; Grzanka, D.; Smolińska, M.; Antosik, P.; Piesiaków, M.L.; et al. HA PEGylated filler in association with an infrared energy device for the treatment of facial skin aging: 150 day follow-up data report. *Pharmaceuticals* **2022**, *15*, 1355. [CrossRef]
48. Jurairattanaporn, N.; Amornpetkul, W.; Rutnin, S.; Vachiramon, V. The effect of combined hyaluronic acid filler injection and radiofrequency treatment: A clinic histological analysis. *J. Cosmet. Dermatol.* **2023**, *22*, 798–803. [CrossRef]
49. Rohrich, R.J.; Schultz, K.P.; Chamata, E.S.; Bellamy, J.L.; Alleyne, B. Minimally Invasive Approach to Skin Tightening of the Face and Body: A Systematic Review of Monopolar and Bipolar Radiofrequency Devices. *Plast. Reconstr. Surg.* **2022**, *150*, 771–780. [CrossRef]

cosmetics

MDPI

Article

Assessment of Systemic Safety of *Althaea rosea* Flower Extract for Use in Cosmetics: Threshold of Toxicological Concern and History of Safe Consumption Approaches

Sangwon Gil [1] and Kyung-Min Lim [1,2,*]

[1] College of Pharmacy, Graduate School of Pharmaceutical Sciences, Ewha Womans University, Seoul 03760, Republic of Korea; swkil514@naver.com
[2] Graduate Program in Innovative Biomaterials Convergence, Ewha Womans University, Seoul 03760, Republic of Korea
* Correspondence: kmlim@ewha.ac.kr

Abstract: *Althaea rosea* flower extract (ARFE) is widely used as a food and cosmetic ingredient. However, the systemic safety of ARFE for use in cosmetics has not been confirmed, yet. Here, we adopted the threshold of toxicological concern (TTC) and history of safe food consumption approaches to evaluate the systemic safety of ARFE as a cosmetic ingredient. A systematic literature review identified 48 chemical constituents in ARFE, 92.6% of which are common food components. Through a literature review, 48 chemical constituents of ARFE were identified. To exclude the potential genotoxicity issues, in silico predictions of an in vitro AMES test and additional literature reviews were performed, demonstrating that all the chemical constituents of ARFE have no genotoxicity issues. To evaluate the systemic toxicity of ARFE, a comparison with the dietary intake of ARFE was performed. The daily dietary intake of ARFE through tea products was estimated to be 66.67 mg/kg/day. Since exposure to ARFE through cosmetic use ranges from 0.0045 to 5.380 mg/kg/day, which is far lower than dietary intake, it is unlikely to pose any additional health risk. The TTC approach along with in silico predictions of dermal absorption also revealed that systemic exposure doses (SEDs) of all the chemical constituents are below TTC thresholds, further supporting its systemic safety for use in cosmetics.

Keywords: safety assessment; cosmetics; threshold of toxicological concern approach (TTC); next-generation risk assessment (NGRA); *Althaea rosea* flower extract; hollyhock

1. Introduction

Traditionally, cosmetic safety assessments have relied on conventional animal testing as the sole source of toxicological data. The European Union implemented a stepwise ban on animal testing for cosmetics: finished product testing was banned in 2004, ingredient testing in 2009, and a complete marketing ban (including repeated-dose toxicity) in 2013 (EU No. 1223/2009) [1]. Following this regulatory shift, over 40 countries have established full or partial bans on cosmetics animal testing [2]. Consequently, alternative strategies are required to evaluate the safety of cosmetic ingredients without reliance on animal testing [3].

The safety of cosmetic ingredients must be assessed with respect to local and systemic toxicity [4]. Efforts has been directed to the development of non-animal test methods to evaluate local toxicity. As a result, key local toxicity endpoints including skin irritation, eye irritation, phototoxicity and skin sensitization can be evaluated with officially accepted non-animal based in vitro test methods [5,6]. However, due to the complexity of systemic toxicity endpoints, there is no regulatory accepted non-animal test method to replace conventional animal tests yet.

Next-generation risk assessment (NGRA) methods have emerged as promising approaches in this regard [7]. These include the threshold of toxicological concern (TTC), Read-Across, and Quantitative Structure-Activity Relationship (QSAR) models. Among these approaches, the TTC approach has gained significant attention as a viable option for assessing the safety of cosmetic ingredients, particularly fragrance and botanical extracts [8–15]. The TTC approach systematically evaluates the systemic safety of substances with limited toxicological data based on their chemical structure and systemic exposure dose (SED) [16]. In June 2019, the European Food Safety Authority (EFSA) introduced new guidelines for applying the TTC approach to food safety assessments [17]. In addition to EFSA, the TTC approach is also recommended for the safety assessment of cosmetic ingredients, as stated in the guidance issued by the major authorities like the EU Scientific Committee on Consumer Safety (SCCS) [18], the US Cosmetic Ingredient Review (CIR) [19] and China's National Institutes for Food and Drug Control (NIFDC) [20].

The TTC approach categorizes chemicals into five classifications [21]. For substances identified as potentially DNA-reactive mutagens or carcinogens, the TTC value is set at 0.0025 μg/kg·bw/day. For organophosphates and carbamates, the TTC value is 0.3 μg/kg·bw/day, and according to the Cramer classification system, which estimates the theoretical toxicity of a compound based on its chemical structure, TTC values are set to 30 μg/kg·bw/day, 9 μg/kg·bw/day, and 1.5 μg/kg·bw/day for Cramer Class I (low toxicity), Cramer Class II (moderate toxicity), and Cramer Class III (high toxicity), respectively. Substances with exposure levels below the TTC threshold are considered unlikely to pose a significant health risk. Yang et al. [9] refined these TTC values through considering over 1000 chemicals used in cosmetics and the EU SCCS adopted 2.3 μg/kg·bw/day for Cramer Class II and III, and 46 μg/kg·bw/day for Cramer Class I [18].

The TTC approach has been successfully employed for the systemic safety evaluation of botanical extracts used in cosmetics, owing to the generally low abundance of bioactive constituents in them [8,13,14,22]. Another approach to ensure the safety of a botanical extract used in cosmetics without animal experiments is to prove its history of safe use as food, medicine or other related product types [19,23]. Especially, by demonstrating the history of safe food consumption of a botanical extract and its equivalence with the food materials with respect to species, source, process, composition, exposure and usage, mandatory systemic toxicological test data for safety assessment can be exempted. Since many botanical extracts used in cosmetics are used as food materials, this approach may be effective in ensuring the systemic safety of a botanical extract used in cosmetics

Althaea rosea, commonly called hollyhock, is a perennial ornamental plant that is included in the *Malvaceae* family [24–26]. The root and flowers of *Althaea rosea* have been used as folk medicine and herbal tea in Asian countries for many years. Several studies reported the pharmacological and biological effects of *Althaea rosea* such as antioxidant, anti-inflammatory, immunomodulatory, anti-urolithiatic and anti-cancer activities [27–32]. Recently, the cosmetic function of *Althaea rosea* flower extract (ARFE) was discovered [33], paving the way for the use of ARFE in cosmetics. ARFE contains various bioactive phytochemicals including anthocyanins and flavonoids [28]. However, it is yet unclear whether ARFE is safe to use in cosmetics.

The aim of this study is to evaluate the systemic safety of ARFE for its use in cosmetics, using the TTC method and food consumption assessment. For the TTC approach, full chemical constituents' identification using instrumental analysis is necessary. Fortunately, the chemical constituents of *Althaea rosea* flower powder were recently reported [25,28,34]. We used this information to apply the TTC approach and food consumption assessment in an effort to provide a useful example of the systemic safety assessment methods for botanical extracts used in cosmetics.

2. Materials and Methods

2.1. Literature Search Strategy and Selection Criteria for ARFE Constituent Analysis

We searched PubMed, the Web of Science and Google Scholar with search terms ("*Althaea rosea*" OR "hollyhock") AND (extract OR flower). We identified 84 papers with relevant information. Three papers had detailed information on the chemical constituents of *Althaea rosea* flower powder [25,28,34], which we used to identify the ARFE constituents.

2.2. Classification of Chemical Compounds in A. rosea Flower Extract into Cramer Class

SMILES codes for analysis were obtained from PubChem (https://pubchem.ncbi.nlm.nih.gov/, accessed on 1 June 2025) and input into ChemTunes® (https://mn-am.com/products/chemtunestoxgps/, accessed on 1 June 2025) for TTC calculation. Toxtree package 3.1 was incorporated in ChemTunes® to assign TTC Cramer Classes based on the extended decision tree [35].

2.3. In Silico Prediction of Genotoxicity Assay

The genotoxicity of *A. rosea* flower was evaluated using in silico predictions by inputting the constituents identified through the literature review and their corresponding chemical structures in SMILES format into ChemTunes®.

2.4. Estimation of Systemic Exposure Dosages

The daily systemic exposure levels for cosmetics were estimated using the daily exposure value for each cosmetic product type provided by the Scientific Committee on Consumer Safety (SCCS) [18]. Daily exposure to a cosmetic product category per kg (body weight) was calculated based on a 60 kg adult (female).

Systemic exposure dosages (SEDs, μg/kg/day) of the chemical constituents of ARFE from cosmetic use were calculated using the following equation:

$$\mathbf{SED} = \mathbf{Eproduct} \times \frac{\boldsymbol{C}}{\mathbf{100}} \times \frac{\boldsymbol{E}}{\mathbf{100}} \times \frac{\boldsymbol{B}}{\mathbf{100}} \times \frac{\boldsymbol{DAp}}{\mathbf{100}}$$

Eproduct (mg/kg bw/day): Estimated daily exposure to a cosmetic product category per kg (body weight) based on the applied quantity and frequency of application.

C (%): Concentration of ARFE used in the finished cosmetic product.

E (%): Extraction ratio (ratio of extractants vs. *A. rosea* flower powder).

B (%): Content of the constituent in *A. rosea* flower powder.

DAp (%): Dermal absorption of the component as a percentage.

The usage concentration of *A. rosea* flower in cosmetics (*C* (%)) was referenced from the Chinese Cosmetic Ingredient Regulatory Database (China CosIng [36]).

For ***E*** (%), since it is common to add solvent (water) to a plant at 1:9 or higher for extraction [10], it was assumed to be 10%. ***DAp*** (%) was initially set at 100% as a worst-case scenario of Tier 1 to be conservative, even though it was overly conservative for high-molecular-weight or highly polar constituents such as flavonoid glycosides and mucilages. For the constituents that needed a refined dermal absorption rate, we used the CDC Finite Dose Skin Permeation Calculator, which considers physicochemical properties.

2.5. Food Consumption Assessment

An equivalence of *A. rosea* flower powder used as the food material was examined through comparing the preparation of tea and the manufacture of cosmetic raw material. The safe exposure level was evaluated by comparing the dietary intake level of *A. rosea* L. flower powder through tea with the cosmetic exposure level.

3. Results

3.1. Identification of Chemical Constituents of A. rosea Flower Extract

The chemical composition of *A. rosea* flower powder was recently analyzed and published [28], as presented in Table 1. A total of 51 chemical constituents of *A. rosea* flower powder was identified from the literature review, accounting for 105.24% of *A. rosea* flower extract on mass basis. Among them, 92.6% were classified as common food or nutritional ingredients that are considered safe and assumed to pose no health concerns to consumers. *A. rosea* flower powder has the high contents of fat, proteins, ash, and fiber. *A. rosea* flower powder contains mucilages composed of glucuronic acid, galacturonic acid, rhamnose, and galactose, which are part of high-molecular-weight dietary fiber [37].

Table 1. Chemical constituents of *A. rosea* flower powder [28].

No.	Chemicals	CAS	Chemical Class	Contents (%)
1	Water	7732-18-5		8.67
2	Fat			2.67
3	Proteins			12.1
4	Ash			7.62
5	Carbohydrates (Total carbohydrates)			68.95
6	Hexanoic acid	142-62-1	Fatty Acids	0.00
7	Octanoic acid	124-07-2	Fatty Acids	0.00
8	Decanoic acid	334-48-5	Fatty Acids	0.00
9	Dodecanoic acid	143-07-7	Fatty Acids	0.00
10	Tridecanoic acid	638-53-9	Fatty Acids	0.00
11	Tetradecanoic acid	544-63-8	Fatty Acids	0.02
12	Pentadecanoic acid	1002-84-2	Fatty Acids	0.00
13	Hexadecanoic acid	57-10-3	Fatty Acids	0.39
14	Heptadecanoic acid	506-12-7	Fatty Acids	0.01
15	Octadecanoic acid	57-11-4	Fatty Acids	0.10
16	Arachidic acid	506-30-9	Fatty Acids	0.09
17	Docosanoic acid	112-85-6	Fatty Acids	0.04
18	Tricosanoic acid	2433-96-7	Fatty Acids	0.01
19	Tetracosanoic acid	557-59-5	Fatty Acids	0.02
20	(Z)-tetradec-9-enoic acid	544-64-9	Fatty Acids	0.01
21	(Z)-hexadec-9-enoic acid	373-49-9	Fatty Acids	0.00
22	(Z)-heptadec-10-enoic acid	29743-97-3	Fatty Acids	0.02
23	(Z)-octadec-9-enoic acid	112-80-1	Fatty Acids	0.18
24	(Z)-icos-9-enoic acid	29204-02-2	Fatty Acids	0.01
25	(9Z,12Z)-octadeca-9,12-dienoic acid	60-33-3	Fatty Acids	0.29
26	(9E,12E)-octadeca-9,12-dienoic acid (Linolaidicacid)	506-21-8	Fatty Acids	0.05
27	(9Z,12Z,15Z)-octadeca-9,12,15-trienoic acid	463-40-1	Fatty Acids	0.50
28	(6Z,9Z,12Z)-octadeca-6,9,12-trienoic acid (Gamma-Linolenic acid)	506-26-3	Fatty Acids	0.04
29	11,14-eicosadienoic acid	5598-38-9	Fatty Acids	0.10
30	(5Z,8Z,11Z,14Z)-icosa-5,8,11,14-tetraenoic acid	506-32-1	Fatty Acids	0.06
31	4-Hydroxybenzoic acid	99-96-7	Phenolic Acid	0.00148
32	Gallic acid	149-91-7	Phenolic Acid	0.00081

Table 1. *Cont.*

No.	Chemicals	CAS	Chemical Class	Contents (%)
33	Protocatechuic acid	99-50-3	Phenolic Acid	0.00041
34	Syringic acid	530-57-4	Phenolic Acid	0.00204
35	Caffeic acid (+derivatives)	331-39-5	Phenolic Acid	0.0501
36	Chlorogenic acid	327-97-9	Phenolic Acid	0.02
37	Ferulic acid	1135-24-6	Phenolic Acid	0.00494
38	P-Coumaric acid	501-98-4	Phenolic Acid	0.0359
39	Hydrocinnamic acid	501-52-0	Phenolic Acid	0.0234
40	Cyanidin (Cy-3-O-β-glucopyranoside)	13306-05-3	Anthocyanins	0.0147
41	Malvidin (Mv-3-O-β-glucopyranoside+ Mv-3,5-O-β-glucopyranoside)	10463-84-0 (643-84-5)	Anthocyanins	0.1301
42	Pelargonidin (Pg-3-O-β-glucopyranoside)	134-04-3	Anthocyanins	0.057
43	Petunidin (Pt-3-O-β-glucopyranoside)	13270-60-5	Anthocyanins	0.100
44	Anthocyanin I-III	-	Anthocyanins	0.2395
45	Gallocatechin gallate	5127-64-0	Flavonoid	0.0535
46	Naringin (Naringin glycoside I+II)	10236-47-2	Flavonoid	1.6070
47	Apigenin (Apigenin glycoside II+III)	520-36-5	Flavonoid	0.0171
48	Luteolin (Luteolin glycoside II)	491-70-3	Flavonoid	0.7224
49	Kaempherol (+Kaempherol glycoside II+III)	520-18-3	Flavonoid	0.0258
50	Quercetin (Quercetin glycosides)	117-39-5	Flavonoid	0.1260
51	Rutin			0.0620
	Total (%)			105.24%

The fatty acid constituents of *A. rosea* flower powders include linoleic acid and palmitic acid. In addition, flower oil contained significant amounts of oleic acid and eicosadienoic acid.

Besides these food components, *A. rosea* flower powder contains a total of twenty-one potentially bioactive constituents; nine phenolic acids, five anthocyanins, and seven flavonoids.

3.2. Assessment of Genotoxicity of A. rosea Flower Extract

According to the literature review conducted so far, no publicly available genotoxicity test results have been reported for extracts of *A. rosea* flower. It is essential to exclude the genotoxic potential of botanical extracts, so we searched or predicted in silico the genotoxicity potential of the 20 non-food chemical constituents, except for anthocyanins, the structure of which was not available with ChemTunes®, based on the SMILES format. Even though we could not use an in silico tool to predict the genotoxicity of anthocyanins, it is widely known that anthocyanins protect against DNA damage [38], allowing us to conclude that they do not have genotoxicity potential. As a result, most chemical constituents were concluded as non-genotoxic, except for luteolin (Table 2).

Table 2. Constituents of *Althaea rosea* flower extract (μg/kg/day), in silico and literature-based assessment of genotoxicity and systemic exposure dose estimation.

No.	Cramer Class	Chemical	Classification	Genotoxicity		TTC Threshold (μg/kg bw/Day)	Leave-on Skin and Hair (SCCS) 20.786 mg/kg
				Results	Detail		Realistic (μg/kg bw/Day)
1	I	4-Hydroxybenzoic acid	Phenolic acid	-	AMES	46	0.00615
2	I	Gallic acid	Phenolic acid	-	AMES	46	0.00337
3	I	Protocatechuic acid	Phenolic acid	-	IN SILLICO	46	0.00170
4	I	Syringic acid	Phenolic acid	-	IN SILLICO	46	0.00848
5	I	Caffeic acid	Phenolic acid	-	AMES	46	0.20828
6	I	Ferulic acid	Phenolic acid	-	IN SILLICO	46	0.08314
7	I	P-Coumaric acid	Phenolic acid	-	IN SILLICO	46	0.02054
8	I	Hydrocinnamic acid	Phenolic acid	-	IN SILLICO	46	0.14924
9	II	Chlorogenic acid	Phenolic acid	-	AMES	2.3	0.09728
10	III	Cyanidin	Anthocyanins	-	IN SILLICO	2.3	0.06111
11	III	Malvidin	Anthocyanins	-	IN SILLICO	2.3	0.54085
12	III	Pelargonidin	Anthocyanins	-	IN SILLICO	2.3	0.23696
13	III	Petunidin	Anthocyanins	-	IN SILLICO	2.3	0.41572
14	III	Anthocyanins I-III	Anthocyanins		IN VIVO	2.3	0.99565
15	III	Gallocatechin gallate	Flavonoid	-	IN SILLICO	2.3	0.22241
16	III	Naringenin (Naringin glycoside)	Flavonoid	-	IN SILLICO	2.3	6.68062
17	III	Apigenin (Apigenin glycoside)	Flavonoid	-	AMES	2.3	0.07109
18	III	Luteolin (luteolinglycosideii)	Flavonoid	+	IN SILICO (genotox concern)	2.3	3.00316
19	III	Kaempherol	Flavonoid	-	IN VIVO	2.3	0.10726
20	III	Quercetin (Quercetin glycoside)	Flavonoid	-	IN VIVO	2.3	0.52381
21	III	Rutin	Flavonoid	-	IN VIVO (Carcino-genesis study)	2.3	0.25775

To assess the genotoxicity issue of luteolin, we reviewed relevant genotoxicity studies and in vitro experimental data. Numerous studies have reported conflicting results regarding its genotoxic potential. For example, the literature review revealed that several studies have demonstrated that these constituents may actually protect cells against genetic damage. Taj and Nagarajan reported that luteolin protected cells in micronucleus assays and reduced chromosomal aberrations in rats [39]. Moreover, luteolin derived from *Clerodendrum cyrtophyllum* Turcz leaves exhibited no genotoxic effect towards HepG2 cells but rather showed antioxidant activity by increasing cell viability [40]. Also, dietary flavonoids including quercetin, luteolin and genistein reduce oxidative DNA damage and lipid peroxidation in vivo [41,42]. Recently, Fernando et al. demonstrated that luteolin at 2.5 μg/mL does not induce DNA damage but rather protects against H_2O_2-induced DNA damage in lung fibroblasts with COMET and phosphor-H2A.X assays [43], further supporting the absence of the genotoxicity of luteolin.

3.3. *Systemic Exposure Dosages of the Chemical Constituents of A. rosea Flower Extract from the Use in Cosmetics*

According to the ChinaCosING database, ARFE (CAS. No. 90045-76-4) was used at concentrations of up to 2% in 2021 [36]. More recently, in 2024, registered products have been reported to contain ARFE at concentrations ranging from 0.0001% to 0.005%. We assumed the worst-case scenario considering that ARFE is used at 2%. Using these data and the content levels in Table 1, we estimated systemic exposure dosages (SEDs, μg/kg/day) of the chemical constituents of ARFE from cosmetic use (see Section 2.3).

For our calculations, we defined C (%) as 2% and set Eproduct at 207.86 mg/kg/day (207,860 μg/kg/day, reflecting the daily use of leave-on type cosmetics: body lotion, facial cream, hand cream, deodorant and hair styling products) (SCCS, 2023) [18].

For E (%), since it is common to add a solvent (water) to a plant at 1:9 or higher for extraction [10], it was assumed to be 10%. DAp (%) was initially set at 100% as a worst-case scenario of Tier 0. The resulting SED values are presented in Table 2.

3.4. *The Safety Assessment Using the TTC Approach*

Twenty non-food constituents with structure information were classified into Cramer Classes using ChemTunes™ based on the ToxTree 3.1 algorithm. Interestingly, anthocyanins were classified as Cramer Class III due to the presence of phenolic hydroxyl groups and heterocyclic moieties although they are regarded as safe food constituents. There were eight Cramer Class I, one Class II and twelve Class III chemical constituents in ARFE.

The SEDs for most of chemical constituents by their use in leave-on products were below their respective TTC thresholds.

However, naringin and luteolin were estimated to exceed their respective TTC thresholds. For these high-molecular-weight or highly polar constituents, a Tier 0 assumption of 100% Dap is overly conservative. Therefore, we further refined the dermal absorption using in silico prediction tools, EpiSuite™ and the CDC Finite Dose Skin Permeation Calculator (Table 3). By applying Tier 1 in silico refinement, SEDs of naringin (naringin glycoside) and luteolin were estimated to be lower than their TTC thresholds, reflecting that ARFE is safe to use in cosmetics up to 2%.

Table 3. In silico refinement of systemic exposure dose estimation of constituents of *A. rosea* flower extract (μg/kg/day).

Chemical (Formula)	M.W.	LogKow	Melting Point	Boiling Point	Vapor Pressure (mmHg, 25 °C)	Applied Conc. (μg/mL)	Applied Amount μg/cm²	24 h Cumulative Permeation (%)	Tier 0 SED μg/kg/Day	Tier 2 SED μg/kg/Day
Naringin ($C_{27}H_{32}O_{14}$)	580.5	−0.52	349.84	843.85	7.00×10^{-24}	32	0.435	>1.00	6.681	>0.067
Luteolin ($C_{15}H_{10}O_6$)	286.24	2.36	212.22	499.19	2.19×10^{-13}	14	0.190	50.00	3.003	1.500

3.5. *The History of Safe Food Consumption of A. rosea Flower Supporting the Safety of ARFE in Cosmetics*

It has been confirmed through the literature that *A. rosea* flower has been consumed as food since ancient times [24,26]. It is recommended to brew tea with 1–2 teaspoons of dried *A. rosea* flower and consume one or four cups a day. Therefore, the tea preparation method (hot water) and manufacture of raw materials for cosmetics (80 °C hot water extraction) are equivalent. For a 60 kg adult, the dietary intake can be converted to mg/kg/day as follows:

- 1 to 2 teaspoons: One teaspoon of dried flowers weighs about 2 to 3 g.

- Daily Intake: Minimum intake = 2 g, maximum intake = 6 g. Thus, the average intake of dried flowers can be estimated as 4 g per a cup and when two cups are consumed daily.

$$\text{Daily Intake of ARFE (mg/kg/day)} = \frac{4000\text{ mg}}{60\text{ kg}} \times 2 = 133.34\text{ mg/kg/day}$$

Assuming ARFE is used in cosmetics at 0.05–2% according to ChinaCosing, the SED of ARFE from its use in leave-on products [207.86 mg/kg (Eproduct) × 2%/100 (C/100) * 10%/100 (Extraction ratio) × 100%/100 (Dap)] would not be more than 0.42 mg/kg bw/d, which is 317-fold lower than the daily food consumption of ARFE, supporting the safety of the use of ARFE in cosmetics.

4. Discussion

Here, we evaluated the systemic safety of ARFE used in cosmetics using the TTC and history of safe food consumption. The TTC approach was particularly useful in assessing the safety of ARFE as botanical extracts are mostly composed of macro- or micronutrients, with only small amounts of bioactive constituents potentially posing a risk to human health. Indeed, 92.6% of ARFE constituted common food components or nutritional ingredients, which are considered safe and assumed to pose no health concerns to consumers, leaving only 12.6% of the constituents of ARFE to be considered safety-wise. Furthermore, botanical extracts used in cosmetics are generally manufactured as diluents, i.e., soaking and stirring solvents with dried plants by a 9:1 ratio or more, substantially lowering the exposure level of chemical constituents by the use of cosmetics. Indeed, by considering the dilution factor, SEDs of ARFE constituents were reduced to 10%, meaning that only two bioactive constituents exceeded TTC thresholds at Tier 0 estimation.

To further refine the SEDs of the two constituents with SEDs exceeding the TTC threshold, we applied an in silico tool, the CDC Finite Dose Skin Permeation Calculator, which estimated the dermal absorption of chemicals using physicochemical properties. Lipophilicity and molecular weight are important parameters determining the dermal absorption of chemicals. Of note, natural bioactive substances occurring in botanical extracts were generally hydrophilic and of high molecular weights, which limits their dermal absorption further. By applying in silico dermal absorption prediction, the SEDs of two constituents were significantly lowered below the TTC, demonstrating the utility of the in silico approach in the safety assessment of botanical extracts.

There are some limitations to consider in the TTC approach we employed. We identified the constituents of ARFE and their content from the literature, which may not fully account for the unknown constituents of ARFE used in cosmetics. It would be more accurate to conduct an instrumental analysis for ARFE used in cosmetics to obtain this information. However, it is often impractical to conduct analysis since the instrumental analysis is expensive and time-consuming. Moreover, a sufficient amount of ARFE must be used, which is often made difficult by the scarcity of plant materials.

To supplement the TTC approach, we also conducted the history of safe food consumption approach to ensure the safety of ARFE used in cosmetics. From tea products, dried herbs can be extracted with hot water, of which the preparation method is similar to the manufacturing process of ARFE for use in cosmetics, providing the rationale for the adoption of this approach. The dietary intake of *A. rosea* flower through teas was estimated to be 133.34 mg/kg/day, whereas the exposure to ARFE through cosmetic use, assuming a worst-case scenario with 2% concentration in all products, was estimated at 0.42 mg/kg/day, indicating that no additional health risk over its food consumption was anticipated. Its long history of food consumption may also support the safety of ARFE across different batches and harvests of *A. rosea* flower, which can supplement the TTC approach at least in part. However, the history of food consumption approach does not consider individual chemical constituents of a botanical extract, which may be problematic when trace but hazardous constituents exist. By applying two distinct approaches, we can effectively and more credibly ensure the systemic safety of ARFE used in cosmetics.

When considering the safety of ARFE in cosmetic applications, its historical use as a food product suggests a degree of safety. However, the different exposure routes—oral versus dermal—introduce critical uncertainty. Flavonoids, including anthocyanins, are major components of ARFE. When consumed orally, these compounds exhibit poor bioavailability (typically 1–2%) [44] and undergo extensive metabolism in the intestines and liver [45]. This metabolic transformation can significantly alter their intrinsic toxicity profiles. In contrast, while the dermal application of flavonoids in cosmetics also leads to limited absorption [46], they encounter substantially less metabolism compared to oral routes. This difference in metabolic processing means that the safety profile of ingested ARFE may not directly mirror that of topically applied ARFE. Therefore, safety assessors must carefully consider these distinct exposure and metabolic pathways when evaluating the dermal safety of ARFE in cosmetic products.

5. Conclusions

Collectively, we demonstrated that TTC and history of safe food consumption approaches can be useful in ensuring the systemic safety of the botanical extract ARFE in cosmetics. A limitation of this study is the reliance on literature-derived constituent profiles, which may not capture all unknown and trace components. In addition, the history of safe food consumption approach does not account for the difference in metabolic processing for topically applied ARFE. Further case studies and methodologies to resolve the limitations of these two approaches are necessary to advance the safety assessment of botanical extracts.

Author Contributions: Conceptualization, S.G. and K.-M.L.; methodology, S.G. and K.-M.L.; formal analysis, S.G.; investigation, S.G.; resources, K.-M.L.; data curation, S.G.; writing—original draft preparation, S.G.; writing—review and editing, K.-M.L.; supervision, K.-M.L.; project administration, K.-M.L.; funding acquisition, K.-M.L. All authors have read and agreed to the published version of the manuscript.

Funding: This study was part of the cosmetic safety evaluation project carried out by the Korea Cosmetic Industry Institute (KCII) funded by the Ministry of Health and Welfare and National Research Foundation (NRF) of Korea (2021R1A2C2013347 and 2021R1A6C101A442).

Institutional Review Board Statement: Not applicable.

Informed Consent Statement: Not applicable.

Data Availability Statement: The data presented in this study are available on request from the corresponding author.

Conflicts of Interest: The authors declare no conflict of interest

References

1. Ferreira, M.; Matos, A.; Couras, A.; Marto, J.; Ribeiro, H. Overview of cosmetic regulatory frameworks around the world. *Cosmetics* **2022**, *9*, 72. [CrossRef]
2. Humane Society of International. 45 Countries Have Full or Partial Bans on Cosmetics Animal Testing. Available online: https://www.humaneworld.org/en/issue/animals-in-research (accessed on 19 April 2025).

3. Silva, R.J.; Tamburic, S. A State-of-the-Art Review on the Alternatives to Animal Testing for the Safety Assessment of Cosmetics. *Cosmetics* **2022**, *9*, 90. [CrossRef]
4. Barthe, M.; Bavoux, C.; Finot, F.; Mouche, I.; Cuceu-Petrenci, C.; Forreryd, A.; Chérouvrier Hansson, A.; Johansson, H.; Lemkine, G.F.; Thénot, J.-P. Safety testing of cosmetic products: Overview of established methods and new approach methodologies (NAMs). *Cosmetics* **2021**, *8*, 50. [CrossRef]
5. Nikodinoska, I.; Spohr, C.; Dillon, G.P.; Moran, C.A. Skin and eye irritancy assessment of six lactic acid bacteria strains. *Regul. Toxicol. Pharmacol.* **2023**, *141*, 105406. [CrossRef] [PubMed]
6. Lee, M.; Hwang, J.H.; Lim, K.M. Alternatives to In Vivo Draize Rabbit Eye and Skin Irritation Tests with a Focus on 3D Reconstructed Human Cornea-Like Epithelium and Epidermis Models. *Toxicol. Res.* **2017**, *33*, 191–203. [CrossRef] [PubMed]
7. Dent, M.; Vaillancourt, E.; Thomas, R.; Carmichael, P.; Ouedraogo, G.; Kojima, H.; Barroso, J.; Ansell, J.; Barton-Maclaren, T.; Bennekou, S.H. Paving the way for application of next generation risk assessment to safety decision-making for cosmetic ingredients. *Regul. Toxicol. Pharmacol.* **2021**, *125*, 105026. [CrossRef]
8. Bury, D.; Head, J.; Keller, D.; Klaric, M.; Rose, J. The Threshold of Toxicological Concern (TTC) is a pragmatic tool for the safety assessment: Case studies of cosmetic ingredients with low consumer exposure. *Regul. Toxicol. Pharmacol.* **2021**, *123*, 104964. [CrossRef] [PubMed]
9. Yang, C.; Barlow, S.M.; Jacobs, K.L.M.; Vitcheva, V.; Boobis, A.R.; Felter, S.P.; Arvidson, K.B.; Keller, D.; Cronin, M.T.; Enoch, S. Thresholds of Toxicological Concern for cosmetics-related substances: New database, thresholds, and enrichment of chemical space. *Food Chem. Toxicol.* **2017**, *109*, 170–193. [CrossRef]
10. Mahony, C.; Bowtell, P.; Huber, M.; Kosemund, K.; Pfuhler, S.; Zhu, T.; Barlow, S.; McMillan, D.A. Threshold of toxicological concern (TTC) for botanicals-Concentration data analysis of potentially genotoxic constituents to substantiate and extend the TTC approach to botanicals. *Food Chem. Toxicol.* **2020**, *138*, 111182. [CrossRef]
11. Kawamoto, T.; Fuchs, A.; Fautz, R.; Morita, O. Threshold of Toxicological Concern (TTC) for Botanical Extracts (Botanical-TTC) derived from a meta-analysis of repeated-dose toxicity studies. *Toxicol. Lett.* **2019**, *316*, 1–9. [CrossRef]
12. Nohynek, G.J.; Antignac, E.; Re, T.; Toutain, H. Safety assessment of personal care products/cosmetics and their ingredients. *Toxicol. Appl. Pharmacol.* **2010**, *243*, 239–259. [CrossRef] [PubMed]
13. Jeon, S.; Lee, E.Y.; Hillman, P.F.; Nam, S.J.; Lim, K.M. Safety assessment of Cnidium officinale rhizome extract in cosmetics using the Threshold of Toxicological Concern (TTC) approach. *Regul. Toxicol. Pharmacol.* **2023**, *142*, 105433. [CrossRef]
14. Jeon, S.; Lee, E.Y.; Nam, S.J.; Lim, K.M. Safety assessment of Paeonia lactiflora root extract for a cosmetic ingredient employing the threshold of toxicological concern (TTC) approach. *Regul Toxicol Pharmacol* **2024**, *149*, 105620. [CrossRef]
15. Rana, P.; Pathania, D.; Gaur, P.; Patel, S.K.; Bajpai, M.; Singh, N.T.; Pandey, R.; Shukla, S.V.; Pant, A.B.; Ray, R.S.; et al. Regulatory frameworks for fragrance safety in cosmetics: A global overview. *Toxicol. Res.* **2025**, *41*, 199–220. [CrossRef]
16. Patlewicz, G.; Worth, A.; Yang, C.; Zhu, T. Editorial: Advances and Refinements in the Development and Application of Threshold of Toxicological Concern. *Front. Toxicol.* **2022**, *4*, 882321. [CrossRef] [PubMed]
17. EFSA Scientific Committee; More, S.J.; Bampidis, V.; Benford, D.; Bragard, C.; Halldorsson, T.I.; Hernández-Jerez, A.F.; Hougaard Bennekou, S.; Koutsoumanis, K.P.; Machera, K. Guidance on the use of the Threshold of Toxicological Concern approach in food safety assessment. *EFSA J.* **2019**, *17*, e05708.
18. Bernauer, U.; Bodin, L.; Chaudhry, Q.; Coenraads, P.; Dusinska, M.; Ezendam, J.; Gaffet, E.; Galli, C.; Panteri, E.; Rogiers, V. *SCCS Notes of Guidance for the Testing of Cosmetic Ingredients and Their Safety Evaluation-12th Revision-Final Opinion–SCCS/1647/22-Corrigendum 2*; European Comission: Brussels, Belgium, 2023.
19. Personal Care Products Council CIR Science and Support Committee. Decision Tree for the Safety Assessment of Botanical Cosmetic Ingredients. Available online: https://www.cir-safety.org/sites/default/files/Botanical%20Decision%20Tree%20CIR%20June%202012%202.pdf (accessed on 19 April 2025).
20. NIFDC. Technical Guidelines for Application of Threshold of Toxicological Concern (TTC) Method. Available online: https://www.zmuni.com/en/news/2024-in-review-key-cosmetic-regulatory-updates-in-china/ (accessed on 19 April 2025)
21. Cramer, G.; Ford, R.; Hall, R. Estimation of toxic hazard—A decision tree approach. *Food Cosmet. Toxicol.* **1976**, *16*, 255–276. [CrossRef]
22. Re, T.A.; Mooney, D.; Antignac, E.; Dufour, E.; Bark, I.; Srinivasan, V.; Nohynek, G. Application of the threshold of toxicological concern approach for the safety evaluation of calendula flower (*Calendula officinalis*) petals and extracts used in cosmetic and personal care products. *Food Chem. Toxicol.* **2009**, *47*, 1246–1254. [CrossRef]
23. NIFDC. China Compliant Cosmetic Ingredients: History of Safe Consumption Guidelines—Draft. Available online: https://cisema.com/en/china-compliant-cosmetic-ingredients-history-of-safe-consumption-guidelines-draft-for-comments/ (accessed on 19 April 2025).
24. Lim, T.K. *Edible Medicinal and Non-Medicinal Plants*; Springer: Berlin/Heidelberg, Germany, 2012; Volume 1.

25. Fahamiya, N.; Shiffa, M.; Aslam, M. A comprehensive review on *Althaea rosea* Linn. *J. Pharm. Res.* **2016**, *6*, 6888–6894.
26. Shehzad, M.R.; Hanif, M.A.; Rehman, R.; Bhatti, I.A.; Hanif, A. Hollyhock. In *Medicinal Plants of South Asia*; Elsevier: Amsterdam, The Netherlands, 2020; pp. 381–391.
27. Shah, S.; Akhtar, N.; Akram, M.; Shah, P.A.; Saeed, T.; Ahmed, K.; Asif, H. Pharmacological activity of *Althaea officinalis* L. *J. Med. Plants. Res.* **2011**, *5*, 5662–5666.
28. Castillo-Carrión, M.; Martínez-Espinosa, R.; Pérez-Álvarez, J.Á.; Fernández-López, J.; Viuda-Martos, M.; Lucas-González, R. Nutritional, fatty acids, (poly) phenols and technological properties of flower powders from *Fuchsia hybrida* and *Alcea rosea*. *Foods* **2024**, *13*, 237. [CrossRef] [PubMed]
29. Hage-Sleiman, R.; Mroueh, M.; Daher, C.F. Pharmacological evaluation of aqueous extract of *Althaea officinalis* flower grown in Lebanon. *Pharm. Biol.* **2011**, *49*, 327–333. [CrossRef] [PubMed]
30. Tomoda, M.; Shimizu, N.; Oshima, Y.; Takahashi, M.; Murakami, M.; Hikino, H. Hypoglycemic activity of twenty plant mucilages and three modified products. *Planta Medica* **1987**, *53*, 8–12. [CrossRef] [PubMed]
31. European Medicines Agency. 2016. Available online: https://www.ema.europa.eu/en/documents/herbal-monograph/final-european-union-herbal-monograph-althaea-officinalis-l-radix_en.pdf (accessed on 1 June 2025).
32. European Medicines Agency. 2009. Available online: https://www.ema.europa.eu/en/documents/herbal-report/final-assessment-report-althaea-officinalis-l-radix_en.pdf (accessed on 1 June 2025).
33. Lee, G.; Yeom, A.; Won, K.D.; Park, C.-M.; Joung, M.-S.; Lee, G.Y.; Jeong, C.-S. Evaluation of *Alcea rosea* L. callus extract as a natural cosmetic ingredient. *J. Soc. Cosmet. Sci. Korea* **2018**, *44*, 295–302.
34. Azadeh, Z.; Asgharian, S.; Habtemariam, S.; Lorigooini, Z.; Taheri, A. A review of botanical, phytochemical, and pharmacological properties of *Alcea rosea* L. *Future Nat. Prod.* **2023**, *9*, 88–99.
35. Yang, C.; Rathman, J.F.; Bienfait, B.; Burbank, M.; Detroyer, A.; Enoch, S.J.; Firman, J.W.; Gutsell, S.; Hewitt, N.J.; Hobocienski, B. The role of a molecular informatics platform to support next generation risk assessment. *Comput. Toxicol.* **2023**, *26*, 100272. [CrossRef]
36. ChinaCosIng. Althaea Rosea Flower Extract. Available online: https://zhg.cirs-group.com/cosmetic-ingredient-report/iecic/63e0ae6d83f9bdb57625da7a/%E8%9C%80%E8%91%B5%EF%BC%88ALTHAEA%20ROSEA%EF%BC%89%E8%8A%B1%E6%8F%90%E5%8F%96%E7%89%A9 (accessed on 4 June 2025).
37. Stribling, P.; Ibrahim, F. Dietary fibre definition revisited-The case of low molecular weight carbohydrates. *Clin. Nutr. ESPEN* **2023**, *55*, 340–356. [CrossRef]
38. Khandelwal, N.; Abraham, S.K. Protective effects of common anthocyanidins against genotoxic damage induced by chemotherapeutic drugs in mice. *Planta Medica* **2014**, *80*, 1278–1283. [CrossRef]
39. Taj, S.; Nagarajan, B. Inhibition by quercetin and luteolin of chromosomal alterations induced by salted, deep-fried fish and mutton in rats. *Mutat. Res./Genet. Toxicol.* **1996**, *369*, 97–106. [CrossRef]
40. Li, G.; Zhou, J.; Sun, M.; Cen, J.; Xu, J. Role of luteolin extracted from *Clerodendrum cyrtophyllum* Turcz leaves in protecting HepG2 cells from TBHP-induced oxidative stress and its cytotoxicity, genotoxicity. *J. Funct. Foods* **2020**, *74*, 104196. [CrossRef]
41. Cai, Q.; Rahn, R.O.; Zhang, R. Dietary flavonoids, quercetin, luteolin and genistein, reduce oxidative DNA damage and lipid peroxidation and quench free radicals. *Cancer Lett.* **1997**, *119*, 99–107. [CrossRef] [PubMed]
42. Syeda, R.; Roshan, S. Anti Genotoxic Activity of Luteolin on Cyclophosphamide Induced Genotoxicity in Albino Mice. *J. Popul. Ther. Clin. Pharmacol.* **2022**, *29*, 347–353.
43. Fernando, P.; Ko, D.O.; Piao, M.J.; Kang, K.A.; Herath, H.; Hyun, J.W. Protective effect of luteolin against oxidative stress-mediated cell injury via enhancing antioxidant systems. *Mol. Med. Rep.* **2024**, *30*, 121. [CrossRef]
44. Thilakarathna, S.H.; Rupasinghe, H.P. Flavonoid bioavailability and attempts for bioavailability enhancement. *Nutrients* **2013**, *5*, 3367–3387. [CrossRef] [PubMed]
45. Rosales, T.K.O.; Hassimotto, N.M.A.; Lajolo, F.M.; Fabi, J.P. Nanotechnology as a Tool to Mitigate the Effects of Intestinal Microbiota on Metabolization of Anthocyanins. *Antioxidants* **2022**, *11*, 506. [CrossRef]
46. Nagula, R.L.; Wairkar, S. Recent advances in topical delivery of flavonoids: A review. *J. Control. Release* **2019**, *296*, 190–201. [CrossRef]

pharmaceutics

Article

The Synergistic Influence of Polyflavonoids from *Citrus aurantifolia* on Diabetes Treatment and Their Modulation of the PI3K/AKT/FOXO1 Signaling Pathways: Molecular Docking Analyses and In Vivo Investigations

Mohamed A. Hassan [1,*], Ghada M. Abd Elmageed [2], Ibtehal G. El-Qazaz [2], Doaa S. El-Sayed [3], Lamia M. El-Samad [2] and Heba M. Abdou [2,*]

[1] Protein Research Department, Genetic Engineering and Biotechnology Research Institute (GEBRI), City of Scientific Research and Technological Applications (SRTA-City), New Borg El-Arab City 21934, Egypt

[2] Department of Zoology, Faculty of Science, Alexandria University, Alexandria 21321, Egypt; dedy862000@yahoo.com (G.M.A.E.); ibtehalgomaa@yahoo.com (I.G.E.-Q.); lamya.moustafa@alexu.edu.eg (L.M.E.-S.)

[3] Chemistry Department, Faculty of Science, Alexandria University, Alexandria 21321, Egypt; doaa.saeed@alexu.edu.eg

* Correspondence: madel@srtacity.sci.eg (M.A.H.); dr.heba_abdou3000@yahoo.com (H.M.A.)

Abstract: This study was aimed at probing the modulatory influence of polyflavonoids extracted from *Citrus aurantifolia*, lemon peel extract (LPE-polyflavonoids), on attenuating diabetes mellitus (DM) and its complications. HPLC investigations of the LPE exhibited the incidence of five flavonoids, including diosmin, biochanin A, hesperidin, quercetin, and hesperetin. The in silico impact on ligand-phosphatidylinositol 3-kinase (PI3K) interaction was investigated in terms of polyflavonoid class to explore the non-covalent intakes and binding affinity to the known protein active site. The drug likeness properties and pharmacokinetic parameters of the LPE-polyflavonoids were investigated to assess their bioavailability in relation to Myricetin as a control. Remarkably, the molecular docking studies demonstrated a prominent affinity score of all these agents together with PI3K, implying the potency of the extract to orchestrate PI3K, which is the predominant signal for lessening the level of blood glucose. To verify these findings, in vivo studies were conducted, utilizing diabetic male albino rats treated with LPE-polyflavonoids and other groups treated with hesperidin and diosmin as single flavonoids. Our findings demonstrated that the LPE-polyflavonoids significantly ameliorated the levels of glucose, insulin, glycogen, liver function, carbohydrate metabolizing enzymes, G6Pd, and AGEs compared to the diabetic rats and those exposed to hesperidin and diosmin. Furthermore, the LPE-polyflavonoids regulated the TBARS, GSH, CAT, TNF-α, IL-1β, IL-6, and AFP levels in the pancreatic and hepatic tissues, suggesting their antioxidant and anti-inflammatory properties. In addition, the pancreatic and hepatic GLUT4 and GLUT2 were noticeably increased in addition to the pancreatic p-AKT in the rats administered with the LPE-polyflavonoids compared to the other diabetic rats. Remarkably, the administration of LPE-polyflavonoids upregulated the expression of the pancreatic and hepatic PI3K, AMPK, and FOXO1 genes, emphasizing the efficiency of the LPE in orchestrating all the signaling pathways necessitated to reduce the diabetes mellitus. Notably, the histopathological examinations of the pancreatic and hepatic tissues corroborated the biochemical results. Altogether, our findings accentuated the potential therapeutic role of LPE-polyflavonoids in controlling diabetes mellitus.

Keywords: diabetes mellitus; diosmin; FOXO1; hesperidin; high fructose/streptozotocin; lemon peel extract-polyflavonoids; PI3K/AKT; AMPK

1. Introduction

Millions of people worldwide are predisposed to a critical threat to their lives due to the prevalence of metabolic diseases, particularly diabetes mellitus (DM), which cannot be understated. DM is a chronic endocrine metabolic disorder identified by augmented blood glucose concentrations and engendered by insulin insufficiency (Type 1 diabetes, T1DM) or insulin resistance (IR), as perceived in type 2 diabetes mellitus (T2DM), and the latter type is prevalent globally [1]. Alarmingly, humans suffering from T2DM are prone to various complications, including cardiovascular diseases, kidney failure, and nerve damage [1–4]. In T2DM, there are two leading pathophysiological mechanisms, including IR and pancreatic beta cell failure [5].

The pancreas plays a pivotal role in the development of diabetes mellitus, since it orchestrates several homeostatic functions through its exocrine, acinar, and endocrine islet cells [6]. Furthermore, given the property of the pancreas to produce low levels of antioxidant enzymes, pancreatic β-cells are highly prone to oxidative stress (OS) [7]. This OS originates from profuse free radicals, misfolded proteins, and hyperactivity of the endoplasmic reticulum. Consequently, pancreatic β-cells undergo apoptosis on account of this cellular disturbance, instigating pancreatic dysfunction [8]. Furthermore, fructose interferes with glucose metabolism as a result of its specific metabolic fate and lipogenic features. This disturbance provokes IR, dyslipidemia, and hepatic fibrosis, in addition to cardiac and renal malfunctions [9].

It is believed that the impairment of hepatic insulin signaling is predominantly attributed to oxidative damage [10]. Most importantly, the PI3K/AKT signaling pathway is the crucial insulin pathway in glucose metabolism [11]. This signaling pathway is tied up with glucose uptake by the liver, skeletal muscles, and adipose tissues by activating phosphatidylinositol 3-kinase (PI3K), proceeding along the insulin receptor substrates IRS1 and IRS2. The PI3K pathway is the leading modulator of insulin metabolic actions; therefore, a hindrance to this pathway lessens the physiological properties of insulin, which critically provokes IR. Reductions in the glucose utilization and glycogenesis in the liver are the result of these signaling episodes, while hepatic glucose production, glycogenolysis, and triglyceride (TG) buildup are increased [12]. Moreover, multiple signaling pathways are involved in the etiology of T2DM, including adenosine monophosphate-activated protein kinase/sirtuin (AMPK/SIRT1) and adenosine monophosphate-activated protein kinase/mammalian target of rapamycin (AMPK/mTOR) [13].

Importantly, forkhead box-O1 (FOXO1) is one of the forkhead box protein families, which is considerably produced by β-cells; thus, it affects indispensable cellular pathways, such as cell differentiation, metabolism, and cell cycle arrest. It has been strongly evidenced that FOXO1 governs β-cell replication and differentiation [14]. Furthermore, it has been identified as a major influencing protein in T2DM, due to its efficiency in modulating the insulin sensitivity of target tissues [15]. Along with the growth of DM, there has been increasing interest in phenolic and flavonoid compounds from diverse origins to alleviate DM and its detrimental consequences, due to their inherent property of scavenging the overflow of free radical species [16,17], hampering oxidative stress within pancreatic cells [18,19]. Several medicinal plants have introduced alternative products for impeding DM due to their capacity to modify carbohydrate enzymes and adapt glucose homeostasis [20]. Among the plants containing effective flavonoids for different therapies, the citrus plant (*Citrus aurantifolia*), Rutaceae family, is a widespread fruit crop [21]. Grapefruit, lemon, and orange are the principal (dominant) species of citrus, which comprise phenolic components, particularly flavonoids. Citrus peels, which are considered to be worthy by-products, represent the highest flavonoid percentage in whole fruits [22]. It has been argued that lemons could help to reduce the negative consequences of cardiovascular disease, inflammation, tumors, and other malignant diseases [23]. In addition, hesperidin and diosmin are flavonoids abundant in citrus fruits, where the former belongs to the flavanone class and the latter to the flavone glycoside class. Hesperidin is a crucial active component of aged citrus [24]. Both flavonoids possess promising mechanisms as antioxidant, anti-inflammatory, anti-cancer, neuroprotective, lipid-lowering, and glucose-lowering agents [25,26]. In contrast to the available literature discussing the efficacy of hesperidin and diosmin, no research has probed the efficiency of polyflavonoids extracted from lemon peel extract (LPE-polyflavonoids) in alleviating DM. Moreover, no reports have elucidated the potency of LPE-polyflavonoids in mimicking the insulin mechanism in modulating the PI3K/AKT/FOXO1 signaling pathway.

Herein, this study was aimed at identifying the flavonoids in lemon peel extract using high-performance liquid chromatography (HPLC), followed by investigating their binding affinity to PI3K, employing a molecular docking approach. Furthermore, we evaluated the synergistic influence of LPE-polyflavonoids on controlling diabetes mellitus using rats as an animal model, providing extensive evidence concerning their competency as antidiabetic, antioxidant, and anti-inflammatory agents compared to the impact of either hesperidin or diosmin as standard flavonoids from lemons. Furthermore, the potency of the LPE-polyflavonoids in orchestrating PI3K/AKT/FOXO1 signals was assessed. Furthermore, the histological features of pancreatic and liver tissues were surveyed to substantiate their impact on restoring the cellular architectures compared to those organs in the diabetic rat group.

2. Materials and Methods

2.1. Chemicals and Assay Kits

Streptozotocin (STZ) was provided by Sigma-Aldrich Co., Rockville, MD, USA, while D-Fructose (laevulose) ($C_6H_{12}O_6$) was procured from El-Nasr Pharmaceutical Chemicals Co., Cairo, Egypt. Hesperidin 97% ($C_{28}H_{34}O_{15}$) was purchased from Acros Organics Co., Belgium, WI, USA and diosmin 100% pure extracted from citrus fruit was supplied by Nutrition Greenlife Co., New York, NY, USA.

2.2. Preparation of Lemon Peel Extract (LPE)

Lemon plants, *Citrus aurantifolia* (Family: Rutaceae), were collected from the local market and instantly transported to the lab before being identified in terms of their morphological features. The lemon peel extract (LPE) was prepared using a methanol approach following previously reported procedures with minor adaptations [27]. Briefly, lemon peels were sliced into small pieces before being dried at room temperature. Afterward, the dried peels were powdered employing an electric grinder, and about 0.25 g of powder was then extracted for 8 h in 100 mL of methanol under orbital agitation. The extract was then filtered utilizing a polytetrafluoroethylene (PTFE) membrane (0.45 μm pore size). The extraction procedures were repeated three times and the methanol was evaporated at 30 °C under vacuum by the means of a rotary evaporator. The LPE was then stored at 4 °C for further examination.

2.3. High-Performance Liquid Chromatography (HPLC) Analysis

To determine the flavonoids in the LPE, 30 mg of LPE powder was incubated in 4 mL of 6 M HCl in a water bath at 100 °C for 15 min with stirring. Following cooling, the residues were filtered and then extracted three times using 15 mL of dichloromethane, followed by a concentration of the extract employing a rotary evaporator under low pressure. Afterward, the extract was dissolved in 10 mL of methanol before being filtered using a filter of 0.45 μm. Next, the sample was injected into an HPLC (Agilent 1260, Santa Clara, CA, USA) connected to a 5C18 column (4.6 mm × 150 mm) and DAD detector. The elution of the flavonoids was conducted utilizing a mobile phase comprising acetonitrile:water:trifluoroacetic acid (20:80:0.01 (*v*/*v*/*v*)) (A) and acetonitrile:trifluoroacetic acid (100:0.1 (*v*/*v*)) (B). The gradient profile was adapted from 0 to 40% B for 0–10 min, 40% B for 10–11 min, and 40 to 100% B for 11–12 min. To restore the initial conditions, 100% mobile phase A was utilized for 6 min at the end of each run. The flow rate was adapted to 0.7 mL/min and the measurement was achieved at 260 nm. The flavonoids were identified with respect to reference flavonoids, and their concentrations were evaluated in the LPE sample.

2.4. Molecular Docking Examination of Flavonoids in LPE

Previous studies on medicinal plants have described ligand pocket areas as the leading framework for figuring out the target protein active sites of drugs derived from plants efficiently [28]. Mainly, the docking protocol proceeds via a crucial path for highly effective results [29]. Consequently, the flavonoids detected in the LPE, diosmin (Lig I), biochanin A (Lig II), hesperidin (Lig III), quercetin (Lig IV), and hesperetin (Lig V) were subjected to docking simulation using the Autodock Vina [30,31] software (version 4.2) for its dominant capacity for docking all studied phenolic compounds at the same time under comparable conditions. Furthermore, the chimera software was used for a docking data visualization and analysis [32]. The chosen protein was phosphatidylinositol 3-kinase (PI3K, code ID: 1E90) [33], which was retrieved from the protein data bank website (https://www.rcsb.org/structure/5f0g, accessed on 1 July 2023). Despite the presence of some spectrum ranges of PI3K macromolecule codes, selecting the more in silico-efficient active protein code in our study was considered. The selected target protein was implicated with a reference bioflavonoid-like Myricetin (MYC) lipid-kinase inhibitor. The molecular structures of the main polyphenolic components were downloaded from the PubChem site (https://pubchem.ncbi.nlm.nih.gov/, accessed on 1 July 2023). Prior to commencing the docking process, some key steps were carried out, including the preparation of proteins and ligands in a suitable format. Grid box dimensions, along with the identification of the predicted active sites, were generated. Based on the drug-like control inside the protein pocket, the grid box size was estimated with dimensions of 82 × 92 × 92 Å, with 0.3 Å spacing and grid centers x, y, and z of 10.066, 51.723, and 27.004, respectively, respecting an exhaustiveness of 8 with an energy range equal to 4. The binding affinity mode was considered to be Genetic Algorithm (LGA) [34]. Using the Discovery Studio software (Version 21) (https://www.3ds.com/products-services/biovia/, accessed on 1 July 2023), the results were visualized and the respective figures were exported.

2.5. Physicochemical and Pharmacokinetic Protocol

In silico investigations of the drug-like properties of the polyflavonoids and ADMET parameters could be calculated utilizing the open-source online SwissADME tool [35], which deals with the SMILE format. The theoretical interpretations related to the ADMET properties were previously discussed through pkCSM graph-based signatures [36], predicting the small molecules' efficacy as a promising drug. Furthermore, physicochemical indices can control the Lipinski rules by regulating drug absorption and diffusion.

2.6. In Vivo Investigations

Male Wistar albino rats weighing 160–180 g were utilized in the current investigation. The rats were housed at the animal house in the Experimental Animal Center-Medical Research Institute, Alexandria University, Egypt. The animals were sheltered in standard cages following standard laboratory conditions: photoperiod (12/12 h light/dark cycle) and room temperature (22 °C ± 3 °C). The rats were provided with a basal diet and tap water ad libitum. The handling and treatment of the rats were performed according to the ARRIVE guidelines (https://arriveguidelines.org/, accessed on 1 July 2023) and approval for all the animal procedures was obtained from the Institutional Animal Care and Use Committee at Alexandria University, Egypt (approval number: AU04220212302).

Experimental Design and Induction of Diabetes

In total, 30 rats were allocated into two groups: Group 1 (control group, 6 rats), which was was administered with distilled water via oral gavage once daily for 4 weeks. Group 2: this group included 24 rats for diabetes induction. Specifically, the 24 rats were provided with 10% fructose in their drinking water for two weeks before being intraperitoneally injected with a single dose of STZ (40 mg/kg body weight) to instigate type 2 diabetes mellitus, as previously reported [37,38]. Following three days of injection, blood was collected from the tail vein to evaluate the concentration of glucose, employing a glucometer, and the rats with fasting blood glucose levels above 220 mg/dL were identified as diabetic rats. Accordingly, the diabetic rats were randomly allocated into 4 sub-groups (n = 6): the Fructose-STZ group, Fructose-STZ + hesperidin (100 mg/kg body weight/day, orally) group, Fructose-STZ + diosmin (100 mg/kg body weight/day, orally) group, and Fructose-STZ + LPE (400 mg/kg body weight/day, orally) group. The hesperidin, diosmin, and LPE-polyflavonoids doses were prepared by dissolving them in distilled water.

2.7. Blood Collection and Tissue Preparation

The fasted rats were euthanized by exposure to isoflurane and dissected at the end of the in vivo investigation. Blood samples were collected via cardiac punctures in all the animal groups using a sterile syringe, before being centrifuged at 3000 rpm for 10 min to separate the blood serum. The serum was preserved at −20 °C for further use to evaluate the biochemical parameters. On the other hand, the pancreas and liver were isolated from all the experimental groups and cleaned from blood, before being rinsed with a chilled saline solution. For the biochemical investigations, parts of the pancreatic and hepatic tissues were separately homogenized for 5 min in 2 mL/g tissue of cold buffer (50 mM potassium phosphate pH 7.5, 1 mM EDTA) by the means of a homogenizer (Potter-Elvehjem homogenizer). Afterward, the homogenates were clarified for 15 min at 4000 rpm and 4 °C using a centrifuge (Universal 32 R, Hettich, Tuttlingen, Germany), and the supernatants were preserved at −80 °C. For the molecular analyses, small portions of the pancreas and livers were instantly dried and preserved at −80 °C. For the histological investigations, small slices of the pancreas and liver were fixed in formalin solution (10%).

2.8. Biochemical Analysis

The serum glucose was estimated utilizing a commercial assay kit (Bio-diagnostics Co., Cairo, Egypt). The insulin was evaluated utilizing an enzyme-linked immunosorbent assay (ELISA) kit (Cat. No. EMINS, Thermo Fisher Scientific, Waltham, MA, USA), and the glycogen levels were estimated using the Rat Glucagon calorimetric assay kit (Cat. No. ab282931, Abcam Co., Berlin, Germany). Aspartate aminotransferase (AST), alanine transaminase (ALT), alkaline phosphatase (ALP), total protein, and albumin were evaluated in the serum using standard commercial kits from Bio-diagnostics Co., Cairo, Egypt, following the manufacturer's instructions.

Hepatic hexokinase (HK) was estimated employing an ELISA kit (Cat. No. E4600-100, Biovision Co., Berlin, Germany), hepatic glucokinase activity (GK) was estimated using a kit obtained from MyBioSource Co. (Cat. No. MBS704328, San Diego, CA, USA), hepatic glucose-6-phosphate dehydrogenase (G6Pd) was evaluated utilizing an assay kit purchased from Abcam Co. (Cat. No. ab102529, Berlin, Germany), hepatic Alpha-fetoprotein (AFP) was estimated using a rat AFP ELISA kit (Cat. No. CSB-E08281r; Cusabio Biotech Co., Ltd., Beijing, China), and phospho-AKT was estimated using an ELISA kit obtained from Biovision Co. (Cat. No. K4211-100, Berlin, Germany). All the assays were conducted in six replicates, following the protocols provided by the manufacturer.

2.8.1. Lipid Profile Assay

The total cholesterol was estimated employing a respective kit (Cat. No. MAK043, Sigma-Aldrich Co., Rockville, MD, USA). Furthermore, the triglyceride (TG) level was evaluated utilizing a kit (Cat. No. MAK266, Sigma-Aldrich Co., Rockville, MD, USA). HDL-cholesterol was estimated following the approach proposed by Lopes-Virella et al. [39], while LDL-cholesterol was assessed utilizing a kit supplied by Sigma-Aldrich Co. (Cat. No. MAK045, Rockville, MD, USA).

2.8.2. Assessment of Oxidative Stress Biomarkers

To evaluate the lipid peroxidation in the tissue homogenates, thiobarbituric acid reactive substance (TBARS) was estimated, as described earlier [40], while glutathione (GSH) was assessed spectrophotometrically following the protocol reported by Jollow et al. [41].

The activities of superoxide dismutase (SOD) and catalase (CAT) were estimated following the procedures of Nishikimi et al. and Abei et al., respectively [42,43]. Additionally, the advanced glycation end-products' (AGEs) concentrations were estimated utilizing the OxiSelect™ AGE Competitive ELISA Kit (Cell Biolabs, San Diego, CA, USA). All the investigations were replicated six times.

2.8.3. Evaluation of Inflammatory Biomarkers

To quantify the tumor necrosis factor alpha (TNF-α), interleukin-6 (IL-6), and nuclear factor Kappa (NF-κB), ELISA kits obtained from MyBioSource Co., San Diego, CA, USA were utilized (Cat. No. MBS2507393, Cat. No. MBS726707, and Cat. No. MBS453975, respectively). Additionally, interleukin-1β (IL-1β) was assessed adopting a respective ELISA kit (Cat. No. BMS630, Thermo Fisher Scientific, Waltham, MA, USA).

2.8.4. Glucose Transporter 2 (GLUT2) and GLUT4 and p-AKT

The concentrations of glucose transporter 2 (GLUT2) and glucose transporter 4 (GLUT4) were quantified in the liver and pancreatic tissues using sandwich ELISA kits from MyBioSource Co., San Diego, CA, USA (Cat. No. MBS2025604 and Cat. No. MBS2023267, respectively). Using a similar approach, phospho-AKT (p-AKT) was evaluated in the pancreas utilizing a phospho-AKT ELISA kit (Cat. No. K4211-100, Biovision Co., Berlin, Germany).

2.8.5. Quantitative Real-Time PCR (qRT-PCR) Analyses

The relative mRNA expression levels of the phosphatidylinositol 3-kinase (PI3K), adenosine monophosphate-activated protein kinase (AMPK), and forkhead box-O1 (FOXO1) genes were assessed using qRT-PCR. The RNA was obtained from the pancreas and liver utilizing Trizol Reagent (Thermo Fisher Scientific, Waltham, MA, USA), following the manufacturer's protocol. Subsequently, the qRT-PCR was conducted employing the Rotor-Gene SYBR Green RT-PCR Kit (Qiagen, Montgomery, MD, USA), following the protocols supplied by the manufacturer. Briefly, 2x Rotor-Gene SYBR Green RT-PCR Master Mix, template RNA ($\leq$100 ng/reaction), primers, and RNase-free water were mixed and put on ice. The cDNA synthesis was carried out at 55 °C for 10 min and then at 95 °C for 5 min for RNase deactivation. Afterward, the qRT-PCR was initiated, and the temperature and time for the annealing step were optimized for each reaction according to the primer used in the PCR reaction. The sequences of primers applied for the qRT-PCR are shown in Table 1. The fold difference was computed using the Equation (2) $-\Delta\Delta$ct relative to β-actin, as reported earlier [44,45].

Table 1. The nucleotide sequences of the primers applied in qRT-PCR.

Gene	Forward Primer	Reverse Primer
PI3K	5′-CCTGGTAACTGCAACACTTC-3′	5′-AACGAATTCAAACCTACCCG-3′
FOXO1	5′-TCATCCAATTGGTCTTGTGG-3′	5′-GTGTTTGCCTGTCTACCTTT-3′
AMPK	5′-TGTGAAGATCGGACACTACG-3′	5′-TAACTGCCACTTTATGGCCT-3′
β-Actin	5′-ATGTGGCTGAGGACTTTGATT-3′	5′-ATCTATGCCGTGGATACTTGG-3′

Accession number: NM_001371300.2, XM_039103268.1, NM_023991.1, and XM_039089807.1, respectively.

2.9. *Histopathological Examination*

The pancreatic and liver tissues were fixed by immersing them in 10% neutral-buffered formalin for 48 h. Next, the tissues were dehydrated in increasing concentrations of ethanol for 15 min for each concentration, before being embedded in paraffin wax. Subsequently, they were sectioned (4 μm thickness), prior to being stained with Hematoxylin and Eosin (H&E). The H&E sections were surveyed by the means of a light microscope (Olympus CX31, Tokyo, Japan).

2.10. Statistical Analysis

The statistical analysis was accomplished by the means of Statistical Product and Service Solutions (SPSS), version 25, produced by IBM Software, Inc., Chicago, IL, USA. The data were normalized adopting a Kolmogorov–Smirnov test. Multiple comparisons of the data were performed using a one-way analysis of variance (ANOVA) and Tukey's post hoc test. All the data are shown as means ± standard deviation (SD). Differences were considered significant at $p \leq 0.05$.

3. Results

3.1. HPLC Study on Flavonoid Profiles

The HPLC analysis of the flavonoids in the LPE exhibited the emergence of five flavonoid compounds, including diosmin, biochanin A, hesperidin, quercetin, and hesperetin at different retention times with concentrations of 62.75, 62.35, 59.18, 28.32, and 9.31 mg/kg, respectively, as depicted in Table 2. Furthermore, Figure 1 delineates the chemical structures of the flavonoids identified in the LPE, in accordance with the reference flavonoids.

Table 2. Concentrations of flavonoid compounds in LPE obtained by HPLC analysis.

Flavonoid Compounds	R.T./min	Con. (mg/kg)
Diosmin	22.878	62.75
Biochanin A	17.477	62.35
Hesperidin	6.254	59.18
Quercetin	14.297	28.32
Hesperetin	16.147	9.31

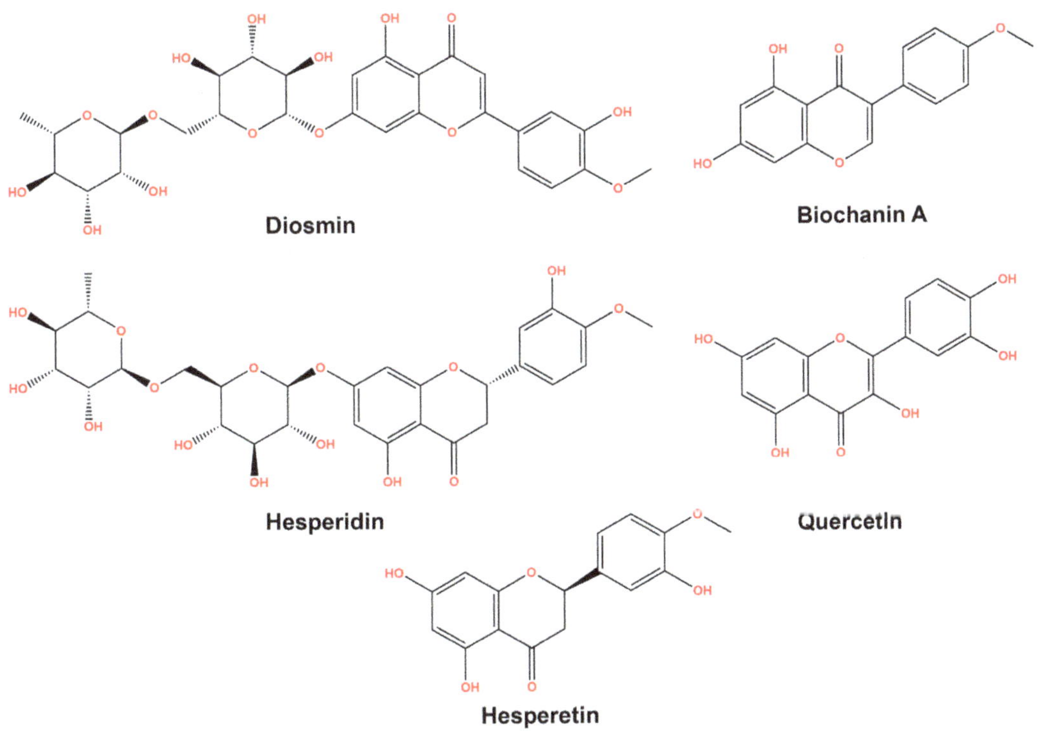

Figure 1. Chemical structures of flavonoid compounds identified in LPE.

3.2. Potential Interaction of the PI3K Protein with LPE-Polyflavonoids

The ligand–protein complex binding mode is best described by a docking simulation analysis with minimum energy generation [46,47]. Proper binding requires protein active site detection, which was depicted for obtaining the site-best score. Figure 2 reveals the results obtained from the docking of the five studied ligands aligned with the best binding energy pose and incorporated with the reference MYC bioactive ligand. The visualized map arranged Lig III (hesperidin) as the best ligand, with a binding energy of −11.1 kcal/mol, followed by Lig I (diosmin), with a value of −10.1 kcal/mol. The score parameter values for Lig IV, Lig V, and Lig II were −9.2, −9.0, and −7.1 kcal/mol, respectively. In more detail, the active sites of PI3K bound to each ligand were mostly different in the binding mode. Based on the sign of the MYC control, Figure 3 discriminates the reference point of the interaction between the control and the binding amino acids of the lipid kinase PI3K. The MYC active site includes some important favorable interactions, particularly conventional H-bonds with the key residues LYS^{833}, ASP^{841}, TYR^{867}, VAL^{882}, and ASP^{964}. Other non-covalent interactions were predicted, such as π-sulfur, π-amide-stacked, and π-alkyl types. Furthermore, the studied polyflavonoids were directed to the same potent pocket, demonstrating significant activity towards PI3K inhibition. It could be perceived from the docking process that Lig III, Lig IV, and Lig V were superimposed on MYS inhibitors, predicting similar behavior in the inhibition process. Additionally, it was observed that Lig III worked with most of the active amino acids in common with the bound-control crystallographic complex.

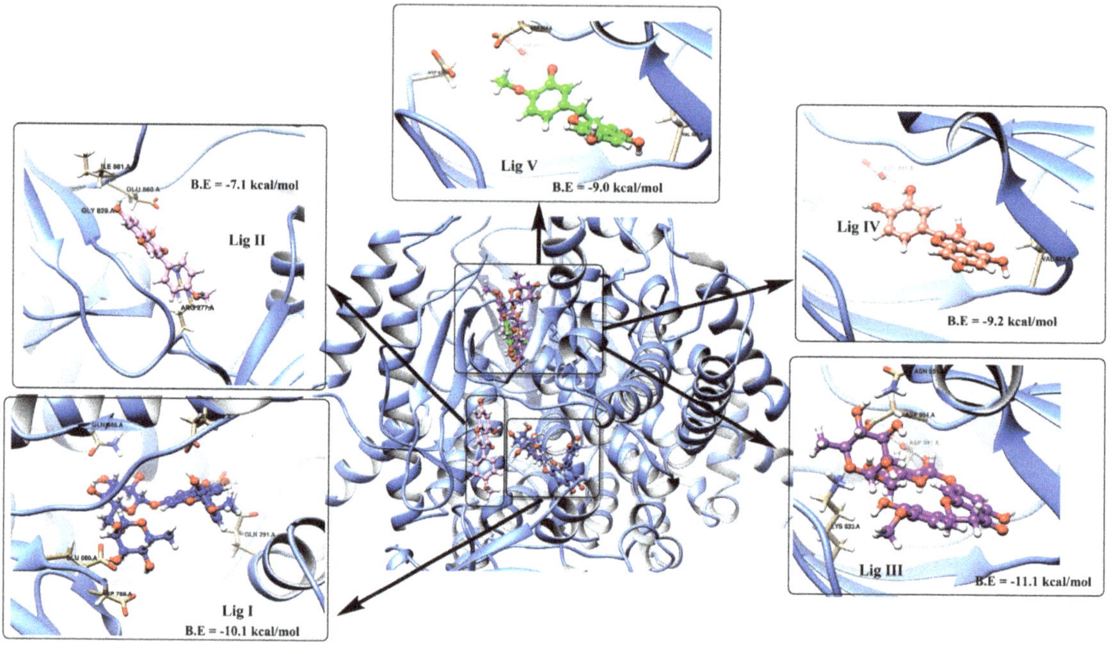

Figure 2. Docking score of the LPE-polyflavonoids ligands detected in the LPE with target protein 1E90, involving the best score value. B.E indicates the value of binding energy.

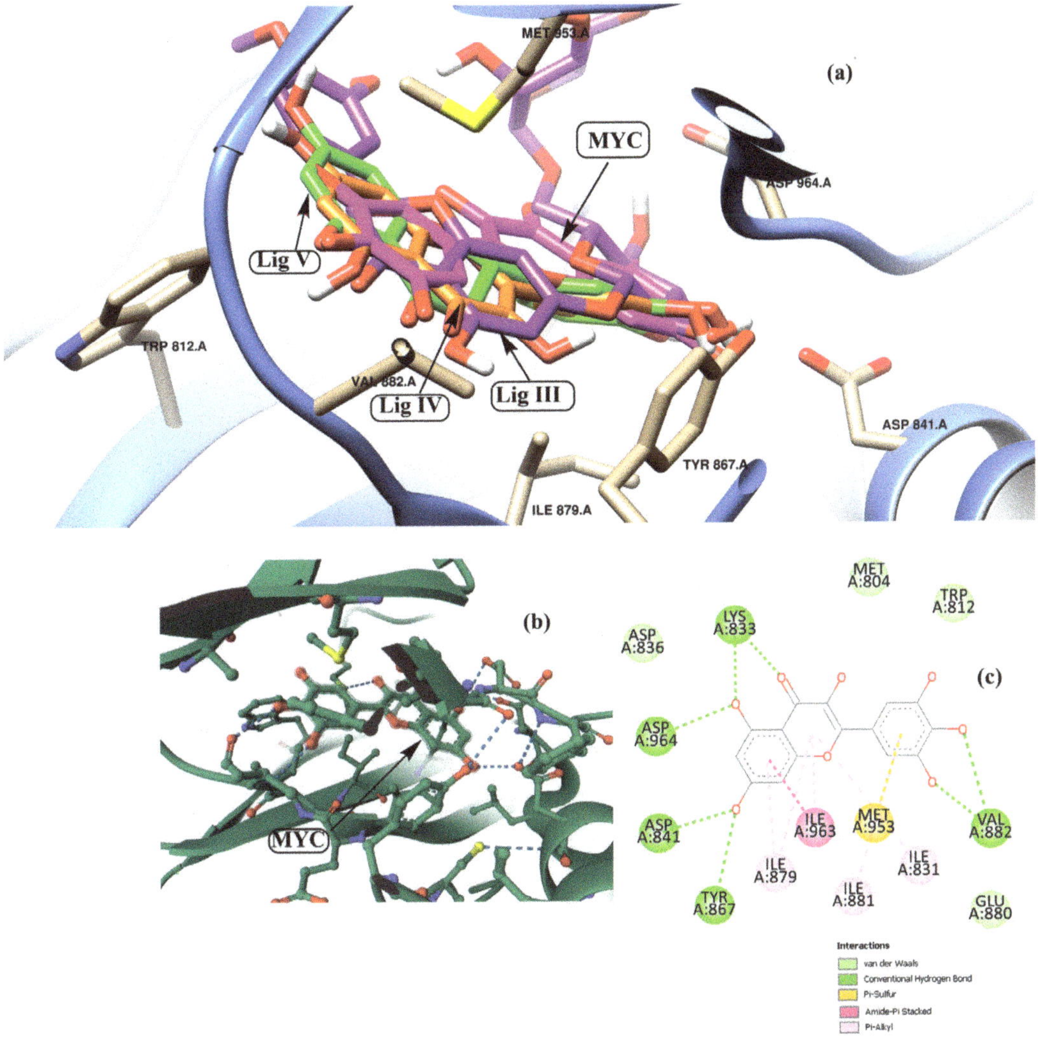

Figure 3. Graphical map describes Lig III, Lig IV, and Lig V superimposed with MYC control (pink color code) in the drug-potent active site (**a**), snapshot [https://www.rcsb.org/3d-view/1E90?preset=ligandInteraction&label_asym_id=B] (accessed on 1 September 2023) of the reported PI3K active sites complex with MYC (**b**), and (**c**) 2D colored interaction map related to MYC active pocket.

Several amino acids that interacted with Lig I and Lig III are delineated in Figures 4 and 5, demonstrating a variety of interaction types, particularly conventional and carbon–hydrogen bonds, where these mainly assisted with further complex stabilization. The majority preferred Lig I and Lig III as potent inhibitors, due to the large number of functional groups in the molecule compared to the other docked ligands. Figures 6–8 represent the docking analyses of Lig II, Lig IV, and Lig V, which revealed lower scores due to the presence of a smaller number of protein–site interactions.

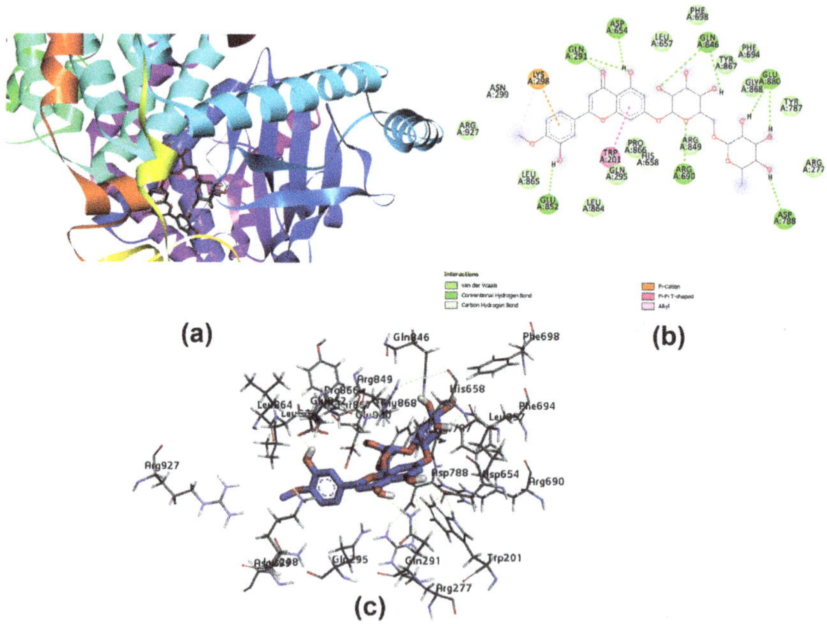

Figure 4. Docking analysis of Lig I with best score pattern: (**a**) 3D map of ligand–receptor docked complex with rainbow ribbon color code, (**b**) 2D interaction map with several 1E90-amino acids, and (**c**) 3D-generated ligand–protein complex.

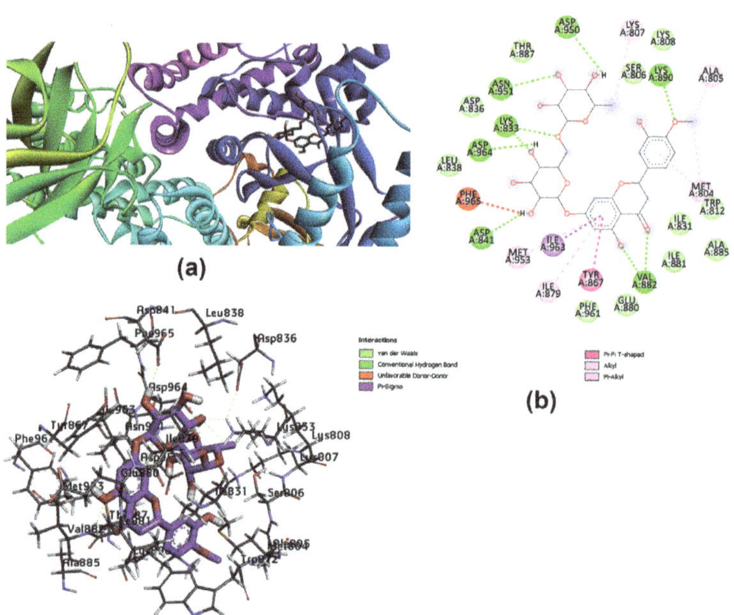

Figure 5. Docking analysis of Lig III with best score pattern: (**a**) 3D map of ligand–receptor docked complex with rainbow ribbon color code, (**b**) 2D interaction map with several 1E90-amino acids, and (**c**) 3D-generated ligand–protein complex.

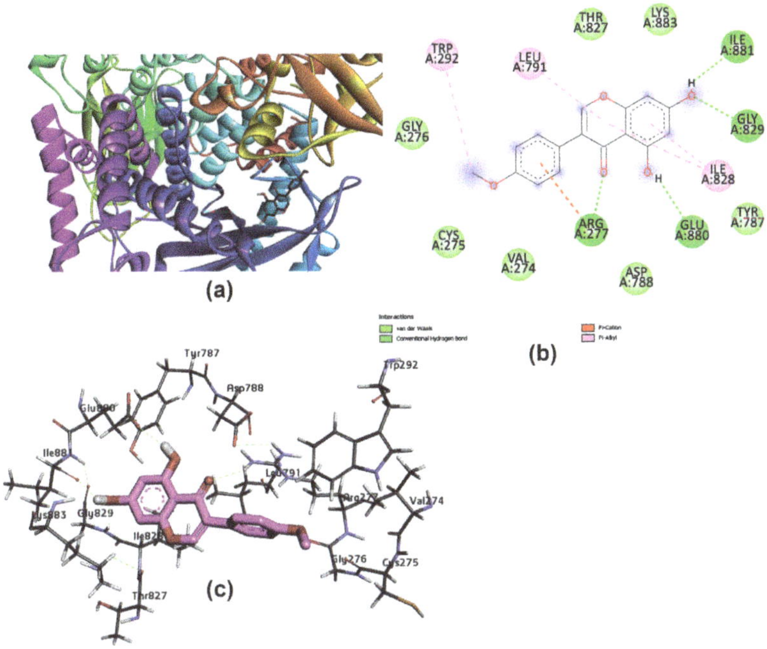

Figure 6. Docking analysis of Lig II with best score pattern: (**a**) 3D map of ligand–receptor docked complex with rainbow ribbon color code, (**b**) 2D interaction map with several 1E90-amino acids, and (**c**) 3D-generated ligand–protein complex.

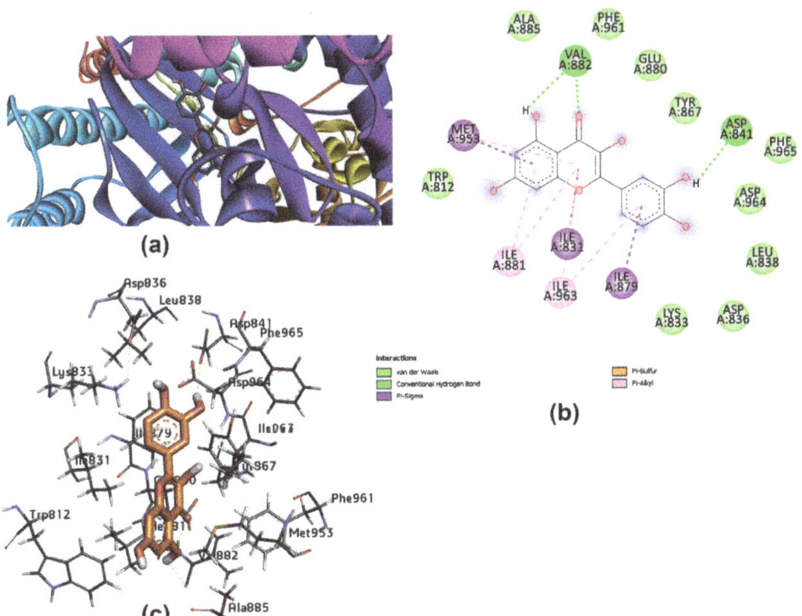

Figure 7. Docking analysis of Lig IV with best score pattern: (**a**) 3D map of ligand–receptor docked complex with rainbow ribbon color code, (**b**) 2D interaction map with several 1E90-amino acids, and (**c**) 3D-generated ligand–protein complex.

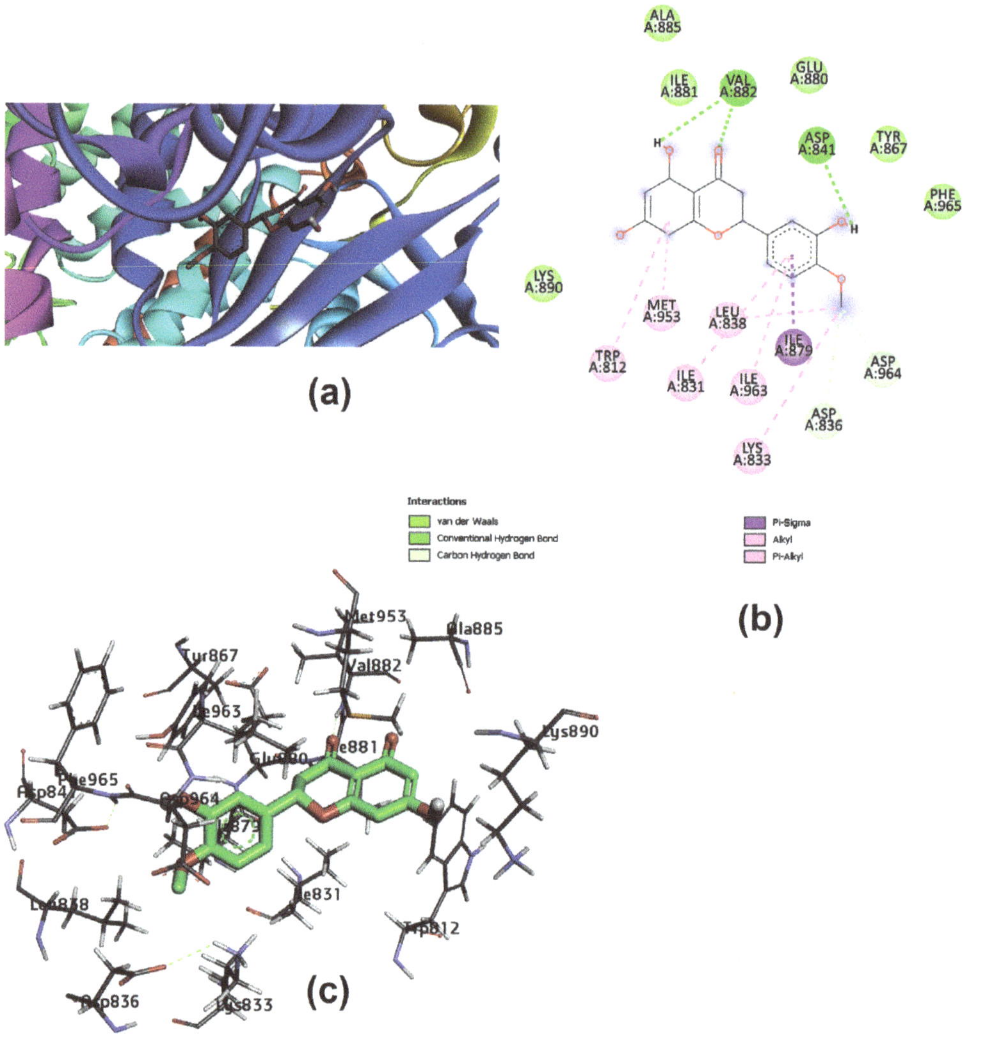

Figure 8. Docking analysis of Lig V with best score pattern: (**a**) 3D map of ligand–receptor docked complex with rainbow ribbon color code, (**b**) 2D interaction map with several 1E90-amino acids, and (**c**) 3D-generated ligand–protein complex.

3.3. Pharmacokinetic Parameters and Drug-Likeness Prediction

The Lipinski rules can assess the drug properties' validity in the studied compounds based on five determinants: a molecular weight of < 500 Da and a high lipophilic property, including a partition coefficient LogP value of < 5, H-bond donors < 5, H-bond acceptors < 10, and a molar refractivity value occurring between 40 and 130. Achieving more than two of these determinants can allow for the studied compound to have drug-like properties [48]. Other filters, such as Ghose, Veber, and Egan, for the studied polyflavonoids substantially help in identifying the bioavailability score [49], which was more significant in Lig II, Lig IV, and Lig V. Table 3 displays the pharmacokinetic parameters and drug-likeness properties elucidated for the studied compounds.

Table 3. Pharmacokinetic parameters and drug-likeness properties of the LPE-polyflavonoids.

	Lig I	Lig II	Lig III	Lig IV	Lig V	MYS Control
Physicochemical properties						
Molecular weight (Da)	608.54	284.26	610.56	302.24	302.28	318.24
Log $P_{o/w}$ (MLOGP)	−3.23	0.77	−3.04	−0.56	0.41	
Number of H-bond acceptors	15	5	15	7	6	8
Number of H-bond donors	8	2	8	5	3	6
Molar refractivity	143.82	78.46	141.41	78.04	78.06	80.06
Number of rotatable bonds	7	2	7	1	2	1
TPSA (Å^2)	238.2	79.9	234.29	131.36		151.59
Pharmacokinetics						
Gastrointestinal (GI) absorption	Low	High	Low	High	High	Low
Blood–brain barrier (BBB) permeant	No	No	No	No	No	No
P-glycoprotein substrate	Yes	No	Yes	No	Yes	No
Log Kp (skin permeation)	−9.91	−5.91	−10.12	−7.05	−6.3	−7.4
Drug likeness						
Log S (ESOL)	−3.51	−3.92	−4.33	−3.16	−3.62	−3.01
Water solubility class	Moderately soluble	Moderately soluble	Moderately soluble	Soluble	Soluble	Soluble
Lipinski rule	No	Yes	No	Yes	Yes	Yes
Ghose	No	Yes	No	Yes	Yes	Yes
Veber	No	Yes	No	Yes	Yes	No
Egan	No	Yes	No	Yes	Yes	No
PAINs	0 alerts	0 alerts	0 alerts	1 alert: catechol_A	0 alerts	1 alert: catechol_A
Bioavailability score	0.17	0.55	0.17	0.55	0.55	0.55
Metabolism						
CYP1A2 inhibitor	No	Yes	No	Yes	Yes	Yes
CYP2C19 inhibitor	No	No	No	No	No	No
CYP2C9 inhibitor	No	No	No	No	No	No
CYP2D6 inhibitor	No	Yes	No	Yes	No	No
CYP3A4 inhibitor	No	Yes	No	Yes	Yes	Yes

Pharmacokinetic properties also deal with the availability of a predicted drug for partitioning in gastrointestinal cells with high proportions, such as Lig II, Lig IV, and Lig V, and low proportions such Lig I and Lig III, which resemble the control effect. The blood–brain barrier (BBB) parameter is an indicator of the proportional distribution of the studied molecules, where the current flavones are undistributed in human brain blood. Metabolism factors can estimate a compound's tendency to inhibit five isoenzymes (CYP1A2, CYP2C19, CYP2C9, CYP2D6, and CYP3A4) related to cytochrome P450 enzymes. This property is a crucial contributor to toxicity and other medication side effects [50]. In this study, it was found that Lig III exhibited no function as an inhibitor for the five isoenzymes compared to the other candidates.

3.4. Impact of LPE-Polyflavonoids on Serum Glucose, Insulin, Glycogen, and Liver Enzymes

The level of glucose and activities of AST, ALT, and ALP were substantially augmented in the Fructose-STZ rats in comparison to the control rats, as portrayed in Figure 9A,D–F. Moreover, it was apparent from the data in Figure 9B,C that the levels of insulin and glycogen were noticeably diminished in the Fructose-STZ group in comparison to the control group. In contrast, the rats administered with hesperidin and diosmin demonstrated remarkable enhancements in terms of glucose, insulin, glycogen, and the performance of their liver enzymes. Interestingly, the rats administered with the LPE manifested significant enhancements in relation to these parameters compared to all the other treated rats, implying a synergistic influence of flavonoids in restoring cellular functions.

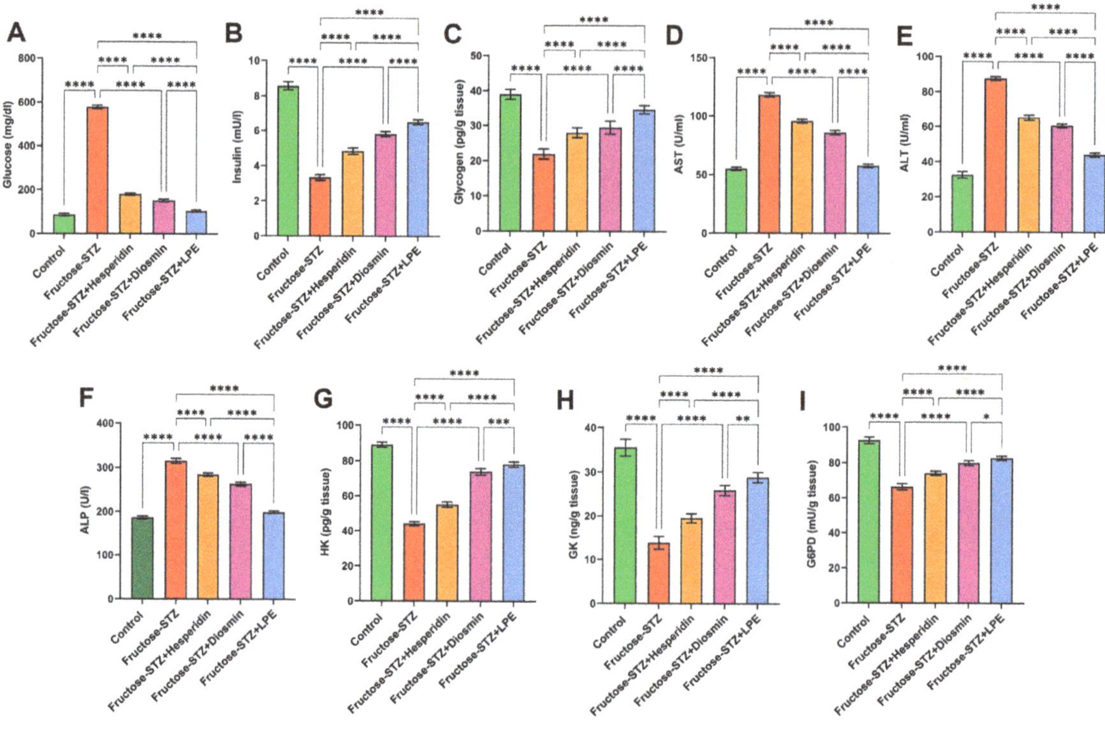

Figure 9. Evaluation of (**A**) glucose, (**B**) insulin, and (**C**) glycogen in the serum of different animal groups. Analysis of liver function enzymes, including (**D**) AST, (**E**) ALT, and (**F**) ALP in the serum of experimental animal groups. Estimation of hepatic carbohydrate-metabolizing enzymes, including (**G**) HK, (**H**) GK, and (**I**) G6PD in animal groups. The results are depicted as mean ± SD when n = 6, which were analyzed following one-way ANOVA associated with multiple comparisons between different groups adopting Tukey's test (**** $p < 0.0001$, *** $p < 0.001$, ** $p < 0.01$, and * $p < 0.05$). Abbreviations: AST: Aspartate aminotransferase; ALT: alanine transaminase; ALP: alkaline phosphatase; HK: hepatic hexokinase; GK: glucokinase; and G6Pd: glucose-6-phosphate dehydrogenase.

3.5. Influence of LPE-Polyflavonoids on Hepatic Carbohydrate-Metabolizing Enzyme Activities

It can be observed from Figure 9G–I that the activities of GK, HK, and G6Pd in the Fructose-STZ group markedly dwindled ($p \leq 0.05$) in comparison to those of the control animals. On the contrary, the treatment with hesperidin and diosmin resulted in remarkable growth in the levels of GK, HK, and G6Pd. Moreover, the rats exposed to the LPE showed significant increases ($p \leq 0.05$) in these enzymes, more so than those treated with either hesperidin or diosmin, which supports the previous findings.

3.6. Effect of LPE-Polyflavonoids on Total Protein and Albumin Levels

The diabetic rats treated with Fructose-STZ showed noticeable diminutions in their total protein and albumin in relation to the control rats, as displayed in Figure 10A,B. In contrast, the rats treated only with hesperidin or diosmin demonstrated marked increases in their concentrations of total protein and albumin. Importantly, the treatment with the LPE caused further enhancements in both factors, with a significant difference ($p \leq 0.05$) in comparison to all the diabetic rats. Altogether, these results corroborate that the treatment with the LPE, including the polyflavonoids, had a vital influence on the treatment of diabetes.

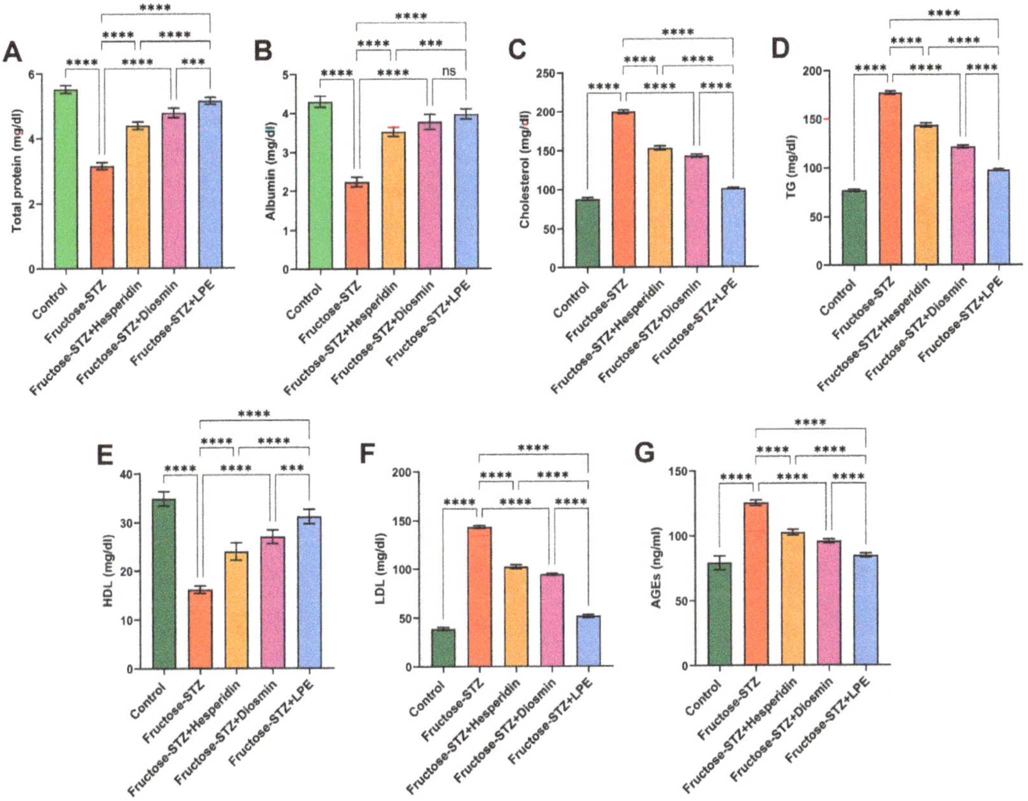

Figure 10. Assessment of (**A**) total protein and (**B**) albumin in the serum of different animal groups. Changes in the serum lipid profiles of experimental groups involving (**C**) Total cholesterol, (**D**) TG, (**E**) LDL-C, and (**F**) LDL of animal groups. Estimation of (**G**) AGEs levels in animal groups. The results are exhibited as mean ± SD when n = 6, which were analyzed following one-way ANOVA associated with multiple comparisons between different groups adopting Tukey's test (**** $p < 0.0001$, *** $p < 0.001$, and ns alludes to non-significant differences). Abbreviations: TG: triglyceride (TG); HDL-cholesterol; LDL-cholesterol; and AGEs: advanced glycation end-products.

3.7. *Influence of LPE-Polyflavonoids on Serum Lipid Profile and AGEs*

Figure 10C–F exhibit that there were obvious trends of remarkable increases in the concentrations of serum total cholesterol, TG, and LDL-C in the Fructose-STZ rats, associated with a substantial decrease in comparison to the control rats. With regard to AGEs, as a fundamental biomarker for proteins and lipids glycated as a consequence of reactions with sugars, the diabetic rats in the Fructose-STZ group evinced a considerable escalation in their AGEs compared to the untreated rats, as presented in Figure 10G. On the contrary, the administration of hesperidin or diosmin to the diabetic rats exposed ameliorative influences in relation to the lipid profiles, associated with a noticeable reduction in AGEs. Most notably, the LPE-treated rats demonstrated higher enhancements in their lipid parameters and AGEs compared to those treated only with one of the flavonoid compounds, reflecting the key role of polyflavonoids in the LPE for managing diabetes and its further deleterious effects.

3.8. Effect of LPE-Polyflavonoids on Oxidative Stress Biomarkers in Pancreatic and Liver Tissues

Concerning the oxidative stress biomarkers in the Fructose-STZ group, the results exhibited noticeable disorders in the TBARS, GSH, and CAT levels. Precisely, the level of lipid peroxidation was markedly amplified through a determination of the TBARS content in both the pancreatic and liver tissues, linked with significant drops in the GSH and CAT levels in relation to the control rats, as shown in Figure 11A–F. Conversely, the application of hesperidin or diosmin counteracted these detrimental impacts, indicating their efficiency in treating diabetes. Remarkably, the LPE-polyflavonoids alleviated the oxidative stress levels in the liver and pancreas, with significant differences in relation to one of the flavonoids, suggesting a synergistic influence of the five flavonoids, which impart the LPE with intrinsic antioxidant attributes.

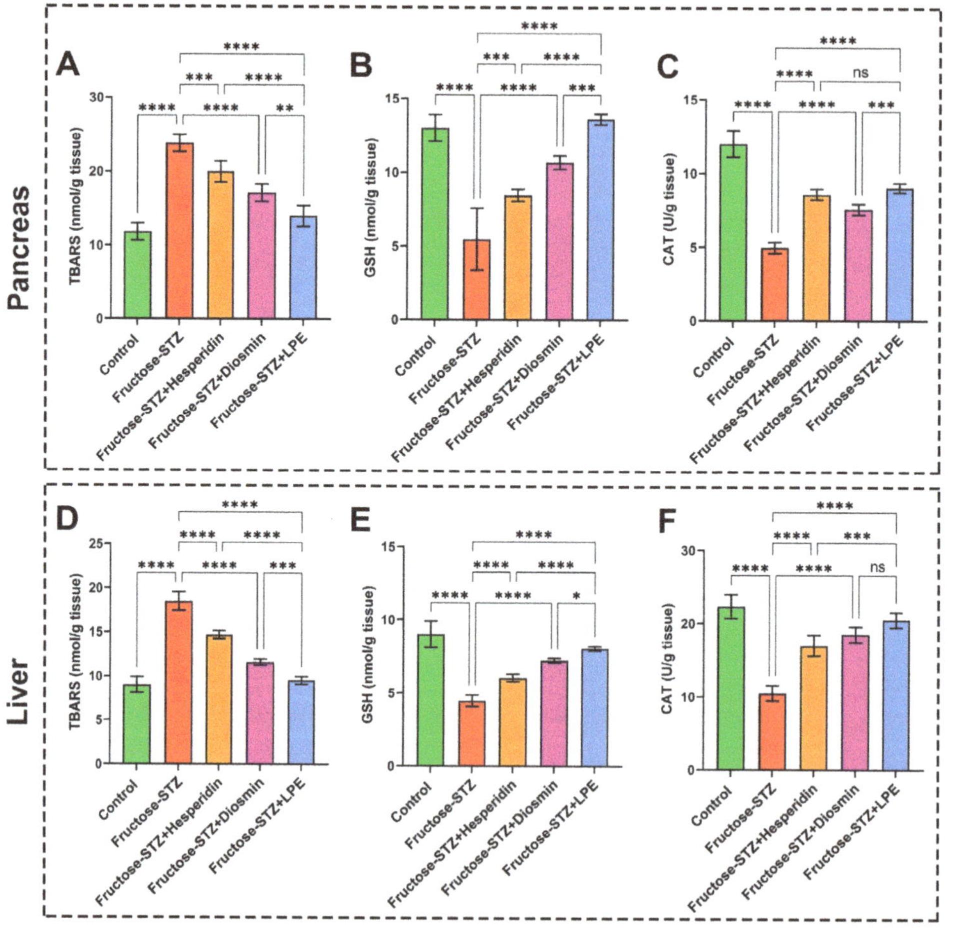

Figure 11. Evaluation of oxidative stress biomarkers and catalase in pancreatic and hepatic tissues of different animal groups. (**A**,**D**) TBARS levels in pancreatic and hepatic tissues, respectively, to appraise the lipid peroxidation. (**B**,**E**) GSH concentrations in pancreatic and hepatic tissues, respectively. (**C**,**F**) CAT activity in pancreatic and hepatic tissues, respectively. The results are shown as mean ± SD when n = 6, which were analyzed following one-way ANOVA associated with multiple comparisons between different groups adopting Tukey's test (**** $p < 0.0001$, *** $p < 0.001$, ** $p < 0.01$, * $p < 0.05$, and ns refers to non-significant differences). Abbreviations: TBARS: thiobarbituric acid reactive substance; GSH: glutathione; and CAT: catalase activity.

3.9. Impact of LPE-Polyflavonoids on Liver and Pancreas Inflammatory Markers, NF-κB, and AFP

The results of the inflammatory markers demonstrated significant augmentations in the TNF-α, IL-1β, and IL-6 in the pancreatic tissues obtained from the Fructose-STZ-treated group in relation to the control animals, as delineated in Figure 12A–C. Moreover, Figure 12D–G illustrate that the concentrations of TNF-α, IL-6, NF-κB, and AFP were markedly heightened in the hepatic tissues of the diabetic animals. In contrast to these results, the rats treated with hesperidin or diosmin showed significant mitigations of inflammatory marker production in both their pancreatic and liver tissues, in addition to a discernible reduction in the NF-κB and AFP levels in their hepatic tissues. Furthermore, the rats administered with the LPE revealed a noticeable modulation of their inflammatory marker levels alongside NF-κB and AFP, better than that in all the diabetic rat groups. These findings substantiate the crucial contribution of the LPE to controlling diabetes and verify the previous results.

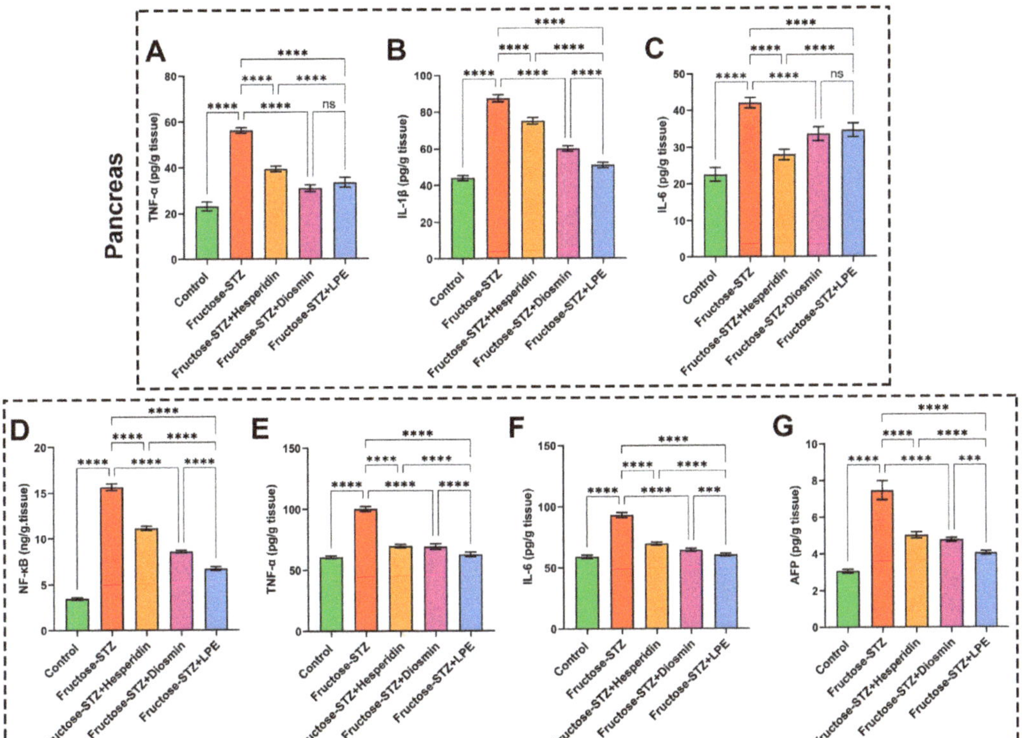

Figure 12. Estimation of inflammatory biomarkers in pancreatic and liver tissues in addition to hepatic Alpha fetoprotein (AFP) from different rat groups. (**A**,**E**) TNF-α concentrations in pancreatic and hepatic tissues, respectively. (**B**) IL-1β in pancreatic homogenates. (**C**,**F**) IL-6 levels in pancreatic and hepatic tissues, respectively. (**D**) NF-κB and (**G**) AFP in hepatic tissues of animal groups. The findings are exhibited as mean ± SD when n = 6, which were analyzed following one-way ANOVA associated with multiple comparisons between different groups adopting Tukey's test (**** $p < 0.0001$, *** $p < 0.001$, and ns points to a non-significant difference). Abbreviations: TNF- α: tumor necrosis factor alpha; IL-1β: interleukin-1β; IL-6: interleukin-6; NF-κB: nuclear factor kappa B; and AFP: Alpha-fetoprotein.

3.10. Effect of LPE-Polyflavonoids on GLUT4 and GLUT2 in Pancreas and Liver Tissues

It was evident from the GLUT4 and GLUT2 results that the levels of both parameters in the pancreases and livers obtained from the Fructose-STZ-treated rats were remarkably lower than those observed in the control group, as presented in Figure 13A,B,F,G. Compared to the Fructose-STZ group, the treatments with hesperidin, diosmin, and the LPE gave rise to notable increases ($p \leq 0.05$) in GLUT4 and GLUT2, with the highest levels seen in the LPE group.

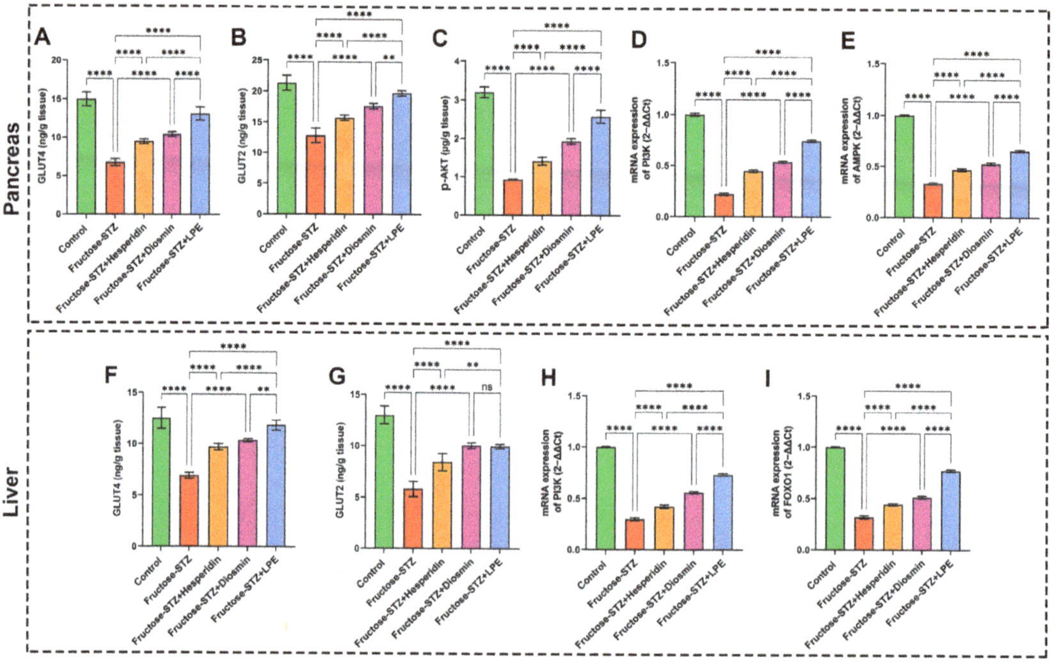

Figure 13. Determination of (A) and (F) GLUT4 levels in pancreatic and hepatic tissues, respectively. (**B**,**G**) GLUT2 concentrations in pancreatic and hepatic tissues, respectively. (**C**) p-AKT concentration in pancreatic tissues of different rat groups. (**D**,**H**) mRNA expression of PI3K in pancreatic and hepatic tissues, respectively. (**E**) mRNA expression of AMPK in the pancreas and (**I**) mRNA expression of FOXO1 in liver. The results are shown as mean ± SD when n = 6, which were analyzed following one-way ANOVA associated with multiple comparisons between different groups adopting Tukey's test (**** $p < 0.0001$, ** $p < 0.001$, and ns indicates a non-significant difference). Abbreviations: GLUT4: Glucose transporter 4; GLUT2: Glucose transporter 2; p-AKT: phospho-Akt or protein kinase B; PI3K: phosphatidylinositol 3-kinase; AMPK: adenosine monophosphate-activated protein kinase; and FOXO1: forkhead box-O.

3.11. Effect of LPE-Polyflavonoids on p-AKT in the Pancreas and Gene Expressions of PI3K, AMPK, and FOXO1 in Pancreatic and Hepatic Tissues

In the Fructose-STZ-treated group, it was distinct that diabetes instigated a significant diminution ($p \leq 0.05$) in the pancreatic p-AKT, alongside notable downregulations of the pancreatic PI3K and AMPK signals compared to the control rats, as illustrated in Figure 13C–E. Additionally, it can be perceived from the results in Figure 13H,I that the gene expressions of hepatic PI3K and FOXO1 were markedly reduced in the untreated diabetic rats. Nevertheless, the administration of hesperidin, diosmin, and the LPE to the diabetic rats substantially promoted the level of the p-AKT signal and upregulated the mRNA expressions of PI3K and AMPK in the pancreas tissues and PI3K and FOXO1 in the liver tissues. Moreover, it was discernible that the treatment with the LPE-polyflavonoids markedly promoted the expressions of various diabetic modulators, higher than those treated with one flavonoid agent. Collectively, these findings emphasize the pivotal functions of flavonoids, including LPE-polyflavonoids, as countermeasures to oxidative stress and other complications derived from diabetes by orchestrating the expressions and functions of PI3K, p-AKT, and FOXO1.

3.12. Histopathological Analysis of Pancreatic and Liver Tissues

To provide further support for the previous findings, histological sections of the pancreases and livers of the rats in all the experimental groups were investigated, as presented in Figures 14 and 15. It can be perceived from the H&E sections of the pancreatic tissues dissected from the control rats that the pancreas exposed a regular cellular architecture associated with typical large Langerhans islets and normal pyramidal-shaped cells "pancreatic acini", as depicted in Figure 14A. In contrast, Figure 14B illustrates that the untreated diabetic rats demonstrated aberrations in their cellular structures of Langerhans islets, including a retraction and distortion of the islets correlated with a lack of B cells, cytoplasmic vacuolation, and the pyknotic nuclei of several islet cells. Prominently, the diabetic rats treated with hesperidin, diosmin, and the LPE restored the major pancreatic architecture close to that of the control group, as portrayed in Figure 14C–E.

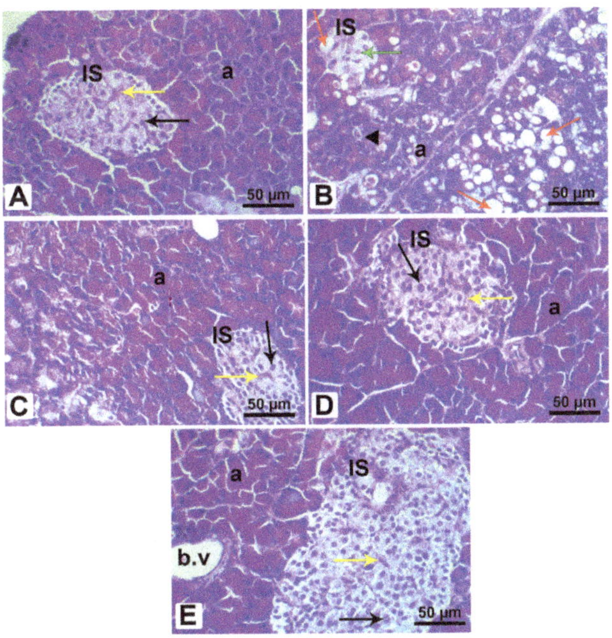

Figure 14. Photomicrographs of pancreatic sections from male rats stained with H&E. (**A**) Section of the control pancreas, exhibiting normal pancreatic structure, typical pancreatic acini with pyramidal-shaped cells and rounded nuclei (a), regular pale-stained islet of Langerhans (IS) with typical ß cells (black arrow) and a blood vessel (yellow arrow). (**B**) Pancreas from Fructose-STZ-treated rats, revealing a shrunken, distorted islet of Langerhans with loss of ß cells (green arrow), emergence of vacuolated cytoplasm (red arrow), acinar cells with pyknotic nuclei (black arrowhead), and incidence of fat infiltration (red arrow). (**C**) Pancreatic sections of diabetic rats treated only with hesperidin, (**D**) pancreatic sections of diabetic rats treated only with diosmin, and (**E**) pancreatic sections of diabetic rats treated with LPE-polyflavonoids, demonstrating restoration of the regular pancreatic architecture, with particular improvement in rats treated with LPE-polyflavonoids, including typical pancreatic acini (a), a regular pale-stained islet of Langerhans (IS) with regular ß cells (black arrow) and blood vessels (yellow arrow).

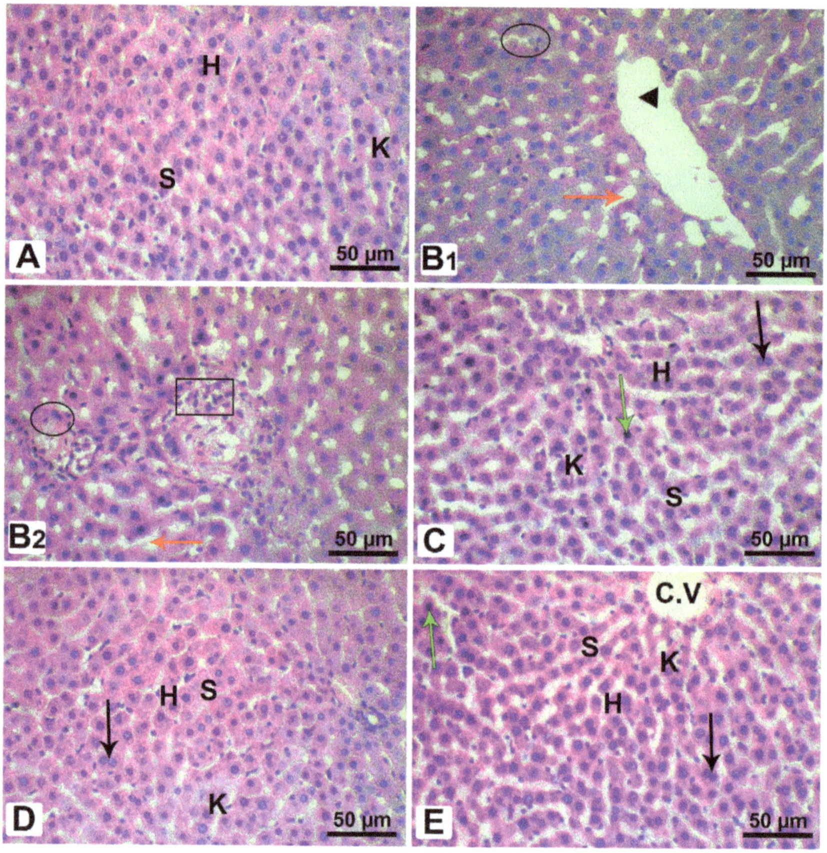

Figure 15. Photomicrographs of hepatic sections from male rats stained with H&E. (**A**) Section of the control liver from control rats, revealing typical histological structure, including the regular pattern of hepatocytes (H), typical blood sinusoids (S), and Kupffer cells (K). (**B1**,**B2**) Liver sections from Fructose-STZ-treated rats, showing dilation of the central vein (head arrow) and dilation of the blood sinusoids (red arrow), pyknotic nuclei (black circle), and accumulation of inflammatory cells among hepatocytes (black square). (**C**) Liver sections of diabetic rats treated only with hesperidin, (**D**) liver sections of diabetic rats administered only with diosmin, and (**E**) liver sections of diabetic rats treated with LPE-polyflavonoids, demonstrating ameliorations in the histological structures associated with the appearance of the central vein (C.V.), regular organization of hepatocytes (H), typical blood sinusoids (S), and Kupffer cells (K), which could be perceived in the control rats, except for the emergence of few pyknotic nuclei (green arrow) and binucleated cells (black arrow). Compared to all groups of diabetic rats, it could be discerned that there was substantial restoration of liver architecture in rats treated with LPE-polyflavonoids.

Considering the histopathological inspection of the livers, it is evident from Figure 15A that the control rats manifested a standard hepatic architecture, involving sinusoids and the central vein. On the other hand, the hepatic tissues from the Fructose-STZ rats revealed a dilation of the central vein and blood sinuses and the necrosis of hepatocytes with nuclear pyknosis, in addition to extensive inflammatory cells among the hepatocytes, as shown in Figure 15B1,B2. In contrast to these findings, the rats administered with hesperidin, diosmin, and the LPE exhibited a reinstatement of histological features, except for the emergence of a few pyknotic nuclei and binucleated cells, as represented in Figure 15C–E.

4. Discussion

It is imperative to implement comprehensive strategies that focus on the impediments, early detection, and effective management of T2DM and its deleterious consequences. Among the various therapies for hampering diabetes and its complications, phenolic and flavonoid compounds have been receiving profuse attention due to their antioxidant and anti-diabetic properties [18,51,52]. In our study, we investigated the efficiency of LPE-polyflavonoids in alleviating DM and enhancing the pancreatic and liver functions by regulating the expressions of PI3K, p-AKT, and FOXO1 as key modulators for controlling DM. The results manifested a prevalence of diosmin, biochanin A, hesperidin, quercetin, and hesperetin in the LPE in different amounts.

Considering the docking simulation of the designed polyflavonoids with the lipid-modifying enzyme PI3K, it was illustrated that most MYS-binding residues were active with the studied ligands, especially for Lig III. The genetic algorithm docking proceeded via a binding score generation for the designed polyflavonoids to conclude the binding energy values. A lower binding energy score initiates a higher affinity for inhibition. Accordingly, Lig III showed a lower energy value (−11.1 kcal/mol). Furthermore, the predicted docked ligand-active site for Lig III, Lig IV, and Lig V was congruent with the key site of the reference MYC inhibitor. To analyze the 2D map interactions for all the studied ligands and compare the results with MYC, it was projected that Lig III and Lig IV interacted with most key residues of the control. Such residues, which interact with some non-covalent interactions, were LYS^{833}, ASP^{964}, ASP^{841}, VAL^{882}, MET^{953}, and ILE^{879}. Assuming the importance of the pharmacokinetic and drug-likeness properties [53], ADME parameters were provided to find that the polyflavonoids had drug-like properties in relation to the MYC inhibitors. Despite the bioavailability prediction, with a score of 0.55 for Lig III, the Lipinski rule detected three unachievable factors of the total five indices. Molecular weight (Mwt = 610.56 Da) is the main factor involved in evaluating drug-like properties. Furthermore, the pan-assay interferences (PAINS) reported no alert for most of the studied ligands. This factor was detected for Lig IV and the control in the same examination [54]. Previous investigations have reported that the binding of phenol and flavonoid compounds to proteins adapts their conformational structure, engendering alterations in their intrinsic protein properties and functions, such as solubility and isoelectric point [19]. Altogether, the molecular docking studies highlighted the potency of all five flavonoid agents in binding to the p-AKT protein, which mimics the insulin mechanism to inhibit the augmentation of glucose in the blood.

To emphasize the prospective application and evaluation of the extent to which the LPE-polyflavonoids thwarted DM, in vivo investigations, alongside extensive evaluations of various biochemical factors, were conducted. The in vivo results indicated a significant elevation in the level of glucose, with a reduction in the insulin level in the Fructose-STZ group, which could have been related to increased oxidative stress as a result of the progression of diabetes, followed by an oxidative degradation of glycated proteins [55]. Furthermore, it was believed that β-cells survived the STZ toxicity and became dysfunctional, resulting in high blood glucose levels and a subsequent toxicity of glucose [56]. Moreover, poor GLUT2-mediated glucose sensing is a key factor in the emergence of glucose intolerance and a manifestation of β-cell malfunction [57].

Our findings evinced that the treatment of the diabetic rats with the LPE-polyflavonoids led to improvements in their glucose and insulin levels, greater than those in the rats administered with a single agent of hesperidin and diosmin. It was identified that diosmin and hesperidin possess anti-diabetic properties, which are exerted primarily through glucose metabolism regulation [58,59]. Additionally, diosmin nanoparticle therapy modulated the glucose and insulin levels, which could have been linked to its anti-hyperglycemic activity by boosting insulin synthesis from the existing pancreatic β-cells [60].

Recently, it was reported that hyperglycemia and diabetes could be instigated by deficiencies in glycogen synthesis [61]. Correspondingly, our results exhibited a significant diminution in relation to the glycogen content in the diabetic rats, which matched those observed in earlier studies by Oyedemi et al. [62]. These could be explained by a reduction in the ability of the liver to induce glycogen synthase, which augments postprandial glycogen storage, causing a drop in glycogen levels. Sundaram et al. found that insulin promoted hepatic intracellular glycogen deposition by activating glycogen synthase and suppressing the glycogen phosphorylase performances in diabetic rats [63]. Correspondingly, in the current study, the rats doped with the LPE-polyflavonoids manifested a remarkable promotion in their glycogen content compared to the rats treated only with either the hesperidin or diosmin, suggesting a synergistic influence of polyflavonoids on stimulating glycogen synthase. Likewise, previous studies have demonstrated that treatment with only hesperidin modulates insulin secretion, which, in turn, enhances glycogen production [64].

To provide further insights into the activities of liver enzymes, the present results exposed elevated levels of the AST, ALT, and ALP enzymes in the Fructose-STZ-treated rats. These results may be ascribed to hepatic IR, OS, and inflammation associated with excess triglyceride deposition in liver cells, which incite hepatic cell damage and a successive discharge of liver enzymes into the bloodstream [11]. Conversely, the diabetic rats treated with hesperidin, diosmin, and the LPE-polyflavonoids revealed reductions in their liver function enzymes, particularly in the animals treated with the LPE, signifying the hindrance of hepatic impairments, as previously demonstrated [65]. These results agree with the findings from previous reports, which manifested the potency of hesperidin and diosmin in maintaining the integrity of plasma membranes, boosting the restoration of these hepatic enzyme levels [66].

To expand on the performance of other hepatic enzymes, the diabetic group in the current research illustrated noticeable reductions in their activities of HK, GK, and G6Pd. It is well known that HK is a rate-limiting enzyme in the glycolytic process and has a pivotal function in the first phosphorylation of glucose during glycolysis. Moreover, it has a paramount role in the glucose homeostasis in the liver and the progression of DM [67]. Prior investigations have exposed a decrease in the HK enzyme activities in several target tissues during the impaired intake of glucose in diabetic rats [68]. This clearly proposes that the increased blood glucose level could have been correlated with a decrease in the activities of these hepatic enzymes in the diabetic rats [69]. On the other hand, the application of the LPE-polyflavonoids significantly improved HK, GK, and G6Pd compared to those in the hesperidin- and diosmin-treated rats. It has been postulated that the modulating influence of hesperidin in hepatic and extra-hepatic tissues is attributed to the promotion of insulin secretion by existing β-cells [64]. Furthermore, diosmin improved the liver G6PD activity, along with an increase in insulin synthesis. Therefore, the elevation in nicotinamide adenine diphosphate synthesis and reduction in OS could be reported as being due to an enhancement in the glucose influx into the pentose monophosphate shunt [70].

It is worth mentioning that a reduction in albumin levels is not only linked to amplified glucose levels, but also increases vulnerability to DM [71]. Our results thus exhibited a reduction in the albumin concentrations in the diabetic rats. In contrast, the treatment with hesperidin, diosmin, and the LPE-polyflavonoids increased these albumin levels, with the highest level being observed in the case of the polyflavonoids. Hesperidin and diosmin were found to adjust the level of albumin, as a result of their protection against hepatic injury [72,73].

Importantly, the elevations in the cholesterol, TG, and LDL-C levels with a diminution in the HDL-C levels in the diabetic rats in this study indicated insulin deficiency, an excessive mobilization of chylomicrons, and VLDL, leading to hypertriglyceridemia, as has been previously elucidated [74]. On the contrary, the LPE-polyflavonoids treatment ameliorated the entire lipid profile to a greater extent than that of hesperidin and diosmin, stemming from the positive influence of the flavonoid compounds being used together. Prior investigations have accentuated the capacity of hesperidin to preclude cholesterol production and absorption, in addition to the efficacy of diosmin in reducing cholesterol and TG. This is likely related to the effect of flavonoids on inhibiting the activities of both hepatic HMG-CoA reductase and acyl CoA cholesterol acyl-transferase [75].

By studying the oxidative parameters in both the pancreatic and liver tissues, our findings demonstrated disorders in the diabetic rats, identified by elevations in the AGEs and TBARS associated with increases in GSH and CAT, implying an overabundance of free radicals in the tissues, which further provoked OS. Likewise, Yazdi et al. reported that hyperglycemia incited OS and subsequent disruptions in the carbohydrate, protein, and lipid metabolism [76]. Furthermore, Vlassara and Uribarri showed that OS is a dominant cause of impairment in pancreatic β-cells in T2DM, accompanied by a generation of AGEs [77]. Conversely, the treatment with hesperidin, diosmin, and the LPE-polyflavonoids significantly lessened the levels of AGEs and TBARS and evidently improved the GSH and CAT levels in the pancreas and liver. These performances stem from the intrinsic antioxidant properties of flavonoids, which function in tissues either by direct radical scavenging or by boosting the cellular antioxidant defense mechanisms [78,79].

Concerning the inflammatory biomarkers, the marked augmentations of the pancreatic TNF-α, IL-1β, and IL-6 and hepatic NF-κB, TNF-α, IL-1β, IL-6, and AFP in the Fructose-STZ-treated rats could be imputed to excessive glucose concentrations, which triggered oxidative-mechanism-mediated NF-κB, prompting pro-inflammatory cytokine expansion [80]. These cytokines not only instigate NADPH oxidase-generated ROS, but also impair the integrity of pancreatic β-cells. Moreover, an increase in TNF-α has been associated with the activation of NF-κB and NO production, leading to the stimulation of the mitochondrial apoptosis pathways and β-cell dysfunction via the hindering of insulin synthesis [81]. These findings match those observed in earlier investigations, which manifested that the AFP expression was substantially amplified in T2DM throughout the early period of diabetes [82].

In contrast to these findings, the supplementation with hesperidin, diosmin, and the LPE-polyflavonoids outstandingly regulated the production of inflammatory biomarkers and AFP compared to that in the Fructose-STZ-diabetic group, with a particular improvement in the presence of polyflavonoids. Prior investigations have shown that treatment with only diosmin revealed an anti-inflammatory effect by attenuating the TNF-α, NF-κB, and IL-1β production [73].

Furthermore, GLUT2 has a crucial role in glucose transportation inside and outside of the hepatocytes [83]. Our results elucidated that the pancreatic and hepatic levels of GLUT4 and GLUT2 were significantly lowered in the diabetic rats. Similarly, Xu et al. [83] demonstrated that STZ-diabetic animals had diminished levels of hepatic GLUT2 and GLUT4. Thus, an inhibition of the GLUT gene levels is one of the leading causes for hyperglycemia in the diabetic state, as a result of a reduction in glucose uptake [84]. On the contrary, the hesperidin-, diosmin-, and LPE-polyflavonoids-treated groups manifested high levels of GLUT2 and GLUT4 in comparison to the diabetic group. These observations are in accordance with Ali et al. [85], who found that the administration of flavonoids to diabetic rats, including hesperidin and quercetin, improved their insulin sensitivity, modulating their GLUT2 levels, while Hsu et al. [86] reported that diosmin treatment reversed a reduction in GLUT4 expression. Accordingly, they concluded that diosmin serves as a positive modulator of glucose metabolism and an inhibitor of hepatic gluconeogenesis in the insulin-deficient case [86].

In our current results, we found that the gene expressions of p-AKT, PI3K, AMPK, and FOXO1 in the pancreases and PI3K and FOXO1 in the livers of the Fructose-STZ group were significantly downregulated. However, the supplementation of hesperidin, diosmin, and the LPE-polyflavonoids remarkably orchestrated the PI3K, p-AKT, FOXO1, and AMPK gene expressions in comparison to those in the diabetic rats. Similar observations were previously demonstrated by Qiu et al. [87], who reported that diabetic rats exhibited lower gene expressions of PI3K, p-AKT, and FOXO1 in their pancreatic and liver tissues. It is well known that the mechanism of insulin fundamentally relies on the sequential activations and correlated interactions of the PI3K, p-AKT, and FOXO1 signals to lessen the glucose in blood [88]. Furthermore, the PI3K/AKT signaling pathways are specifically implicated in hepatic glucose storage and uptake. Moreover, FOXO1 is a key transcription factor downstream of these insulin signaling pathways. Besides the promotion of PI3K by insulin, nuclear exclusion occurs due to the subsequent phosphorylation of AKT and FOXO1. Kinyua et al. [89] indicated that FOXO1 translocation from the nucleus to the cytoplasm inhibits the expressions of FOXO1 gluconeogenic targets, phosphoenolpyruvate carboxylase (PEPCK), and glucose-6-phosphatase (G-6pase). Considering the pivotal role of AMPK, it has previously been postulated that AMPK in T2DM coordinates the anabolic and catabolic processes, in addition to regulating metabolic homeostasis, which is linked to inflammation and oxidative stress [90].

Intriguingly, the rats treated with hesperidin, diosmin, and the LPE-polyflavonoids showed an upregulation of their PI3K, p-AKT, FOXO1, and AMPK gene expressions. These results may be explained by the capacity of flavonoids and their metabolites to immediately exert an amendment in the cellular system via various signaling pathways or via their interaction with intracellular signaling cascades, such as AMPK and the p-AKT signaling pathways [91]. It can be inferred from these findings that the treatment with the LPE-polyflavonoids showed greater expressions of the previous genes compared to a single flavonoid, implying their potency in orchestrating the expression of the genes responsible for decreasing the glucose level in blood, which mimics the interaction of insulin within cells.

To verify the previous findings, we inspected the histopathological features of the pancreatic and liver tissues harvested from all the studied rat groups. As anticipated, various alterations were perceived in the diabetic rats, such as a degeneration of their pancreatic islets with an irregular outline and vacuolated cytoplasm and other anomalies, which were chiefly related to glucose toxicity and lipotoxicity [92]. Prior investigations have described that diabetic rats exhibit shrinkage in the size of their pancreatic islets and degranulated β-cells, giving rise to hyperglycemia due to an inadequate β-cell mass. These can be elucidated as a consequence of diabetes progression [55]. Conversely, the treatment of the diabetic rats with hesperidin, diosmin, and the LPE-polyflavonoids displayed manifestations of β-cells regeneration and enhancement in the pancreatic tissues, which could be derived from the antioxidant attributes of the flavonoids. Moreover, Ghorbani et al. [93] justified these modifications as being due to the prophylactic mechanisms of flavonoids toward β-cell survival and their capacity to alleviate OS, which eventually thwarts the caspase cascade and DNA damage, protecting pancreatic cells against autophagy, apoptosis, or necrosis.

Our histopathological results, in relation to the hepatic tissues in the diabetic rats, were in agreement with those observed by Abdelmageed et al. [5], who provided evidence for the appearance of hepatic and aortic lesions in diabetic rats on account of hyperglycemia, hyperinsulinemia, elevated OS, and hyperlipidemia. Furthermore, we detected a loss in the hepatocyte architecture alongside hepatic necrosis, which is in line with the results of Huang et al. [94]. Moreover, the hepatic steatosis observed in T2DM is likely to be related to a lipid metabolism disorder associated with an abnormal lipid deposition in hepatocytes [95]. Significantly, the rats treated with the LPE-polyflavonoids showed remarkable ameliorations in their hepatic tissues compared to those treated with one of the studied flavonoids. Notably, hesperidin has previously shown a marked improvement in the liver by raising cellular antioxidants [64]. In addition, the diosmin obviously improved the hepatic tissue by restoring its typical structure and architecture, with a centrally positioned vein and undamaged hepatocytes [25]. Considering all of this evidence presented in our study, we could suggest the potential metabolic pathway of LPE-polyflavonoids in attenuating type 2 diabetes mellitus and its deleterious consequences, which mirrors the action of insulin, as portrayed in Figure 16.

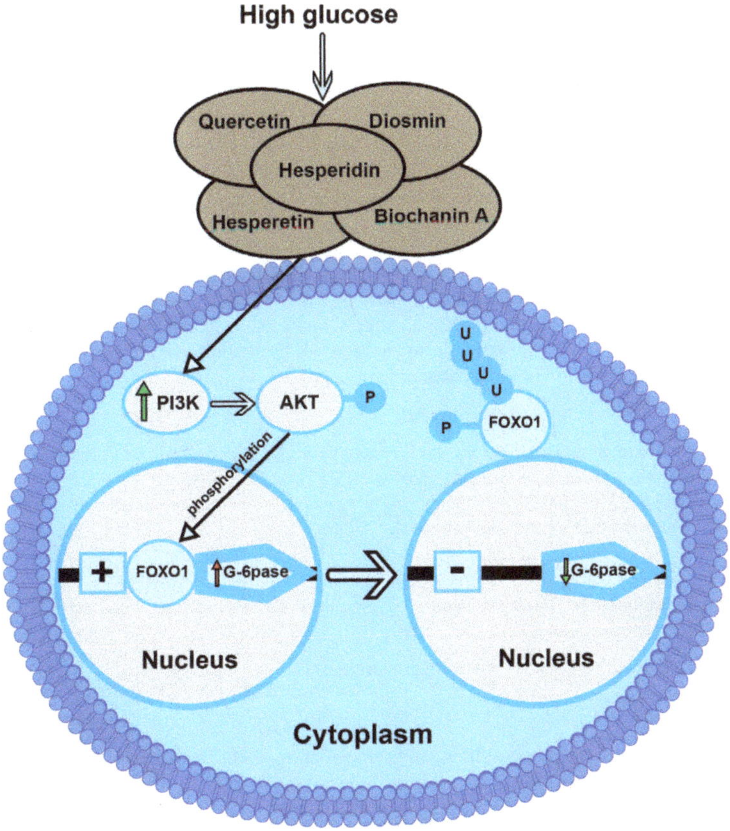

Figure 16. A schematic illustration of the potential mechanism of LPE-polyflavonoids based on the evidence presented in our current study, which demonstrates the respective signaling pathways of LPE-polyflavonoids in mitigating T2DM through diminishing the rates of gluconeogenesis and glycogenolysis. Abbreviations: PI3K: phosphatidylinositol 3-kinase; p-AKT: phospho-Akt or protein kinase B; FOXO1: forkhead box-O; G-6pase: glucose 6-phosphatase; P: phosphorylation; and U: Ubiquitination.

5. Conclusions

In conclusion, this study highlighted, for the first time, the curative and synergistic influences of the polyflavonoids in lemon peel extract (LPE-polyflavonoids) on governing the glucose level in the bloodstream of diabetic rats and restoring the morphology and activity of pancreatic and hepatic cells, in comparison to diabetic rats treated with a single flavonoid agent. Interestingly, superimposed polyflavonoid structures were observed for hesperidin, quercetin, and hesperetin through the identification of active sites. Furthermore, the docked complex binding affinity supported the robust interaction present in hesperidin (−11.2 kcal/mol). The common key residues in the lipid kinase enzyme PI3K explained the favorable inhibition of the LPE-polyflavonoids. Furthermore, the physicochemical and pharmacokinetic properties were studied and demonstrated significant drug-likeness behavior for hesperidin with a low GI absorption, no BBB parameter effect, and no PAINS alert. Most importantly, the in vivo investigations manifested that the LPE-polyflavonoids revealed an antidiabetic, antioxidant, and anti-inflammatory efficiency by regulating the insulin, glucose, glycogen levels, liver function, and oxidative and inflammatory biomarkers. These properties hindered oxidative stress and its respective complications within the pancreatic and hepatic cells, including inflammatory and cellular impairments. Significantly, the LPE-polyflavonoids manifested a coordination of the PI3K, AKT, AMPK, and FOXO1 signaling pathways in the restoration of the histopathological alterations in the pancreas and liver tissues. Altogether, LPE-polyflavonoids could be implemented to manage diabetes mellitus and its consequences. Therefore, future investigations could focus on loading LPE-polyflavonoids onto different biopolymers with various formulations as drug carriers and delivering them to intended target sites, which could sustain the activity of the extract until its release into the selected site.

Author Contributions: M.A.H.: conceptualization, project administration, investigation, data acquisition, formal analysis, statistical analysis, validation, writing—original draft, reviewing and editing the manuscript, finalizing the manuscript. G.M.A.E.: investigation. I.G.E.-Q.: investigation. D.S.E.-S.: formal analysis, writing—original draft. L.M.E.-S.: formal analysis. H.M.A.: conceptualization, project administration, investigation, formal analysis, writing—original draft. All authors have read and agreed to the published version of the manuscript.

Funding: This research received no external funding.

Institutional Review Board Statement: All the experimental and animal handling procedures were performed according to the guidelines approved by Alexandria University Institutional Animal Care and Use Committee (ALEXU-IACUC), a member of the International Council for Laboratory Animal Science (ICLAS) (Approval number: AU04220212302).

Informed Consent Statement: Not applicable.

Data Availability Statement: The datasets generated during the current study are available from the corresponding authors upon reasonable request.

Conflicts of Interest: The authors declare no conflict of interest.

Abbreviations

T2DM: Type 2 diabetes mellitus; PI3K: phosphatidylinositol 3-kinase; p-AKT: phospho-Akt or protein kinase B; AMPK: adenosine monophosphate-activated protein kinase; SIRT1: sirtuin; mTOR: mammalian target of rapamycin; FOXO1: forkhead box-O1; GK: glucokinase; HK: hepatic hexokinase; G6Pd: glucose-6-phosphate dehydrogenase; AGEs: advanced glycation end-products; AST: Aspartate aminotransferase; ALT: alanine transaminase; ALP: alkaline phosphatase; CAT: catalase activity; GSH: glutathione; TBARS: thiobarbituric acid reactive substance; TNF- α: tumor necrosis factor alpha; IL-1β: interleukin-1β; IL-6: interleukin-6; NF-κB: nuclear factor kappa B; AFP: alpha-fetoprotein; and MAPK: mitogen activated protein kinases.

References

1. Xu, Y.; Wang, L.; He, J.; Bi, Y.; Li, M.; Wang, T.; Wang, L.; Jiang, Y.; Dai, M.; Lu, J.; et al. Prevalence and control of diabetes in Chinese adults. *JAMA* **2013**, *310*, 948–959. [CrossRef] [PubMed]
2. Forouhi, N.G.; Wareham, N.J. Epidemiology of diabetes. *Medicine* **2022**, *50*, 638–643. [CrossRef]
3. Hassan, M.A.; Tamer, T.M.; Rageh, A.A.; Abou-Zeid, A.M.; Abd El-Zaher, E.H.F.; Kenawy, E.-R. Insight into multidrug-resistant microorganisms from microbial infected diabetic foot ulcers. *Diabetes Metab. Syndr. Clin. Res. Rev.* **2019**, *13*, 1261–1270. [CrossRef] [PubMed]
4. Hassan, M.A.; Abd El-Aziz, S.; Elbadry, H.M.; El-Aassar, S.A.; Tamer, T.M. Prevalence, antimicrobial resistance profile, and characterization of multi-drug resistant bacteria from various infected wounds in North Egypt. *Saudi J. Biol. Sci.* **2022**, *29*, 2978–2988. [CrossRef]
5. Abdelmageed, M.E.; Shehatou, G.S.; Abdelsalam, R.A.; Suddek, G.M.; Salem, H.A. Cinnamaldehyde ameliorates STZ-induced rat diabetes through modulation of IRS1/PI3K/AKT2 pathway and AGEs/RAGE interaction. *Naunyn-Schmiedeberg's Arch. Pharmacol.* **2019**, *392*, 243–258. [CrossRef] [PubMed]
6. Ahlqvist, E.; Storm, P.; Käräjämäki, A.; Martinell, M.; Dorkhan, M.; Carlsson, A.; Vikman, P.; Prasad, R.B.; Aly, D.M.; Almgren, P. Novel subgroups of adult-onset diabetes and their association with outcomes: A data-driven cluster analysis of six variables. *Lancet Diabetes Endocrinol.* **2018**, *6*, 361–369. [CrossRef]

7. El-Samad, L.M.; Hassan, M.A.; Basha, A.A.; El-Ashram, S.; Radwan, E.H.; Abdul Aziz, K.K.; Tamer, T.M.; Augustyniak, M.; El Wakil, A. Carboxymethyl cellulose/sericin-based hydrogels with intrinsic antibacterial, antioxidant, and anti-inflammatory properties promote re-epithelization of diabetic wounds in rats. *Int. J. Pharm.* **2022**, *629*, 122328. [CrossRef]
8. Qin, S.; Hou, D.X. Multiple regulations of Keap1/Nrf2 system by dietary phytochemicals. *Mol. Nutr. Food Res.* **2016**, *60*, 1731–1755. [CrossRef]
9. Nakayama, T.; Kosugi, T.; Gersch, M.; Connor, T.; Sanchez-Lozada, L.G.; Lanaspa, M.A.; Roncal, C.; Perez-Pozo, S.E.; Johnson, R.J.; Nakagawa, T. Dietary fructose causes tubulointerstitial injury in the normal rat kidney. *Am. J. Physiol. Renal. Physiol.* **2010**, *298*, F712–F720. [CrossRef]
10. Gerber, P.A.; Rutter, G.A. The Role of Oxidative Stress and Hypoxia in Pancreatic Beta-Cell Dysfunction in Diabetes Mellitus. *Antioxid. Redox Signal.* **2017**, *26*, 501–518. [CrossRef]
11. Abdelmageed, M.E.; Shehatou, G.S.G.; Suddek, G.M.; Salem, H.A. Protocatechuic acid improves hepatic insulin resistance and restores vascular oxidative status in type-2 diabetic rats. *Environ. Toxicol. Pharmacol.* **2021**, *83*, 103577. [CrossRef] [PubMed]
12. Titchenell, P.M.; Quinn, W.J.; Lu, M.; Chu, Q.; Lu, W.; Li, C.; Chen, H.; Monks, B.R.; Chen, J.; Rabinowitz, J.D. Direct hepatocyte insulin signaling is required for lipogenesis but is dispensable for the suppression of glucose production. *Cell Metab.* **2016**, *23*, 1154–1166. [CrossRef] [PubMed]
13. Li, Q.; Jia, S.; Xu, L.; Li, B.; Chen, N. Metformin-induced autophagy and irisin improves INS-1 cell function and survival in high-glucose environment via AMPK/SIRT1/PGC-1α signal pathway. *Food Sci. Nutr.* **2019**, *7*, 1695–1703. [CrossRef]
14. Cheng, Z. The FoxO—Autophagy axis in health and disease. *Trends Endocrinol. Metab.* **2019**, *30*, 658–671. [CrossRef] [PubMed]
15. Gu, L.; Ding, X.; Wang, Y.; Gu, M.; Zhang, J.; Yan, S.; Li, N.; Song, Z.; Yin, J.; Lu, L. Spexin alleviates insulin resistance and inhibits hepatic gluconeogenesis via the FoxO1/PGC-1α pathway in high-fat-diet-induced rats and insulin resistant cells. *Int. J. Biol. Sci.* **2019**, *15*, 2815. [CrossRef] [PubMed]
16. Tamer, T.M.; Sabet, M.M.; Alhalili, Z.A.H.; Ismail, A.M.; Mohy-Eldin, M.S.; Hassan, M.A. Influence of Cedar Essential Oil on Physical and Biological Properties of Hemostatic, Antibacterial, and Antioxidant Polyvinyl Alcohol/Cedar Oil/Kaolin Composite Hydrogels. *Pharmaceutics* **2022**, *14*, 2649. [CrossRef]
17. Tamer, T.M.; ElTantawy, M.M.; Brussevich, A.; Nebalueva, A.; Novikov, A.; Moskalenko, I.V.; Abu-Serie, M.M.; Hassan, M.A.; Ulasevich, S.; Skorb, E.V. Functionalization of chitosan with poly aromatic hydroxyl molecules for improving its antibacterial and antioxidant properties: Practical and theoretical studies. *Int. J. Biol. Macromol.* **2023**, *234*, 123687. [CrossRef]
18. Nabil-Adam, A.; Ashour, M.L.; Tamer, T.M.; Shreadah, M.A.; Hassan, M.A. Interaction of *Jania rubens* Polyphenolic Extract as an Antidiabetic Agent with α-Amylase, Lipase, and Trypsin: In Vitro Evaluations and In Silico Studies. *Catalysts* **2023**, *13*, 443. [CrossRef]
19. Lakey-Beitia, J.; Burillo, A.M.; La Penna, G.; Hegde, M.L.; Rao, K. Polyphenols as potential metal chelation compounds against Alzheimer's disease. *J. Alzheimer's Dis.* **2021**, *82*, S335–S357. [CrossRef]
20. Sundaram, R.; Naresh, R.; Shanthi, P.; Sachdanandam, P. Modulatory effect of green tea extract on hepatic key enzymes of glucose metabolism in streptozotocin and high fat diet induced diabetic rats. *Phytomedicine* **2013**, *20*, 577–584. [CrossRef]
21. Mehl, F.; Marti, G.; Boccard, J.; Debrus, B.; Merle, P.; Delort, E.; Baroux, L.; Raymo, V.; Velazco, M.I.; Sommer, H. Differentiation of lemon essential oil based on volatile and non-volatile fractions with various analytical techniques: A metabolomic approach. *Food Chem.* **2014**, *143*, 325–335. [CrossRef]
22. Singh, B.; Singh, J.P.; Kaur, A.; Singh, N. Phenolic composition, antioxidant potential and health benefits of citrus peel. *Food Res. Int.* **2020**, *132*, 109114. [CrossRef]
23. Tag, H.M.; Kelany, O.E.; Tantawy, H.M.; Fahmy, A.A. Potential anti-inflammatory effect of lemon and hot pepper extracts on adjuvant-induced arthritis in mice. *J. Basic Appl. Zool.* **2014**, *67*, 149–157. [CrossRef]
24. Zhang, M.; Zhu, J.; Zhang, X.; Zhao, D.G.; Ma, Y.Y.; Li, D.; Ho, C.T.; Huang, Q. Aged citrus peel (chenpi) extract causes dynamic alteration of colonic microbiota in high-fat diet induced obese mice. *Food Funct.* **2020**, *11*, 2667–2678. [CrossRef] [PubMed]
25. El-Dakhly, S.M.; Salama, A.A.A.; Hassanin, S.O.M.; Yassen, N.N.; Hamza, A.A.; Amin, A. Aescin and diosmin each alone or in low dose-combination ameliorate liver damage induced by carbon tetrachloride in rats. *BMC Res. Notes* **2020**, *13*, 259. [CrossRef] [PubMed]
26. Parhiz, H.; Roohbakhsh, A.; Soltani, F.; Rezaee, R.; Iranshahi, M. Antioxidant and anti-inflammatory properties of the citrus flavonoids hesperidin and hesperetin: An updated review of their molecular mechanisms and experimental models. *Phytother. Res.* **2015**, *29*, 323–331. [CrossRef] [PubMed]
27. Mcharek, N.; Hanchi, B. Maturational effects on phenolic constituents, antioxidant activities and LC-MS/MS profiles of lemon (*Citrus limon*) peels. *J. Appl. Bot. Food Qual.* **2017**, *90*, 1–9.
28. Abchir, O.; Yamari, I.; Nour, H.; Daoui, O.; Elkhattabi, S.; Errougui, A.; Chtita, S. Structure-Based Virtual Screening, ADMET analysis, and Molecular Dynamics Simulation of Moroccan Natural Compounds as Candidates α-Amylase Inhibitors. *ChemistrySelect* **2023**, *8*, e202301092. [CrossRef]
29. Yamari, I.; Abchir, O.; Mali, S.N.; Errougui, A.; Talbi, M.; Kouali, M.E.; Chtita, S. The anti-SARS-CoV-2 activity of novel 9, 10-dihydrophenanthrene derivatives: An insight into molecular docking, ADMET analysis, and molecular dynamics simulation. *Sci. Afr.* **2023**, *21*, e01754. [CrossRef]
30. Eberhardt, J.; Santos-Martins, D.; Tillack, A.F.; Forli, S. AutoDock Vina 1.2.0: New Docking Methods, Expanded Force Field, and Python Bindings. *J. Chem. Inf. Model.* **2021**, *61*, 3891–3898. [CrossRef]

31. Tamer, T.M.; Zhou, H.; Hassan, M.A.; Abu-Serie, M.M.; Shityakov, S.; Elbayomi, S.M.; Mohy-Eldin, M.S.; Zhang, Y.; Cheang, T. Synthesis and physicochemical properties of an aromatic chitosan derivative: In vitro antibacterial, antioxidant, and anticancer evaluations, and in silico studies. *Int. J. Biol. Macromol.* **2023**, *240*, 124339. [CrossRef] [PubMed]
32. Yang, Z.; Lasker, K.; Schneidman-Duhovny, D.; Webb, B.; Huang, C.C.; Pettersen, E.F.; Goddard, T.D.; Meng, E.C.; Sali, A.; Ferrin, T.E. UCSF Chimera, MODELLER, and IMP: An integrated modeling system. *J. Struct. Biol.* **2012**, *179*, 269–278. [CrossRef] [PubMed]
33. Walker, E.H.; Pacold, M.E.; Perisic, O.; Stephens, L.; Hawkins, P.T.; Wymann, M.P.; Williams, R.L. Structural Determinants of Phosphoinositide 3-Kinase Inhibition by Wortmannin, LY294002, Quercetin, Myricetin, and Staurosporine. *Mol. Cell* **2000**, *6*, 909–919. [CrossRef] [PubMed]
34. Morris, G.M.; Goodsell, D.S.; Halliday, R.S.; Huey, R.; Hart, W.E.; Belew, R.K.; Olson, A.J. Automated docking using a Lamarckian genetic algorithm and an empirical binding free energy function. *J. Comput. Chem.* **1998**, *19*, 1639–1662. [CrossRef]
35. Daina, A.; Michielin, O.; Zoete, V. SwissADME: A free web tool to evaluate pharmacokinetics, drug-likeness and medicinal chemistry friendliness of small molecules. *Sci. Rep.* **2017**, *7*, 42717. [CrossRef]
36. Pires, D.E.V.; Blundell, T.L.; Ascher, D.B. pkCSM: Predicting Small-Molecule Pharmacokinetic and Toxicity Properties Using Graph-Based Signatures. *J. Med. Chem.* **2015**, *58*, 4066–4072. [CrossRef]
37. Hassan, R.M.; Elsayed, M.; Kholief, T.E.; Hassanen, N.H.M.; Gafer, J.A.; Attia, Y.A. Mitigating effect of single or combined administration of nanoparticles of zinc oxide, chromium oxide, and selenium on genotoxicity and metabolic insult in fructose/streptozotocin diabetic rat model. *Environ. Sci. Pollut. Res. Int.* **2021**, *28*, 48517–48534. [CrossRef]
38. Álvarez-Almazán, S.; Solís-Domínguez, L.C.; Duperou-Luna, P.; Fuerte-Gómez, T.; González-Andrade, M.; Aranda-Barradas, M.E.; Palacios-Espinosa, J.F.; Pérez-Villanueva, J.; Matadamas-Martínez, F.; Miranda-Castro, S.P.; et al. Anti-Diabetic Activity of Glycyrrhetinic Acid Derivatives FC-114 and FC-122: Scale-Up, In Silico, In Vitro, and In Vivo Studies. *Int. J. Mol. Sci.* **2023**, *24*, 12812. [CrossRef]
39. Lopes-Virella, M.F.; Stone, P.; Ellis, S.; Colwell, J.A. Cholesterol determination in high-density lipoproteins separated by three different methods. *Clin. Chem.* **1977**, *23*, 882–884. [CrossRef]
40. Arafat, E.A.; El-Sayed, D.S.; Hussein, H.K.; Flaven-Pouchon, J.; Moussian, B.; El-Samad, L.M.; El Wakil, A.; Hassan, M.A. Entomotherapeutic Role of Periplaneta americana Extract in Alleviating Aluminum Oxide Nanoparticles-Induced Testicular Oxidative Impairment in Migratory Locusts (*Locusta migratoria*) as an Ecotoxicological Model. *Antioxidants* **2023**, *12*, 653. [CrossRef]
41. Jollow, D.; Mitchell, J.; Zampaglione, N.A.; Gillette, J. Bromobenzene-induced liver necrosis. Protective role of glutathione and evidence for 3, 4-bromobenzene oxide as the hepatotoxic metabolite. *Pharmacology* **1974**, *11*, 151–169. [CrossRef] [PubMed]
42. Nishikimi, M.; Rao, N.A.; Yagi, K. The occurrence of superoxide anion in the reaction of reduced phenazine methosulfate and molecular oxygen. *Biochem. Biophys. Res. Commun.* **1972**, *46*, 849–854. [CrossRef] [PubMed]
43. Abei, H. Catalase. In *Methods of Enzymatic Analysis*; Bergmeyer, H.U., Ed.; Verlag Chemie: Weinheim, Germany; Academic Press: New York, NY, USA, 1974.
44. Livak, K.J.; Schmittgen, T.D. Analysis of relative gene expression data using real-time quantitative PCR and the $2^{-\Delta\Delta CT}$ Method. *Methods* **2001**, *25*, 402–408. [CrossRef] [PubMed]
45. El-Samad, L.M.; Bakr, N.R.; El-Ashram, S.; Radwan, E.H.; Abdul Aziz, K.K.; Hussein, H.K.; El Wakil, A.; Hassan, M.A. Silver nanoparticles instigate physiological, genotoxicity, and ultrastructural anomalies in midgut tissues of beetles. *Chem.-Biol. Interact.* **2022**, *367*, 110166. [CrossRef]
46. El Sayed, D.S.; Abdelrehim, E.-S.M. Computational details of molecular structure, spectroscopic properties, topological studies and SARS-Cov-2 enzyme molecular docking simulation of substituted triazolo pyrimidine thione heterocycles. *Spectrochim. Acta Part A Mol. Biomol. Spectrosc.* **2021**, *261*, 120006. [CrossRef]
47. Cañizares-Carmenate, Y.; Mena-Ulecia, K.; Perera-Sardiña, Y.; Torrens, F.; Castillo-Garit, J.A. An approach to identify new antihypertensive agents using Thermolysin as model: In silico study based on QSARINS and docking. *Arab. J. Chem.* **2019**, *12*, 4861–4877. [CrossRef]
48. Ouassaf, M.; Belaidi, S.; Mogren Al Mogren, M.; Chtita, S.; Ullah Khan, S.; Thet Htar, T. Combined docking methods and molecular dynamics to identify effective antiviral 2, 5-diaminobenzophenonederivatives against SARS-CoV-2. *J. King Saud Univ.-Sci.* **2021**, *33*, 101352. [CrossRef]
49. Abchir, O.; Daoui, O.; Nour, H.; Yamari, I.; Elkhattabi, S.; Errougui, A.; Chtita, S. Exploration of Cannabis constituents as potential candidates against diabetes mellitus disease using molecular docking, dynamics simulations and ADMET investigations. *Sci. Afr.* **2023**, *21*, e01745. [CrossRef]
50. Daoui, O.; Elkhattabi, S.; Chtita, S.; Elkhalabi, R.; Zgou, H.; Benjelloun, A.T. QSAR, molecular docking and ADMET properties in silico studies of novel 4,5,6,7-tetrahydrobenzo[D]-thiazol-2-Yl derivatives derived from dimedone as potent anti-tumor agents through inhibition of C-Met receptor tyrosine kinase. *Heliyon* **2021**, *7*, e07463. [CrossRef]
51. Musarra-Pizzo, M.; Ginestra, G.; Smeriglio, A.; Pennisi, R.; Sciortino, M.T.; Mandalari, G. The Antimicrobial and Antiviral Activity of Polyphenols from Almond (*Prunus dulcis* L.) Skin. *Nutrients* **2019**, *11*, 2355. [CrossRef]
52. Loureiro, G.; Martel, F. The effect of dietary polyphenols on intestinal absorption of glucose and fructose: Relation with obesity and type 2 diabetes. *Food Rev. Int.* **2019**, *35*, 390–406. [CrossRef]

53. Abchir, O.; Daoui, O.; Belaidi, S.; Ouassaf, M.; Qais, F.A.; ElKhattabi, S.; Belaaouad, S.; Chtita, S. Design of novel benzimidazole derivatives as potential α-amylase inhibitors using QSAR, pharmacokinetics, molecular docking, and molecular dynamics simulation studies. *J. Mol. Model.* **2022**, *28*, 106. [CrossRef] [PubMed]
54. Wu, A.T.H.; Lawal, B.; Wei, L.; Wen, Y.-T.; Tzeng, D.T.W.; Lo, W.-C. Multiomics Identification of Potential Targets for Alzheimer Disease and Antrocin as a Therapeutic Candidate. *Pharmaceutics* **2021**, *13*, 1555. [CrossRef] [PubMed]
55. Gopalan, D.H.; Vani, M.; Manikandan, S.; Vijayakumar, V. An immunohistochemical and biochemical evidences of pancreatic β-cell regeneration in type 2 diabetic rats after treated with Aegle marmelos leaf extract and Aegeline. *Int. J. Res. Pharm. Sci.* **2021**, *12*, 868–876. [CrossRef]
56. Hahn, M.; van Krieken, P.P.; Nord, C.; Alanentalo, T.; Morini, F.; Xiong, Y.; Eriksson, M.; Mayer, J.; Kostromina, E.; Ruas, J.L.; et al. Topologically selective islet vulnerability and self-sustained downregulation of markers for β-cell maturity in streptozotocin-induced diabetes. *Commun. Biol.* **2020**, *3*, 541. [CrossRef] [PubMed]
57. Berger, C.; Zdzieblo, D. Glucose transporters in pancreatic islets. *Pflug. Arch.* **2020**, *472*, 1249–1272. [CrossRef] [PubMed]
58. Peng, P.; Jin, J.; Zou, G.; Sui, Y.; Han, Y.; Zhao, D.; Liu, L. Hesperidin prevents hyperglycemia in diabetic rats by activating the insulin receptor pathway. *Exp. Ther. Med.* **2021**, *21*, 53. [CrossRef] [PubMed]
59. Gandhi, G.R.; Vasconcelos, A.B.S.; Wu, D.-T.; Li, H.-B.; Antony, P.J.; Li, H.; Geng, F.; Gurgel, R.Q.; Narain, N.; Gan, R.-Y. Citrus Flavonoids as Promising Phytochemicals Targeting Diabetes and Related Complications: A Systematic Review of In Vitro and In Vivo Studies. *Nutrients* **2020**, *12*, 2907. [CrossRef]
60. Om, H.; El-Naggar, M.E.; El-Banna, M.; Fouda, M.M.G.; Othman, S.I.; Allam, A.A.; Morsy, O.M. Combating atherosclerosis with targeted Diosmin nanoparticles-treated experimental diabetes. *Investig. New Drugs* **2020**, *38*, 1303–1315. [CrossRef]
61. Mohamed, A.H.; Abd El-Hameed, M.N.; Khamis, H.H.; Sharkawy, G.K.; Samir, A.E.-L.H. Antidiabetic and anti-inflammatory effects of two Fabaceae extracts against streptozotocin induced diabetic impairment in male rats. *World J. Adv. Res. Rev.* **2020**, *6*, 012–029. [CrossRef]
62. Oyedemi, S.O.; Nwaogu, G.; Chukwuma, C.I.; Adeyemi, O.T.; Matsabisa, M.G.; Swain, S.S.; Aiyegoro, O.A. Quercetin modulates hyperglycemia by improving the pancreatic antioxidant status and enzymes activities linked with glucose metabolism in type 2 diabetes model of rats: In silico studies of molecular interaction of quercetin with hexokinase and catalase. *J. Food Biochem.* **2020**, *44*, e13127. [CrossRef] [PubMed]
63. Sundaram, R.; Shanthi, P.; Sachdanandam, P. Effect of tangeretin, a polymethoxylated flavone on glucose metabolism in streptozotocin-induced diabetic rats. *Phytomedicine* **2014**, *21*, 793–799. [CrossRef] [PubMed]
64. Sundaram, R.; Nandhakumar, E.; Haseena Banu, H. Hesperidin, a citrus flavonoid ameliorates hyperglycemia by regulating key enzymes of carbohydrate metabolism in streptozotocin-induced diabetic rats. *Toxicol. Mech. Methods* **2019**, *29*, 644–653. [CrossRef] [PubMed]
65. Rabee, A.; Bennasir, H. Hesperidin an antioxidant flavonoid prevents carbon tetrachloride-induced hepatic toxicity in male albino rats. *J. Innov. Pharm. Biol.* **2018**, *5*, 127–132.
66. Ali, F.E.; Azouz, A.A.; Bakr, A.G.; Abo-Youssef, A.M.; Hemeida, R.A. Hepatoprotective effects of diosmin and/or sildenafil against cholestatic liver cirrhosis: The role of Keap-1/Nrf-2 and P38-MAPK/NF-κB/iNOS signaling pathway. *Food Chem. Toxicol.* **2018**, *120*, 294–304. [CrossRef]
67. Saravanan, R.; Vengatash Babu, K.; Ramachandran, V. Effect of Rebaudioside A, a diterpenoid on glucose homeostasis in STZ-induced diabetic rats. *J. Physiol. Biochem.* **2012**, *68*, 421–431. [CrossRef]
68. Vinayagam, R.; Jayachandran, M.; Chung, S.S.M.; Xu, B. Guava leaf inhibits hepatic gluconeogenesis and increases glycogen synthesis via AMPK/ACC signaling pathways in streptozotocin-induced diabetic rats. *Biomed. Pharmacother.* **2018**, *103*, 1012–1017. [CrossRef]
69. Wang, K.; Wang, H.; Liu, Y.; Shui, W.; Wang, J.; Cao, P.; Wang, H.; You, R.; Zhang, Y. Dendrobium officinale polysaccharide attenuates type 2 diabetes mellitus via the regulation of PI3K/Akt-mediated glycogen synthesis and glucose metabolism. *J. Funct. Foods* **2018**, *40*, 261–271. [CrossRef]
70. Pari, L.; Srinivasan, S. Antihyperglycemic effect of diosmin on hepatic key enzymes of carbohydrate metabolism in streptozotocin-nicotinamide-induced diabetic rats. *Biomed. Pharmacother.* **2010**, *64*, 477–481. [CrossRef]
71. Chang, D.C.; Xu, X.; Ferrante, A.W.; Krakoff, J. Reduced plasma albumin predicts type 2 diabetes and is associated with greater adipose tissue macrophage content and activation. *Diabetol. Metab. Syndr.* **2019**, *11*, 14. [CrossRef]
72. Abdel-Sttar, A.R.; Khalaf, M.M.; Aboyoussef, A.M.; Abosaif, A. Ameliorative effect of hesperidin on carbon tetrachloride induced liver fibrosis in rats. *Int. J. Pharm. Pharm. Sci.* **2017**, *9*, 45–51. [CrossRef]
73. Tahir, M.; Rehman, M.U.; Lateef, A.; Khan, R.; Khan, A.Q.; Qamar, W.; Ali, F.; O'Hamiza, O.; Sultana, S. Diosmin protects against ethanol-induced hepatic injury via alleviation of inflammation and regulation of TNF-α and NF-κB activation. *Alcohol* **2013**, *47*, 131–139. [CrossRef] [PubMed]
74. Omolola, S.-O.; Bukoye, O.H.O. Improvement of Oral Glucose Tolerance and Total Lipid Profile of Diabetic Rats Treated with *Ficus exasperata* Leaf-Based Diet. *Covenant J. Phys. Life Sci.* **2019**, *7*, 2354–3574.
75. Abdel-Reheim, M.A.; Messiha, B.A.S.; Abo-Saif, A.A. Hepatoprotective effect of diosmin on iron-induced liver damage. *Int. J. Pharmacol.* **2017**, *13*, 529–540. [CrossRef]
76. Yazdi, H.B.; Hojati, V.; Shiravi, A.; Hosseinian, S.; Vaezi, G.; Hadjzadeh, M.A. Liver Dysfunction and Oxidative Stress in Streptozotocin-Induced Diabetic Rats: Protective Role of Artemisia Turanica. *J. Pharmacopunct.* **2019**, *22*, 109–114. [CrossRef]

77. Vlassara, H.; Uribarri, J. Advanced glycation end products (AGE) and diabetes: Cause, effect, or both? *Curr. Diabetes Rep.* **2014**, *14*, 453. [CrossRef]
78. Eraslan, G.; Sarica, Z.S.; Bayram, L.C.; Tekeli, M.Y.; Kanbur, M.; Karabacak, M. The effects of diosmin on aflatoxin-induced liver and kidney damage. *Environ. Sci. Pollut. Res. Int.* **2017**, *24*, 27931–27941. [CrossRef]
79. Caglayan, C.; Kandemir, F.M.; Darendelioğlu, E.; Küçükler, S.; Ayna, A. Hesperidin protects liver and kidney against sodium fluoride-induced toxicity through anti-apoptotic and anti-autophagic mechanisms. *Life Sci.* **2021**, *281*, 119730. [CrossRef]
80. Keane, K.N.; Cruzat, V.F.; Carlessi, R.; de Bittencourt, P.I., Jr.; Newsholme, P. Molecular Events Linking Oxidative Stress and Inflammation to Insulin Resistance and β-Cell Dysfunction. *Oxid. Med. Cell Longev.* **2015**, *2015*, 181643. [CrossRef]
81. Hanchang, W.; Khamchan, A.; Wongmanee, N.; Seedadee, C. Hesperidin ameliorates pancreatic beta-cell dysfunction and apoptosis in streptozotocin-induced diabetic rat model. *Life Sci.* **2019**, *235*, 116858. [CrossRef]
82. Uslu, S.; Alaca, N.; Kilic, K.D.; Uysal, A.; Kurtel, H. The effects of aerobic exercise frequencies on liver fibrosis, alpha-fetoprotein and cytokeratin 19 in experimental type 2 diabetes-induced rats: An immunohistochemistry study. *Biotech. Histochem.* **2018**, *93*, 615–622. [CrossRef] [PubMed]
83. Xu, P.; Xiao, J.; Chi, S. Piperlongumine attenuates oxidative stress, inflammatory, and apoptosis through modulating the GLUT-2/4 and AKT signaling pathway in streptozotocin-induced diabetic rats. *J. Biochem. Mol. Toxicol.* **2021**, *35*, 1–12. [CrossRef] [PubMed]
84. Al-Shaqha, W.M.; Khan, M.; Salam, N.; Azzi, A.; Chaudhary, A.A. Anti-diabetic potential of Catharanthus roseus Linn. and its effect on the glucose transport gene (GLUT-2 and GLUT-4) in streptozotocin induced diabetic wistar rats. *BMC Complement. Altern. Med.* **2015**, *15*, 379. [CrossRef] [PubMed]
85. Ali, A.M.; Gabbar, M.A.; Abdel-Twab, S.M.; Fahmy, E.M.; Ebaid, H.; Alhazza, I.M.; Ahmed, O.M. Antidiabetic Potency, Antioxidant Effects, and Mode of Actions of Citrus reticulata Fruit Peel Hydroethanolic Extract, Hesperidin, and Quercetin in Nicotinamide/Streptozotocin-Induced Wistar Diabetic Rats. *Oxid. Med. Cell Longev.* **2020**, *2020*, 1730492. [CrossRef]
86. Hsu, C.C.; Lin, M.H.; Cheng, J.T.; Wu, M.C. Antihyperglycaemic action of diosmin, a citrus flavonoid, is induced through endogenous β-endorphin in type I-like diabetic rats. *Clin. Exp. Pharmacol. Physiol.* **2017**, *44*, 549–555. [CrossRef] [PubMed]
87. Qiu, F.; Wang, J.; Liu, H.-Y.; Zhang, Y.-Q. Mulberry bark alleviates effect of STZ inducing diabetic mice through negatively regulating FoxO1. *Evid.-Based Complement. Altern. Med.* **2019**, *2019*, 2182865. [CrossRef] [PubMed]
88. Hao, F.; Kang, J.; Cao, Y.; Fan, S.; Yang, H.; An, Y.; Pan, Y.; Tie, L.; Li, X. Curcumin attenuates palmitate-induced apoptosis in MIN6 pancreatic β-cells through PI3K/Akt/FoxO1 and mitochondrial survival pathways. *Apoptosis* **2015**, *20*, 1420–1432. [CrossRef]
89. Kinyua, A.W.; Ko, C.M.; Doan, K.V.; Yang, D.J.; Huynh, M.K.Q.; Moh, S.H.; Choi, Y.-H.; Kim, K.W. 4-hydroxy-3-methoxycinnamic acid regulates orexigenic peptides and hepatic glucose homeostasis through phosphorylation of FoxO1. *Exp. Mol. Med.* **2018**, *50*, e437. [CrossRef]
90. Ruderman, N.B.; Xu, X.J.; Nelson, L.; Cacicedo, J.M.; Saha, A.K.; Lan, F.; Ido, Y. AMPK and SIRT1: A long-standing partnership? *Am. J. Physiol. Endocrinol. Metab.* **2010**, *298*, E751–E760. [CrossRef]
91. Panda, S.; Kar, A. Antithyroid effects of naringin, hesperidin and rutin in l-T4 induced hyperthyroid rats: Possible mediation through 5′ DI activity. *Pharmacol. Rep.* **2014**, *66*, 1092–1099. [CrossRef]
92. Ekperikpe, U.S.; Owolabi, O.J.; Olapeju, B.I. Effects of Parkia biglobosa aqueous seed extract on some biochemical, haematological and histopathological parameters in streptozotocin induced diabetic rats. *J. Ethnopharmacol.* **2019**, *228*, 1–10. [CrossRef] [PubMed]
93. Ghorbani, A.; Rashidi, R.; Shafiee-Nick, R. Flavonoids for preserving pancreatic beta cell survival and function: A mechanistic review. *Biomed. Pharmacother.* **2019**, *111*, 947–957. [CrossRef] [PubMed]
94. Huang, X.L.; He, Y.; Ji, L.L.; Wang, K.Y.; Wang, Y.L.; Chen, D.F.; Geng, Y.; OuYang, P.; Lai, W.M. Hepatoprotective potential of isoquercitrin against type 2 diabetes-induced hepatic injury in rats. *Oncotarget* **2017**, *8*, 101545–101559. [CrossRef] [PubMed]
95. Hou, B.; Zhao, Y.; Qiang, G.; Yang, X.; Xu, C.; Chen, X.; Liu, C.; Wang, X.; Zhang, L.; Du, G. Puerarin Mitigates Diabetic Hepatic Steatosis and Fibrosis by Inhibiting TGF-beta Signaling Pathway Activation in Type 2 Diabetic Rats. *Oxid. Med. Cell Longev.* **2018**, *2018*, 4545321. [CrossRef]

pharmaceutics

MDPI

Article

Formulation of Polymeric Nanoparticles Loading Baricitinib as a Topical Approach in Ocular Application

Negar Beirampour [1], Paola Bustos-Salgado [1,2], Núria Garrós [1], Roya Mohammadi-Meyabadi [1,2], Òscar Domènech [1,2], Joaquim Suñer-Carbó [1,2], María José Rodríguez-Lagunas [3], Garyfallia Kapravelou [4], María Jesús Montes [5], Ana Calpena [1,2] and Mireia Mallandrich [1,2,*]

1 Departament de Farmàcia i Tecnologia Farmacèutica, i Fisicoquímica, Facultat de Farmàcia i Ciències de l'Alimentació, Universitat de Barcelona, Av. Joan XXIII 29-31, 08028 Barcelona, Spain; nbeirabe7@alumnes.ub.edu (N.B.); paola2006@hotmail.com (P.B.-S.); rmohammo31@alumnes.ub.edu (R.M.-M.); jsuner@ub.edu (J.S.-C.); anacalpena@ub.edu (A.C.)

2 Institut de Nanociència i Nanotecnologia, Universitat de Barcelona (UB), 08028 Barcelona, Spain

3 Department of Biochemistry and Physiology, Faculty of Pharmacy and Food Sciences, University of Barcelona, Av. Joan XXIII, 08028 Barcelona, Spain; mjrodriguez@ub.edu

4 Department of Physiology, Institute of Nutrition and Food Technology (INyTA), Biomedical Research Center (CIBM), Universidad de Granada, 18100 Granada, Spain; kapravelou@ugr.es

5 Department de Biologia, Sanitat i Medi Ambient, Facultat de Farmàcia i Ciències de l'Alimentació, Universitat de Barcelona, Av. Joan XXIII 29-31, 08028 Barcelona, Spain; mjmontes@ub.edu

* Correspondence: mireia.mallandrich@ub.edu

Abstract: Topical ocular drug delivery faces several challenges due to the eye's unique anatomy and physiology. Physiological barriers, tear turnover, and blinking hinder the penetration of drugs through the ocular mucosa. In this context, nanoparticles offer several advantages over traditional eye drops. Notably, they can improve drug solubility and bioavailability, allow for controlled and sustained drug release, and can be designed to specifically target ocular tissues, thus minimizing systemic exposure. This study successfully designed and optimized PLGA and PCL nanoparticles for delivering baricitinib (BTB) to the eye using a factorial design, specifically a three-factor at five-levels central rotatable composite 2^{3+} star design. The nanoparticles were small in size so that they would not cause discomfort when applied to the eye. They exhibited low polydispersity, had a negative surface charge, and showed high entrapment efficiency in most of the optimized formulations. The Challenge Test assessed the microbiological safety of the nanoparticle formulations. An ex vivo permeation study through porcine cornea demonstrated that the nanoparticles enhanced the permeability coefficient of the drug more than 15-fold compared to a plain solution, resulting in drug retention in the tissue and providing a depot effect. Finally, the in vitro ocular tolerance studies showed no signs of irritancy, which was further confirmed by HET-CAM testing.

Keywords: baricitinib; poly(lactic-co-glycolic acid) nanoparticles; poly(ε-caprolactone) nanoparticles; transcorneal permeation; ocular tolerance; ocular delivery

1. Introduction

The eye is an exceptional organ of complexity. It is vulnerable to different disorders and diseases. Among these is uveitis, an autoimmune condition where abnormal immune responses target various structures within the eye [1]. These diseases require a systematic and multidisciplinary management strategy aimed at controlling inflammation, preserving visual function, and preventing further damage to ocular structures. Corticosteroids, such as prednisolone or dexamethasone, are commonly used as the initial therapeutic intervention to rapidly suppress inflammation [2]. In addition to corticosteroids, other immunosuppressive agents may be prescribed, such as methotrexate, azathioprine, mycophenolate mofetil, and cyclosporine. These medications work by modulating the immune response and preventing the production of inflammatory cells or chemicals [3]. In extreme cases, surgical interventions such as corneal transplants or retinal surgeries may be necessary, though their success can be compromised by immunological allograft rejection [4,5]. Research continues to focus on preventing graft rejection and improving graft survival [6,7].

Baricitinib (BTB), a Janus kinase (JAK) inhibitor, selectively inhibits Janus kinases 1 and 2, reducing inflammation, cellular activation, and the proliferation of key immune cells, thereby mitigating disease symptoms. BTB, administered orally, has shown good outcomes in treating refractory rheumatoid arthritis and atopic dermatitis [8]. In 2020, Yutaka Kaneko et al. examined the use of topical BTB for treating uveitis and dry eye disease, showing that BTB eye drops effectively reduced ocular inflammation and improved symptoms [9].

However, the corneal epithelium and tear fluid act as barriers, limiting drug penetration into the eye. These results in low bioavailability for ophthalmic drugs in aqueous solutions, exacerbated by rapid drug elimination due to reflex blinking and tear drainage, making it difficult to achieve therapeutic drug levels [10]. This presents an opportunity for nanomedicine, which can directly deliver drugs or genes to maximize therapeutic effect and minimize side effects. Nanoparticles (NPs) have shown promise in ocular drug delivery for immunosuppressive drugs. For example, Tacrolimus-loaded NPs have shown promise in treating dry eye syndrome and graft-versus-host disease [11,12], and cyclosporine A encapsulated in NPs, such as poly (lactic-co-glycolic acid) (PLGA) or chitosan NPs, have been studied to improve efficacy and prolong release for treating dry eye disease and ocular surface inflammation [13]. Further, Sirolimus-loaded NPs have been explored for corneal graft rejection and proliferative vitreoretinopathy [14], and Dexamethasone-loaded NPs have been investigated for treating uveitis and macular edema [15]. These NP-based systems offer advantages like improved drug stability, sustained release, and optimized bioavailability, which can reduce dosing interval. Enhancing the local delivery of immunosuppressive drugs aims to increase therapeutic efficacy while minimizing systemic side effects [16,17]. Further research is needed to optimize formulations and evaluate the safety and efficacy of NPs in ophthalmic applications.

This study aims to develop advanced BTB NPs using PLGA and polycaprolactone (PCL) for ocular delivery to manage inflammatory or immunological ocular diseases. The BTB NP formulations were optimized using a factorial design. The formulations were characterized in terms of physicochemical properties and morphology. The capacity of BTB to penetrate the cornea was evaluated through an ex vivo permeation study. The tolerability of the NPs was assessed via the HET-CAM test, corneal transparency, and histological analysis. Additionally, a microbiological study on the optimized formulations during their storage and possible use was conducted.

2. Materials and Methods

2.1. Materials

The BTB bulk powder ingredient was supplied by Henrikang Biotech Co., Ltd. (Xi'an, China). PLGA [poly (lactic-co-glycolic acid)] (Resomer® RG 503H) was acquired from Boehringer Ingelheim (Ingelheim, Germany) and Poloxamer 188 (Lutrol® F68) was acquired from BASF (Ludwigshafen, Germany). PCL [poly (ε-caprolactone)], with an average molecular weight of ~14,000 Da, and Acetone were acquired from Sigma–Aldrich (Barcelona, Spain). Transcutol® was gifted by Gattefossé (Barcelona, Spain). The purified water used in all experiments was obtained from the Milli-Q® Plus System (Millipore Corporation, Burlington, MA, USA) lab supply.

2.2. Biological Materials

The study utilized corneas sourced from the remains of pigs (a crossbreed of Landrace and Large White, weighing between 20 and 25 kg) that had been previously used in surgical procedures at a university. The Ethics Committee of Animal Experimentation at the University of Barcelona approved the use of these corneas (code 10619, approved on 10 January 2019). Immediately after the animals were slaughtered, the eyes were removed for corneal tissue dissection and then stored in a solution of artificial aqueous humor for transport to the laboratory. The corneas were then debrided and prepared for permeation experiments.

2.3. Factorial Design

To develop the NPs, a factorial design was conducted considering three-factor at five-levels-central rotatable composite design 2^{3+} star (Tables 1 and 2). As dependent variables, physicochemical properties, such as average particle size (Z-ave) and polydispersity index (PDI), were chosen according to the central composite design matrix generated by Statgraphics® Plus 5.1 software (Statpoint Technologies, Inc., Warrenton, VA, USA).

Table 1. Factors and levels in experimental design. Polymer (PCL and PLGA) and surfactant (Poloxamer P188) amounts and aqueous phase volumes.

	Factors					
Variables Levels	Polymer Amount (mg)	Surfactant Amount (mg)	Aqueous Phase Volume (mL)	Polymer Amount (mg)	Surfactant Amount (mg)	Aqueous Phase Volume (mL)
	PCL	PCL	PCL	PLGA	PLGA	PLGA
Low (−1)	50	200	38	75	150	50
Centered (0)	60	240	44	95	200	60
High (+1)	70	280	50	115	250	70
Axial (−0.39)	46	185	36.5	72	134	47
Axial (+1.39)	74	295	51.5	118	266	73

Table 2. Resulting formulations of the 2^{3+} star central composite rotatable factorial design based on three levels (low (0), central (0.5) and high (1)) for each factor: amount of Polymer (PLGA or PCL), amount of Surfactant (P-188), and the volume of the aqueous phase.

	Levels of Factors		
Formulations	Polymer Amount (mg)	Surfactant Amount (mg)	Aqueous Phase Volume (mL)
F1	0	0	0
F2	1	0	0
F3	0	1	0
F4	1	1	0
F5	0	0	1
F6	1	0	1
F7	0	1	1
F8	1	1	1
F9	0.5	0.5	0.5
F10	0.5	0.5	0.5
F11	−0.39	0.5	0.5
F12	1.39	0.5	0.5
F13	0.5	−0.39	0.5
F14	0.5	1.39	0.5
F15	0.5	0.5	−0.39
F16	0.5	0.5	1.39
F17	0.5	0.5	0.5
F18	0.5	0.5	0.5
F19	0.5	0.5	0.5

A total of 19 experiments for estimation of the pure error sum of squares were performed (Table 2). The observed experimental responses resulted from the individual influence and interactions of the three independent variables. To identify the significance of the effects and the interactions between them, analysis of variance (ANOVA) was performed for each parameter and p values pf less than 0.05 were considered to be statistically significant.

2.4. Preparation of the Selected Nanoparticles Loading BTB

To prepare the optimized NPs (B-PLGA-95, B-PLGA-75 and B-PCL NPs), the solvent displacement technique was used. A series of organic solutions was prepared by dissolving one the polymers (50 mg of PCL, 95 mg of PLGA, or 75 mg of PLGA) in 10 mL of acetone with 10 mg of BTB. Simultaneously, 200 mg of surfactant (P188) was dissolved in water adjusted to pH 5.5. The volumes of water were determined according to the formulations presented in Table 3. The previously prepared organic solution was added dropwise to the aqueous solution under moderate magnetic stirring at room temperature (25 °C). Then, the organic solvent evaporated overnight at room temperature (around 22 ± 2 °C) under ambient pressure (around 1 atmosphere). These conditions are based on standard laboratory procedures for evaporation techniques to remove organic solvents from nanoparticle suspensions.

Table 3. Composition of the selected formulations, each containing 10 mg of BTB.

Formulation	Polymer Amount (%)	Aqueous Phase (%)	Surfactant Amount P188 (%)
B-PLGA-95	0.157	99.511	0.332
B-PLGA-75	0.107	99.609	0.284
B-PCL	0.100	99.502	0.398

B-PLGA-95: prepared with 95 mg of poly(lactic-co-glycolic) acid; B-PLGA-75: prepared with 75 mg of poly(lactic-co-glycolic) acid; B-PCL: prepared with 50 mg of poly(ε-caprolactone); P188: Poloxamer 188.

2.5. Physicochemical Characterization of Nanoparticles

The physicochemical properties of the NPs were evaluated by measuring the pH, vesicle size (Z-ave), polydispersity index (PDI), and zeta potential (ZP). The Z-ave and PDI properties were determined at 25 °C using a Nano ZS Zetasizer® (Malvern Instruments, Malvern, UK) via dynamic light scattering. ZP measures the surface charge of cells by evaluating the electrokinetic potential; this characterizes the electrical potential of the double layer on the cell surface and serves as an indirect measure of physical stability. The pH was measured at room temperature using a Micro pH 2001 pH meter (Crison Instruments SA, Alella, Spain) [18]. All measurements were conducted in triplicate and are reported as mean ± standard deviation.

2.5.1. Encapsulation Efficiency (EE)

The encapsulation efficiency (EE%) was determined using an indirect method. Initially, 0.5 mL of nanoparticle suspension was placed into a centrifugal filter unit (Amicon Ultra, Millipore, Billerica, MA, USA) and centrifuged at 11,000 r.p.m. (12,000× g) at 4 °C for 15 min. The filtered solution, containing non-encapsulated BTB, was then analyzed using high-performance liquid chromatography (HPLC) [19]. The HPLC system utilized a Chromatograph Waters Alliance 2695 with a Fluorescence Jasco FP-1520 detector (Waters, Milford, CT, USA), employing an excitation wavelength of 310 nm and an emission wavelength of 390 nm. A Symmetry C18 column (3.5 μm, 4.6 × 75 mm) was used with a mobile phase consisting of Ammonium Formate (10 mM, pH 7) and ACN (75:25 v/v) delivered at a flow rate of 1 mL/min under isocratic elution conditions. With an injection volume of 10 μL, the quantification of BTB was validated within a range of 0.06–1 μL/min. Finally, the encapsulation efficacy (%) was calculated by dividing the amount of non-encapsulated BTB from the total amount of BTB in the NPs and multiplying by 100, as described in (Equation (1)):

$$EE\% = (A_{non\text{-}e}/A_{total}) \times 100 \quad (1)$$

$A_{non\text{-}e}$ is the amount of BTB non-encapsulated and A_{total} is the total amount of BTB in the NPs.

2.5.2. Morphological Study of Nanoparticles

The transmission electron microscopy (TEM) was used for the morphological and structural examination of NPs. First, one drop of each NP solution was adsorbed into carbon-coated copper grids for 1 min. Then, they were stained with UranyLess EM Stain (EMS) for 1 min. The excess liquid was manually blotted from the edge of the grids. The sample was observed in a JEOL 1010 (JEOL Inc., Peabody, MA, USA) microscope equipped with a tungsten filament. Finally, images were acquired at 120 kV at room temperature with a 1376 × 1024 pixels CCD Mega view camera [20].

2.5.3. Extensibility of Nanoparticles

To assess the spreading behavior of the NPs, a 50 μL volume of each formulation was added to a steel plate circle with a diameter of 10 cm, with a glass plate weighing 26 g being placed on top. Increasing standard weight pieces (0, 10, 20, 50, 100 g) were added on top of the upper glass plate for 1 min each. The diameter of the spread area was recorded for each formulation and tested in triplicate at room temperature. The resulting increase in surface area (mm^2) of the NPs was plotted as a function of the increasing weights applied [21]. The two-site binding parabola model described in Equation (2) was best fitted to the formulations measured.

$$Y = A + B \times X + C \times X^2, \tag{2}$$

A, B and C are coefficients, X the weight in g, and Y the extensibility in cm^2.

2.5.4. Rheological Characterization

Rheological measurements were performed with a Haake Rheostress® 1 rheometer (Thermo Fisher Scientific, Karlsruhe, Germany) connected to a thermostatic circulator Thermo® Haake Phoenix II + Haake C25P. Steady-state measurements were addressed with a plate-and-cone geometry (C60/2°Ti: 60 mm diameter, 2° angle). The shear stress (τ) was measured as a function of the shear rate (γ). Viscosity curves ($\eta = f(\gamma)$) and flow curves ($\tau = f(\gamma)$) were recorded at 25 ± 0.1 °C. The shear rate ramp program included a 3 min ramp-up period from 0 to 100 s^{-1}, a 1 min constant shear rate period at 100 s^{-1}, and a 3 min ramp-down from 100 to 0 s^{-1}. Representative mathematical models were fit to flow curves as the best descriptive model. Selection of the best fitting was based on the correlation coefficient (observed vs. predicted) and chi-square value. Steady-state viscosity (η, m·Pa s) was determined from the constant shear section at 100 s^{-1}.

2.5.5. Stability of Nanoparticles

The physicochemical stability of NPs was assessed at 25 °C by dynamic light scattering analysis (DLS) with Z-sizer (Malvern Instruments, Malvern, UK) profiles. The optimized NPs were divided into two batches and stored at different temperature conditions according to the established guidelines at 5 ± 3 °C and 25 ± 2 °C, with a relative humidity of 60% ± 5% [22]. The values of Z-Ave and PDI were recorded at 0 d, 7 d and 30 d after the NP preparation.

2.6. Antimicrobial Challenge Test

During the development phase of a pharmaceutical formulation for ophthalmic use, it must be demonstrated that the formulation provides adequate protection against adverse effects that may occur from microbial contamination or growth during storage and use [19,23]. The Challenge Test involved the deliberate contamination of the formulations, in their final containers, with a determined inoculum of suitable microorganisms (Table 4). All the microbial strains (*Pseudomonas aeruginosa*, *Staphylococcus aureus*, *Candida albicans* and *Aspergillus brasiliensis*) used in this study were acquired from the collection at the microbiology laboratory of the Faculty of Pharmacy and Food Sciences, University of Barcelona. Tryptone Soya Agar (TSA; Ref: CM0131B, Oxoid, UK) was inoculated with bacteria and incubated for 18–24 h at 37 ± 2 °C. *C. albicans* and *A. brasiliensis* were grown on Sabouraud Dextrose Agar (SDA; Ref: CM0041B, Oxoid, UK) at 25 °C for 48 h and 5 d, respectively. For the collection of subcultures and preparation of bacterial and fungus suspensions that will be used to inoculate the formulations, a sterile saline test solution (NaCl 0.9% *w/v*) was applied to reach the final concentration of 10^8 CFU/mL (CFU = colony forming units). In the case of the A. *brasiliensis* culture, a sterile saline test solution containing 0.05% (*v/v*) of polysorbate 80 was used.

Table 4. Microorganisms used to conduct the microbiological tests.

Test Microorganisms			Microorganism Concentration	
			B-PCL NPs (CFU/mL)	**B-PLGA-95 and B-PLGA-75 NPs (CFU/mL)**
Bacteria	*Pseudomonas aeruginosa*	ATCC 9027 (NCIMB 8626, CIP 82.118)	1.5×10^8	2.3×10^8
	Staphylococcus aureus	ATCC 6538, (NCTC 10788, NCIMB 9518)	2.1×10^8	2.3×10^8
Fungus	*Candida albicans*	ATCC 10231 (NCPF 3179, IP 48.72)	1×10^8	1×10^8
	Aspergillus brasiliensis	ATCC 16404 (IMI 149007, IP 1431.83)	2×10^7	2×10^7

Tests were conducted with each microorganism individually. CFU: colony forming units.

Each NP formulation was transferred to containers in triplicate and individually inoculated with 1% of each microbial test suspension separately in order to make the final concentrations in the samples match those indicated in the pharmacopoeia (10^5–10^6 CFU/mL) [24]. The NP formulations were kept at room temperature (20–25 °C) and protected from light during the withdrawal of samples from the container (1 mL). At specified time intervals, samples were taken and microorganisms were counted and recorded (Table 5). The samples were neutralized using Beerens (Ref. 02-257 Scharlab, Barcelona, Spain). The average log10 CFU/mL were used to quantify the results. The formulations were found to be adequate if, under the test conditions, there was a significant decrease or, depending on the case, there was no increase in the number of microorganisms after the established times and temperatures (Table 5).

Table 5. Criteria for ophthalmic formulations.

	Log Reduction (CFU/mL)					
	Criteria	**6 h**	**24 h**	**7 d**	**14 d**	**28 d**
Bacteria	A	2	3	-	-	NR
	B	-	1	3	-	NI
Fungus	A	-	-	2	-	NI
	B	-	-	-	1	NI

NR: Not found; NI: Not increased.

Criteria A expresses the recommended effectiveness. In justified cases where it is not possible to meet the A criteria (for example, due to an increased risk of adverse reactions), then B criteria should be met.

2.7. Ex Vivo Cornea Permeation Assays

Ex vivo studies were performed using porcine corneas (see Section 2.2 biological material). The experiments were conducted in independent vertical Franz diffusion cells with a diffusional surface area of 0.64 cm^2. We applied 200 μL of NPs, either B-PLGA-95, B-PLGA-75, and B-PCL, in the donor chamber (in sextuplicate per each formulation).

As a control, a solution of BTB in phosphate buffer pH 7.4 was included in the study. Cornea tissues were positioned between the two compartments of a Franz cell with the corneal side facing upwards, towards the donor chamber, with the other side staying in contact with the receptor medium. The donor chamber was covered with a parafilm to prevent evaporation during the study. Phosphate buffered saline (PBS), namely Transcutol® (1:1), was used as the receptor medium. This was necessary due to the low solubility of BTB on Transcutol® [25]. The permeation study was conducted for 6 h at 32 ± 0.5 °C under continuous stirring to maintain sink conditions. At each sampling interval of up to 6 h, a volume 200 µL of receptor medium was withdrawn and an equal volume of fresh Transcutol® was added. Samples were analyzed by HPLC (see Section 2.5.1) with a fluorescence detector to determine the cumulative amount of the drug that was permeated. Figure 1 shows the isolation of the cornea from the eyeball and the preparation of the Franz cells.

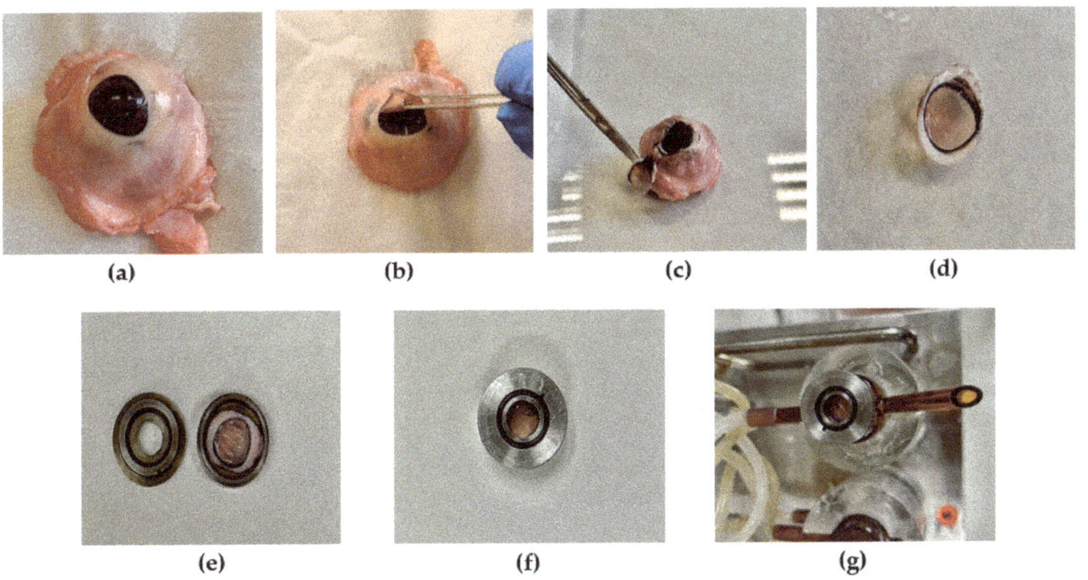

Figure 1. Isolation of the cornea and assembly on the Franz cell with 0.64 cm^2 diffusional area for the ex vivo transcorneal permeation study: (**a**) excised porcine eyeball; (**b**,**c**) isolation of the cornea from the eyeball; (**d**) isolated porcine cornea; (**e**,**f**) fasten of the cornea in the donor compartment; and (**g**) assembly of the donor compartment on the Franz cell.

Following the ex vivo permeation study, the drug retained in the corneas was extracted as follows: after disassembling the corneas from the diffusion cells, the NPs on the corneal surface were eliminated by rinsing them with distilled water. The permeation area was isolated, weighed, pricked with a thin needle, submerged in 1 mL of Transcutol®, and sonicated in an ultrasonic water bath for 15 min. The resultant solution was filtered and quantified by HPLC (see Section 2.5.1) to obtain the amount retained in the tissues (Q_{ret}) according to Equation (3):

$$Q_{ret} = \frac{Q_{ext}}{A \times W \times R} \quad (3)$$

Q_{ext} is the amount of drug retained in the cornea (µg), A is the Franz cell diffusion area (cm^2), W is the weight of the cornea (mg), and R is the recovery of BTB in each tissue [26,27].

Permeation parameters such as the permeability coefficient (K_p, cm/h), flux (J, $\mu g\ h^{-1} cm^{-2}$), partition and diffusion parameters (P_1 and P_2), and the lag time (T_l, h) were calculated from the ex vivo permeation study. The permeation profiles were structured by plotting the cumulative amount of (μg) permeated per surface area of the corneal tissue (cm^2) versus time (h). In the plot, the T_l is the intercept with the *X*-axis (time) and the flux is the slope of the linear part of the permeation profile calculated by linear regression analysis (GraphPad Prism 5 software, Inc., San Diego, CA, USA). The permeability coefficient was calculated through Equation (4), as follows:

$$K_p = \frac{Jss}{C_0} \tag{4}$$

Jss ($\mu g/h/cm^2$) is the flux across the cornea and C_0 ($\mu g/cm^3$) is the initial amount of formulation tested in the donor compartment. It is assumed that, under sink conditions, the drug concentration in the receiver compartment is negligible compared to that in the donor compartment [28]. Other permeation parameters were calculated according to the following equations:

$$P_2 = \frac{1}{6.T_l} \tag{5}$$

where P_2 is the diffusion coefficient and T_l is the lag-time.

$$P_1 = \frac{K_p}{P_2} \tag{6}$$

where P_1 is the partition coefficient, P_2 is the diffusion coefficient, and K_p is the permeability coefficient.

The ratio of the permeability coefficients between NPs and the BTB solution led to calculating the permeation enhancement ratio (*ER*) with Equation (7) [29], as follows:

$$ER = \frac{Kp_{\ nanoparticle}}{Kp_{\ solution}} \tag{7}$$

One method used to assess the integrity of the cornea is to determine the percentage levels of corneal hydration (HL), which was obtained through Equation (8), as follows:

$$\text{HL (\%)} = (1 - W_b/W_a) \times 100 \tag{8}$$

Upon completion of the permeation study, the corneas were recovered from the Franz cells and weighed; this weight determined the wet corneal weight (W_a). The corneas were then desiccated until a constant weight was reached. After desiccation at 80–100 °C for 6 h, samples were weighed again to determine the dry corneal weight (W_b) [30].

2.8. *In Vitro Tolerance Study*

2.8.1. Hen's Egg Test on the Chorioallantoic Membrane (HET-CAM)

The ocular potential irritancy of NPs was determined by observing the adverse changes that occurred in the chorioallantoic membrane (CAM) of fertilized eggs after exposure to the tested NPs. Eggs were incubated for 9 d, after which non-viable or defective eggs were discarded. The shell on the air cell portion was removed and the inner membrane was extracted after being washed and hydrated with PBS to reveal the CAM. Then, 300 μL of the test NPs were applied to the CAM and reactions were observed within 300 s. One egg was treated with 0.1 N sodium hydroxide as the positive control and another with 0.9% NaCl solution as the negative control. The endpoints to be observed were bleeding (bleeding from vessels), vascular lysis (breakdown of blood vessels), and coagulation (intravascular and extravascular protein denaturation) [30,31]. The severity of vascular damage observed in the chorioallantoic membrane is an indication of the product's potential to damage mucosal membranes in vivo, assuming that there is a correlation between the severity of damage and the speed at which it occurs [26]. The time of the appearance and the intensity of any reactions that occur within 5 min were recorded according to the INVITTOX protocol [26]. Then, the eye irritation score (*IS*) was calculated using Equation

(9) and the degree of potential irritation for each formulation was established based on the criteria defined by the INVITTOX protocol (Table 6).

$$IS = \frac{301 - H}{300} \cdot 5 + \frac{301 - L}{300} \cdot 7 + \frac{301 - C}{300} \cdot 9 \quad (9)$$

Table 6. Criteria for classification of the severity of irritation [30].

Irritation Score	Irritation Classification
0–0.9	Non-Irritant
1–4.9	Slight Irritant
5–8.9	Moderate Irritant
9–21	Severe Irritant

H is the time (s) to the beginning of hemorrhage, *L* is the time (s) to lysis, and *C* is the time (s) to coagulation.

2.8.2. Cornea Transparency

The corneas were carefully isolated from the eyeball with a scalpel and rinsed with PBS pH 7.4. One eye from the same donor served as a negative control while the other eye was exposed to either ethanol (positive control) or one of the NPs. Three animals were included per group, which were the following: (i) ethanol; (ii) B-PLGA-95; (iii) B-PLGA-75 and (iv) B-PCL. The corneas were mounted on the Franz cells, and 2 drops of the following solutions were applied to the cornea: physiological saline solution as negative control, ethanol as positive control, and the colloidal suspensions of B-PLGA-95, B-PLGA-75 and B-PCL NPs. The exposure time was 2 h, and 1–2 drops of physiological saline solution was applied to the cornea every 5–10 min to simulate the tear clearance. After two hours, the corneas were demounted from the Franz cells and evaluated for transparency by measuring the transmittance with a Nanodrop® spectrophotometer TM 2000 (Thermo® Fisher Scientific, Waltham, MA, USA). Samples were cleaned with saline solution to remove any contaminants and carefully dried under a stream of N_2. The transmittance of each sample was measured in the range of 380–780 nm in at least 3 different regions of the cornea near its center to evaluate any changes in corneal transparency caused by the NPs.

2.8.3. Cornea Histology Study

To determine if the formulations had an impact on the structure of the cornea, the same procedure used for the cornea transparency evaluation was followed. For the histological examination, the three different NPs formulations (B-PLGA-95, B-PLGA-75 and B-PCL) were applied to the corneal tissues while distilled water served as the negative control. The tissues were exposed to the formulations for 3 h, followed by processing for hematoxylin and eosin staining [31]. To prepare the tissues for histology, the corneas were immersed in 4% buffered paraformaldehyde solution for 24 h to fix the tissues, then dehydrated with increasing concentrations of ethanol. Finally, the corneas were embedded in paraffin, cut into sections of 5 µm, and stained with hematoxylin and eosin; DPX was used as the mounting medium (Sigma–Aldrich). The histological samples were analyzed under a microscope Olympus BX41 and camera Olympus XC50; (Olympus, Hamburg, Germany) in a blinded fashion to evaluate any changes in corneal structure caused by the nanoparticle formulations.

3. Results

3.1. Factorial Design

A factorial design was conducted to optimize the development of NPs with two polymers, PLGA and PCL. The evaluation of the effects considered three factors (amount of polymer, amount of surfactant, and volume of the aqueous phase) and used the responses (Z-ave and PDI) to select the optimized NP composition from a total of 19 formulations with each polymer. Table 7 shows the mean particle size and polydispersity index for PLGA-based NPs and Table 8 reports the results obtained for the responses for the PCL-based NPs.

Table 7. Responses for each PLGA-based formulation: mean particle size (Z-ave) and polydispersity index (PDI).

	Levels of Factors			Responses	
Formulations	**PLGA Amount (mg)**	**P188Amount (mg)**	**Aqueous Phase Volume (mL)**	**Z-Ave ± SD (nm)**	**PDI ± SD**
F1	75	150	50	206.65 ± 2.9	0.303 ± 0.016
F2	115	150	50	245.80 ± 2.1	0.177 ± 0.035
F3	75	250	50	201.50 ± 1.8	0.265 ± 0.049
F4	115	250	50	237.55 ± 5.0	0.150 ± 0.001
F5	75	150	70	180.80 ± 4.9	0.213 ± 0.050
F6	115	150	70	209.90 ± 3.0	0.231 ± 0.035
F7	75	250	70	197.25 ± 1.2	0.217 ± 0.053
F8	115	250	70	233.05 ± 1.3	0.166 ± 0.079
F9	95	200	60	98.20 ± 0.0	0.090 ± 0.056
F10	95	200	60	118.70 ± 0.6	0.168 ± 0.017
F11	72	200	60	209.55 ± 1.8	0.103 ± 0.028
F12	118	200	60	288.20 ± 4.5	0.091 ± 0.051
F13	95	134	60	214.75 ± 3.6	0.218 ± 0.022
F14	95	266	60	225.25 ± 3.0	0.117 ± 0.009
F15	95	200	47	217.55 ± 2.2	0.251 ± 0.008
F16	95	200	73	206.40 ± 1.6	0.101 ± 0.038
F17	95	200	60	116.60 ± 0.7	0.184 ± 0.002
F18	95	200	60	104.90 ± 0.0	0.191 ± 0.016
F19	95	200	60	159.75 ± 0.8	0.056 ± 0.010

The highlighted formulation was one of the selected ones for further studies.

Table 8. Responses for each PCL-based formulation: mean particle size (Z-ave) and polydispersity index (PDI).

	Levels of Factors			Responses	
Formulations	**PCL Amount (mg)**	**P188 Amount (mg)**	**Aqueous Phase Volume (mL)**	**Z-Ave ± SD (nm)**	**PDI ± SD**
F1	50	200	38	104.20 ± 1.70	0.137 ± 0.004
F2	70	200	38	138.30 ± 5.94	0.165 ± 0.016
F3	50	280	38	109.35 ± 0.78	0.174 ± 0.006
F4	70	280	38	120.00 ± 6.93	0.247 ± 0.062
F5	50	200	50	127.50 ± 1.98	0.107 ± 0.037
F6	70	200	50	398.65 ± 15.34	0.159 ± 0.047
F7	50	280	50	366.40 ± 4.38	0.160 ± 0.065
F8	70	280	50	339.75 ± 2.76	0.197 ± 0.009
F9	60	240	44	347.50 ± 2.55	0.223 ± 0.015
F10	60	240	44	426.60 ± 6.79	0.029 ± 0.030
F11	46	240	44	403.90 ± 1.56	0.096 ± 0.053
F12	118	240	44	335.75 ± 3.32	0.355 ± 0.066
F13	60	185	44	370.60 ± 4.53	0.158 ± 0.031
F14	60	295	44	349.95 ± 6.43	0.182 ± 0.040
F15	60	240	36.5	401.35 ± 7.85	0.215 ± 0.011
F16	60	240	51.5	374.95 ± 1.06	0.200 ± 0.013
F17	60	240	44	458.10 ± 4.95	0.203 ± 0.027
F18	60	240	44	356.90 ± 2.83	0.181 ± 0.034
F19	60	240	44	309.70 ± 6.08	0.147 ± 0.048

The highlighted formulation was one of the selected ones for further studies.

The mean particle size for the PLGA-based NPs was within the range of 98.20 and 288.20 nm. The polydispersity index (PDI) indicates the degree of uniformity in the size distribution of the NPs. For polymeric NPs, values of PDI of up to 0.2 are considered acceptable. Most of the NPs exhibit PDI values below 0.2, indicating that the obtained NPs were monodisperse colloidal systems [32]. In general, a low amount of surfactant led to higher PDI values. Figure 2 illustrates the effects of several factors (amount of polymer, surfactant, and volume of the aqueous phase) on particle size and PDI for the PLGA-based NPs.

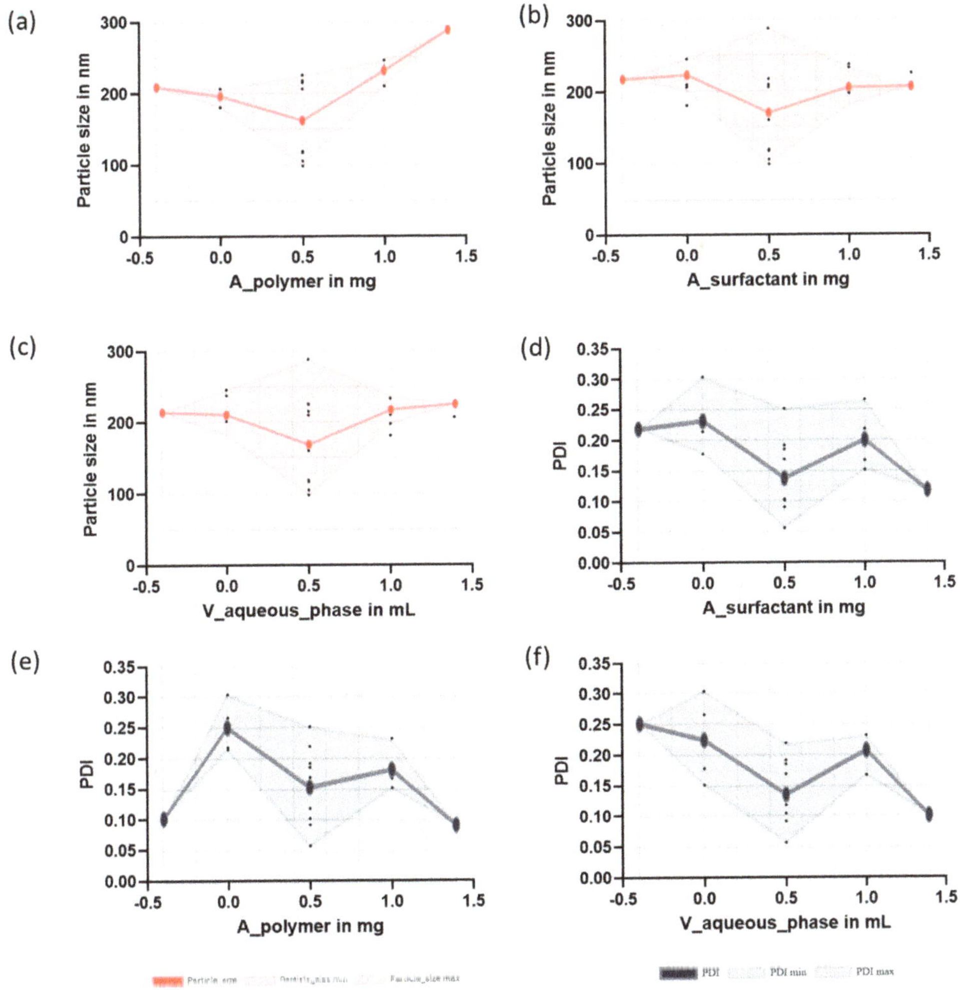

Figure 2. Main effects of the factors on the responses for the PLGA-based NPs. Effects on the particle size (red panels): (**a**) amount of polymer on the particle size; (**b**) volume of the aqueous phase on the particle size; (**c**) amount of surfactant on the particle size. Effects on the PDI (grey panels): (**d**) amount of surfactant on the PDI; (**e**) amount of polymer on the PDI; (**f**) volume of the aqueous phase on the PDI. The *x*-axis is given as levels which are low (0), centered (0.5), high (1), and axial (−0.39; 1.39).

In ocular delivery, the size and distribution of the particles are critical features. Since large particles should irritate and can cause discomfort to the eye [32,33], the optimization of the NPs was focused on obtaining small NPs with narrow distribution sizes [34]. F9 was selected among the 19 formulations prepared, which corresponded to the centered level for the three factors. In addition, a modified F5 formulation was also prepared (with an increased amount of surfactant) taking into account that the central point of surfactant led to small NPs and narrow PDI. The values in Table 2 informed the composition of the modified F5 formulation, which was further characterized and evaluated in subsequent studies.

Figure 3 shows the effects of the factors (amount of polymer, surfactant and aqueous phase) on the responses (particle size and PDI) for the PCL-based NPs. A low amount of polymer and surfactant and a small volume of the aqueous phase led to the smallest particle size. Low PDI values were found for the formulations composed of low amounts of polymer. Based on the results, the selected formulation for PCL-based NPs was F5 (Table 8), which corresponded to the low levels for the three factors (polymer, surfactant, and aqueous phase).

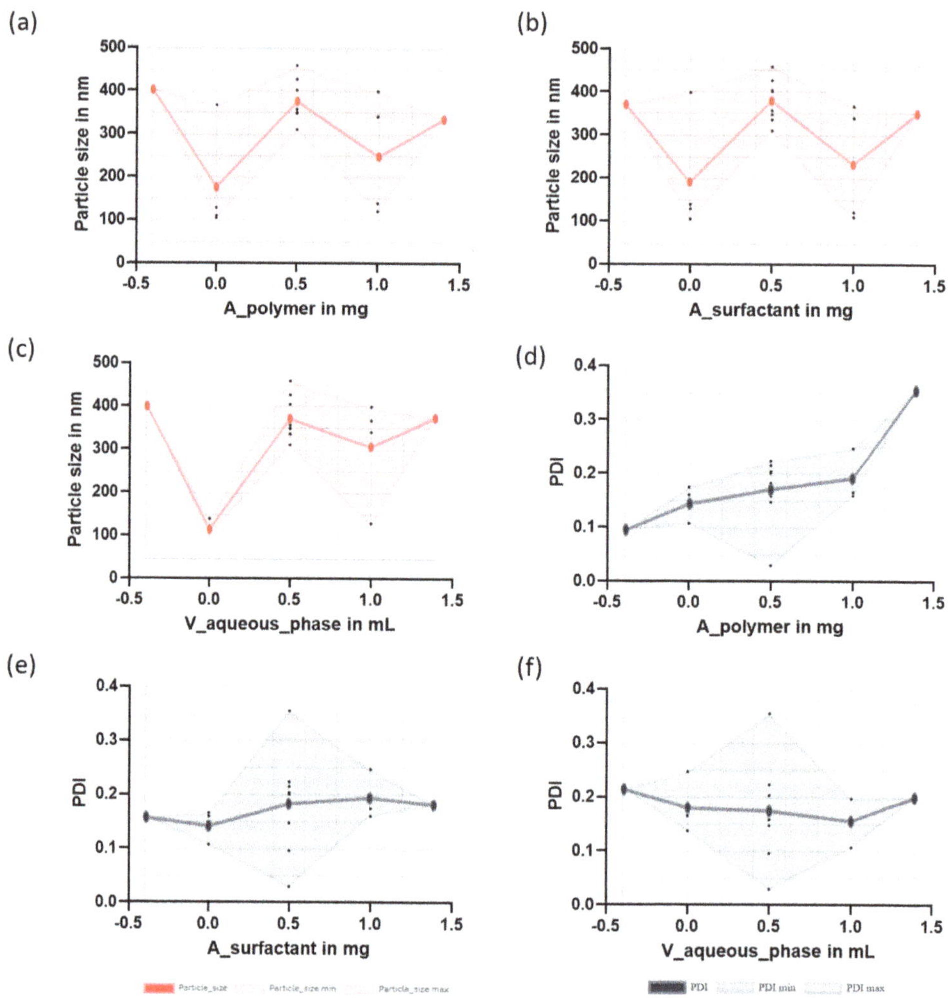

Figure 3. Main effects of the factors on the responses for the PCL-based NPs. Effects on the particle size (red panels): (**a**) amount of polymer on the particle size; (**b**) amount of surfactant on the particle size; (**c**) volume of the aqueous phase on the particle size. Effects on the PDI (gray panels): (**d**) amount of polymer on the PDI; (**e**) amount of surfactant on the PDI; (**f**) volume of the aqueous phase on the PDI. The *x*-axis is given as levels: low (0), centered (0.5), high (1) and axial (−0.39; 1.39).

Thus, further characterization and investigations were conducted on the selected formulations F9-PLGA, modified F5-PLGA and F5-PCL, which corresponded to the formulations indicated in Table 3 (B-PLGA-95, B-PLGA-75 and B-PCL, respectively). Figure 4 shows the suspension of selected NPs.

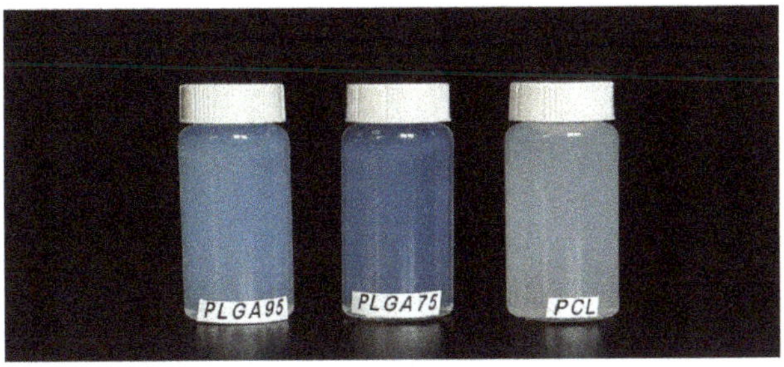

Figure 4. Physical appearance of the colloidal suspension of NPs.

3.2. Nanoparticle Physicochemical Characterization

Particle size is a crucial parameter for ocular drug delivery systems, as it can affect the risk of irritation and discomfort [35]. The particle size of all the formulations that were developed were suitable for ocular administration, featuring diameters smaller than 250 nm (Table 9). Additionally, the polydispersity index (PDI) values for all NPs were less than 0.1, indicating a narrow size distribution [36]. Table 9 presents the physicochemical characteristics of the B-PLGA-95, B-PLGA-75, and B-PCL NPs. Despite the relatively low zeta potential (ZP) values, between −20 and −28 mV, the nanoparticles were shown to be stable for at least one month since they exhibited minor variations in particle size and PDI over time (Section 3.2.5). Zeta potential is a key parameter that influences particle stability; higher absolute ZP values generally indicate better stability due to increased electrostatic repulsion [37]. However, even with ZP values of around −20 mV, the formulations remained stable due to other contributing factors. The narrow size distribution, indicated by low PDI values, and the small particle sizes help maintain stability by reducing the likelihood of particle aggregation [38]. Moreover, all formulations had a pH of approximately 6.8, which is biocompatible with the ocular environment and further supports the stability and safety of these formulations for ocular use [39,40].

Table 9. Physicochemical properties of the optimized NPs: B-PLGA-95, B-PLGA-75, and B-PCL NPs. All measurements have been taken with three replicates ($n = 3$).

Formulation (NPs)	Z-Ave ± SD (nm)	PDI ± SD	ZP (mV) ± SD	pH ± SD	EE (%) ± SD
B-PLGA-95	84.50 ± 0.72	0.10 ± 0.01	−25.70 ± 2.01	6.9 ± 0.3	65.63 ± 3.87
B-PLGA-75	77.65 ± 0.03	0.04 ± 0.02	−28.20 ± 0.14	6.8 ± 0.3	27.39 ± 2.05
B-PCL	138.36 ± 0.92	0.09 ± 0.02	−20.55 ± 0.07	6.8 ± 0.2	77.82 ± 5.33

Z-ave: average particle size; PDI: polydispersity index; ZP: zeta potential; SD: standard deviation; EE: Encapsulation efficiency.

3.2.1. Encapsulation Efficiency

The entrapment efficiency provides an insight into the amount of BTB that is effectively loaded in the NPs. Typically, an excellent drug carrier should have high entrapment efficiency (EE). High EE (above 70%) can increase the efficacy of the drug delivery system and decrease the side effects of the drug [41]. The EE of the B-PLGA-95, B-PLGA-75, and B-PCL NPs were measured in the range of 65.63 ± 3.87, 27.39 ± 2.05, and 77.82 ± 5.33, respectively (Table 9). The highest drug entrapment (77.82 ± 5.33%) was found in the case of B-PCL NPs. The large amount of BTB formulated in this sample is supposed to prevent the drug's diffusion from the polymeric core, thereby enhancing the entrapment of the drug [42].

3.2.2. Morphological Study

The appearance, droplet, and particle size of nanoparticle formulations B-PLGA-95, B-PLGA-75, and B-PCL NPs were evaluated. The NPs were spherical in shape, and the TEM analysis confirmed the size distribution of NPs obtained by dynamic light scattering. As shown in Figure 5, droplets were nearly uniformly distributed in B-PLGA-95 and B-PLGA-75 formulations. However, B-PCL showed larger particles and a less uniform system compared to the other formulations. The results indicated that both the B-PLGA-95 and B-PLGA-75 NPs evaluated in this study were the most suitable for ophthalmic applications [43].

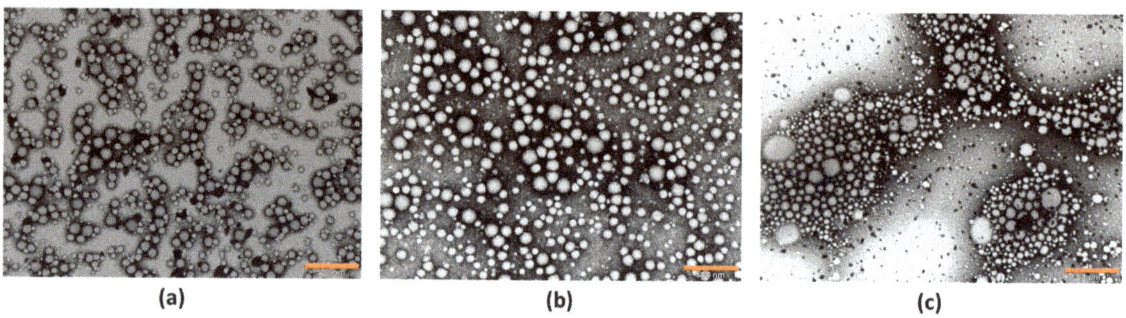

Figure 5. Transmission electron microscopy (TEM) images of NPs of (**a**) B-PLGA-95 NPs, (**b**) B-PLGA-75 NPs, and (**c**) B-PCL NPs. Bar length (**a**,**b**) 500 nm, (**c**) 1 μm.

3.2.3. Extensibility Studies

All of the formulations were in accordance with the two-site binding parabola model (Figure 6). Extensibility values increase with loading weight. The B-PLGA-95 NPs were the most extensible but did not exhibit significant statistical differences with B-PLGA-75 nor B-PCL NPs. Water was used as the reference.

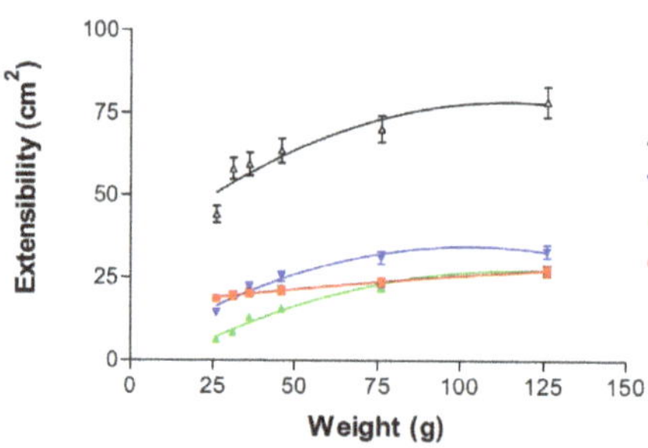

Figure 6. The scheme shows how all NPs follow the same mathematical fit models with their respective parameter values for extensibility (expressed in cm^2 by weight (Mean ± SD) (n = 3)).

Table 10 reports the predicted values of the parameters according to the two-site binding parabola model, which was the mathematical model that best described the extensibility behavior of the NPs.

Table 10. Mathematical fitting of the NPs extensibility, which followed the two-site binding parabola model. Results are reported as the mean of each model parameter and the standard error (Std. Error).

Best-Fit Values	B-PCL	B-PLGA-75	B-PLGA-95	Water
A	16.03	−5.633	2.408	31.99
B	0.125	0.562	0.624	0.812
C	−0.000279	−0.002398	−0.003030	−0.003556
Std. Error				
A	0.456	2.849	3.556	10.720
B	0.016	0.097	0.121	0.365
C	0.000101	0.000629	0.000785	0.002366
Goodness of Fit				
R^2	0.999	0.994	0.988	0.960

3.2.4. Rheological Characterization

Steady-state rheological measurements as a function of shear rate are shown in Figure 7. All formulations exhibited typical Newtonian behavior in which the shear stress to shear rate relationship (flow curve) was linear, with an absence of thixotropy, and the viscosity remained constant. Viscosity values at 100 s^{-1} were 1.22 ± 0.02, 1.03 ± 0.01, and 1.09 ± 0.02 mPa·s for B-PCL, B-PLG-75, and B-PLGA-95 NPs, respectively.

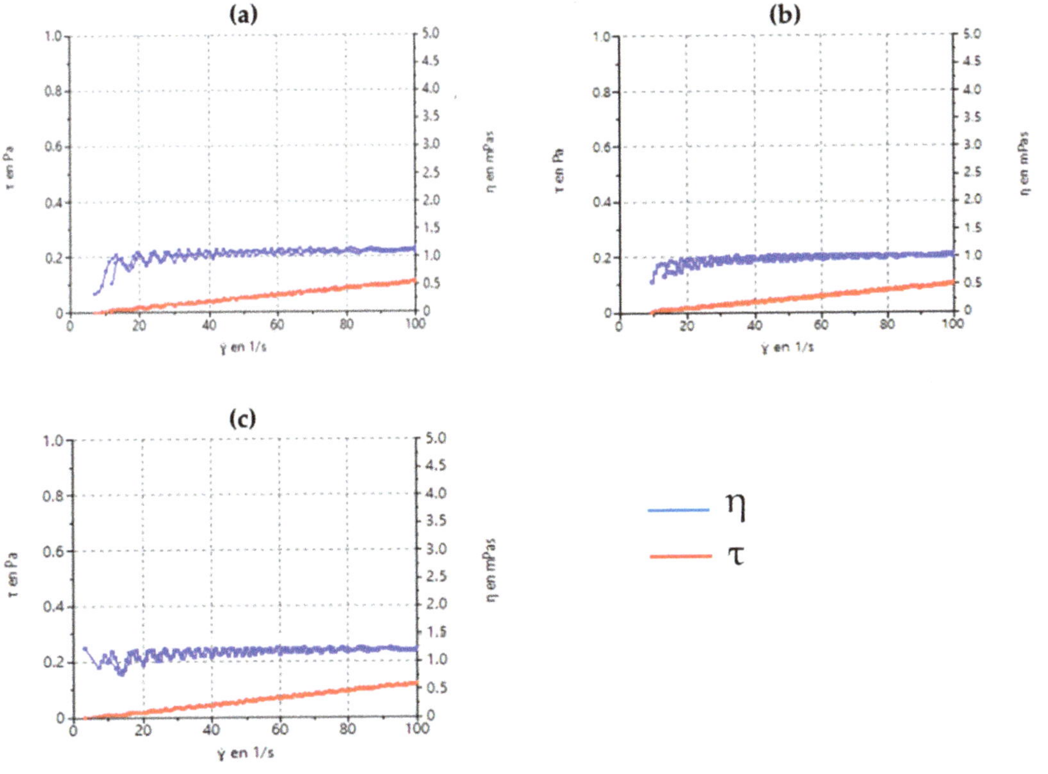

Figure 7. Rheograms of the tested formulations. (**a**) B-PLGA-95 NPs, (**b**) B-PLGA-75 NPs, (**c**) B-PCL NPs. The red line corresponds to the flow curve (τ) while the blue line represents the viscosity curve (η).

3.2.5. Stability of Nanoparticles

The physical stabilities of NPs were evaluated at two different temperatures upon formation (25 and 4 °C) for 30 d (last time point tested). NPs showed no physical changes, such as discoloration during the testing period. A slight increase in the NP size was observed for all the NPs stored at both temperatures, and although these differences were statistically different, they were minor changes that did not indicate particle aggregation. The same was observed for the PDI, i.e., a slight increase in the value over time with statistical differences, but no differences that were relevant to the stability since the NPs showed values of PDI below 0.2. Results are shown in Table 11.

Table 11. Physical stability study of NP formulations after 30 d (n = 3) of storage at 25 and 5 °C. (Mean ± SD).

NPs	Storage Condition (°C)	Z-Ave (nm)			PDI		
		Time Assayed (Days)					
		0	7	30	0	7	30
B-PLGA-95	25	80.56 ± 0.72	81.72 ± 0.22	83.03 ± 2.90	0.03 ± 0.01	0.07 ± 0.02 *	0.07 ± 0.05 *
	4	82.24 ± 0.54	85.48 ± 0.72	91.07 ± 4.85 *	0.02 ± 0.01	0.03 ± 0.01	0.07 ± 0.08 *
B-PLGA-75	25	71.43 ± 0.01	74.75 ± 2.24	81.42 ± 5.42 *	0.05 ± 0.01	0.07 ± 0.02	0.08 ± 0.04
	4	75.23 ± 0.15	77.58 ± 0.24	86.00 ± 5.24 *	0.02 ± 0.01	0.04 ± 0.02	0.06 ± 0.01 *
B-PCL	25	130.36 ± 1.82	134.8 ± 2.12	141.5 ± 5.03 *	0.03 ± 0.01	0.07 ± 0.04	0.12 ± 0.08 *
	4	128.30 ± 1.23	131.30 ± 2.82	152.5 ± 7.79 *	0.06 ± 0.02	0.09 ± 0.07	0.02 ± 0.04 *

* Statistical differences versus Time Assayed = 0 d ($p < 0.01$).

3.3. Antimicrobial Challenge Test

Table 12 shows the results in CFU/mL of the antimicrobial Challenge Test for all the tested formulations. Based on the results, the B-PLGA-95 NP formulation meets criterion B. The B-PLGA-75 NPs met criterion A in terms of logarithmic reduction for both bacterial and fungal inoculum at the indicated times. In the case of the B-PCL formulation, the acceptance criteria were not met. The B-PCL NPs were compatible with *P. aeruginosa* but not with *S. aureus*, *C. albicans*, or *A. brasiliensis*. It would be necessary to add a suitable preservative to prevent growth or limit microbial contamination during storage and use [41].

Table 12. Criteria based on Log reduction for microorganisms by B-PLGA-95, B-PLGA-75, and B-PCL NPs.

Microorganism	Nanoparticle	Log Reduction (CFU/mL)				
		6 h	24 h	7 d	14 d	28 d
P. aeruginosa	B-PLGA-95	1	3	NR		NR
	B-PLGA-75	NR	NR	-	-	NR
	B-PCL	1	NI **	3	-	NR
S. aureus	B-PLGA-95	1	NR *	NR		NR
	B-PLGA-75	NR *	NR	-	-	NR
	B-PCL	-	1	NI (10^5)	-	NR
C. albicans	B-PLGA-95	-	-	NI	1	NI
	B-PLGA-75	-	-	3	NI **	NI
	B-PCL	-	-	NI (10^6)	~1	NI
A. brasiliensis	B-PLGA-95	-	-	1	NI (1)	NI
	B-PLGA-75			NR	NR	NR
	B-PCL	-	-	NI (10^5)	NI (10^5)	NI

* NR: not found; ** NI = no increase.

3.4. Ex Vivo Permeation

The permeation study parameters of the studied NPs are depicted in Table 13. B-PLGA-95 NPs achieved the maximum flux, followed by B-PCL NPs, and B-PLGA-75 NPs. B-PLGA-95 NPs also exhibited the highest permeability K_p. The T_1 parameter indicates the time required for reaching the steady state; this time is inversely proportional to the drug's diffusivity through the cornea. All NPs had a similar T_1 between them. One of the factors, which controls the drug permeability between the formulation and the cornea, is the partition coefficient (P_1). It is directly proportional to the distribution of the drug in the cornea. The greater the value of P_1, the higher the affinity of the drug to the tissue. Thus, once again, B-PLGA-95 NPs were shown to have the highest distribution of NPs. In the case of the diffusion parameter P_2, the results had a similar tendency.

Table 13. Permeation parameters from B-PLGA-95 NPs, B-PLGA-75 NPs, and B-PCL NPs expressed by the mean and standard deviation of six replicates (n = 6). Predicted plasma concentration at the steady-state (C_{ss}); SD = standard deviation; b.l.q. = below the limit of quantification. The statistical analysis consisted of one-way ANOVA test followed by Tukey's Multiple Comparison Test ($p < 0.05$) expressed with letters: ^{A} = statistically significant differences between with B-PLGA-95; ^{B} = statistically significant differences with B-PLGA-75; ^{C} = statistically significant differences with B-PCL.

		B-PLGA-95	B-PLGA-75	B-PCL	B-Solution
Parameters	**Units**	**Mean ± SD**	**Mean ± SD**	**Mean ± SD**	**Mean ± SD**
Flux (J/cm^2)	μg/h/cm^2	20.90 ± 2.47	10.12 ± 1.23 ^{A}	16.12 ± 1.92 ^{A,B}	0.07 ± 0.01 ^{A,B,C}
Permeability coefficient (K_p)	cm/h	0.126 ± 0.014	0.071 ± 0.008 ^{A}	0.081 ± 0.009 ^{A}	0.004 ± 0.000 ^{A,B,C}
Lag Time (T_l)	h	2.90 ± 0.53	2.75 ± 0.37	2.84 ± 0.24	0.83 ± 0.07 ^{A,B,C}
Partition coefficient (P_1)	cm	2.19 ± 0.02	1.17 ± 0.02 ^{A}	1.38 ± 0.01 ^{A,B}	0.03 ± 0.00 ^{A,B,C}
Diffusion coefficient (P_2)	1/h	0.057 ± 0.007	0.061 ± 0.005	0.059 ± 0.005	0.139 ± 0.014 ^{A,B,C}
Amount of BTB in the skin (Q_{ret})	μg/cm^2	3.35 ± 0.29	0.40 ± 0.04 ^{A}	6.09 ± 0.57 ^{A,B}	b.l.q.
Enhancement ratio (ER)	-	27.7	15.6	17.8	-

The NPs composed of PLGA led to lower amounts of baricitinib retained in the cornea than the ones prepared with PCL, being about one-tenth B-PLGA-75 with respect to B-PLGA-95. Thus, when the NPs are exposed to the corneal surface, they form a drug reservoir from which BTB is directly delivered. The drug depot produced with BTB entrapped into NPs guarantees sustained permeation.

The hydration levels of the corneas were measured at the end of the ex vivo transcorneal permeation study to evaluate the integrity of the cornea. The values obtained were 77.1 ± 0.2%, 78.4 ± 0.5%, and 79.0 ± 0.3% for B-PLGA-95, B-PLGA-75, and B-PCL NPs, respectively.

3.5. In Vitro Tolerance Study

3.5.1. Hen's Egg Test on the Chorioallantoic Membrane (HET-CAM)

To determine if the nanoparticle formulations were well tolerated for ocular use, we conducted a HET-CAM in vitro test. The irritation score calculated for positive control was 20.48 ± 1.96. On the contrary, when the different NPs were tested, including B-PLGA-95, B-PLGA-75, and B-PCL formulations with 0.9% NaCl (negative control), no signs of ocular irritancy, such as coagulation, vascular lysis, or hemorrhage, were detected within 5 min. The irritancy score was below 0.1, demonstrating the safety of the NP formulations for ocular use since no reaction occurred during the test period (Figure 8).

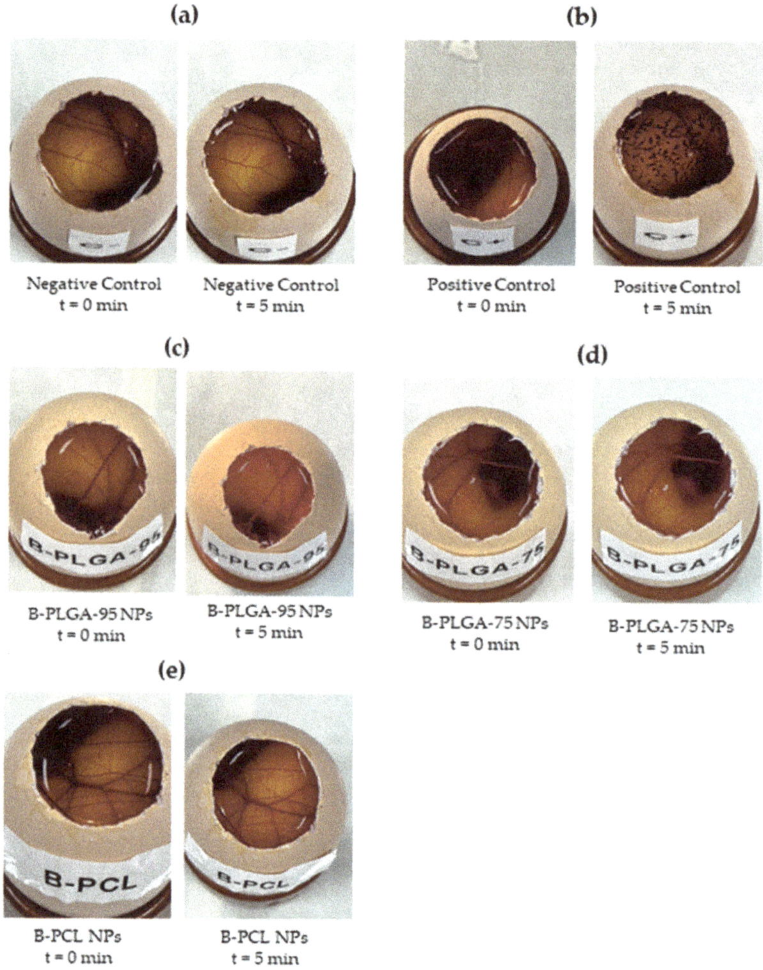

Figure 8. Evaluation of the irritant effect of the formulations by HET-CAM; (**a**) negative control (saline solution), (**b**) positive control (sodium hydroxide solution 0.1 N), (**c**) B-PCL nanoparticles, (**d**) B-PLGA-75 nanoparticles, and (**e**) B-PLGA-95 nanoparticles.

3.5.2. Cornea Transparency

In Figure 9, transmittance values are shown as a function of wavelength for the different treated corneas. As expected, the positive control showed the lowest transmittance values in the range studied. On the one hand, corneas treated with B-PLC NPs or B-PLGA-75 NPs presented a clearness quite similar to the negative control, especially in the range of 380–480 nm. On the other hand, B-PLGA-95 NP-treated corneas showed higher transmittance values than the negative control, suggesting a possible improvement of corneal tissue clarity.

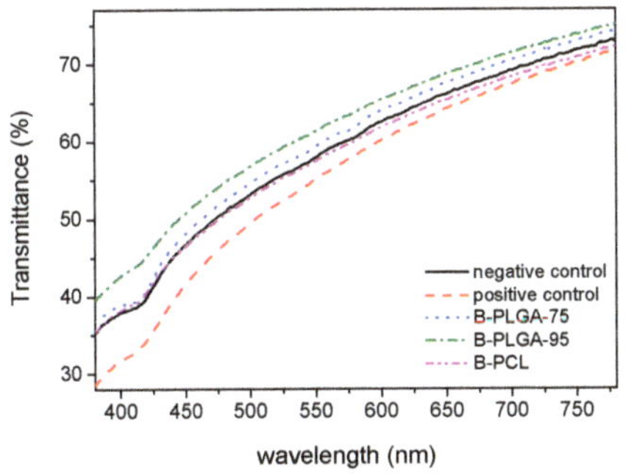

Figure 9. Transmittance values as a function of wavelength in the visible region for the different groups: Ethanol as the positive control, physiological saline solution as the negative control, and the treated corneas with B-PLGA-75, B-PLGA-95, and B-PCL NPs.

3.5.3. Cornea Histological Study

The treatment of the cornea with ethanol induced a disruption of the corneal epithelium that is noticeable in Figure 10b. In negative control conditions, serum did not harm the cornea, as can be seen in Figure 10a. Similarly, the B-PLGA-75 and B-PCL formulations (Figure 10d,e) showed no damage to the corneal epithelium or the substantia propria. In the case of B-PLGA-95 (Figure 10c), the epithelium was not altered; however, the substantia propria showed some dilatations of the tissue that are comparable to the cornea treated with ethanol.

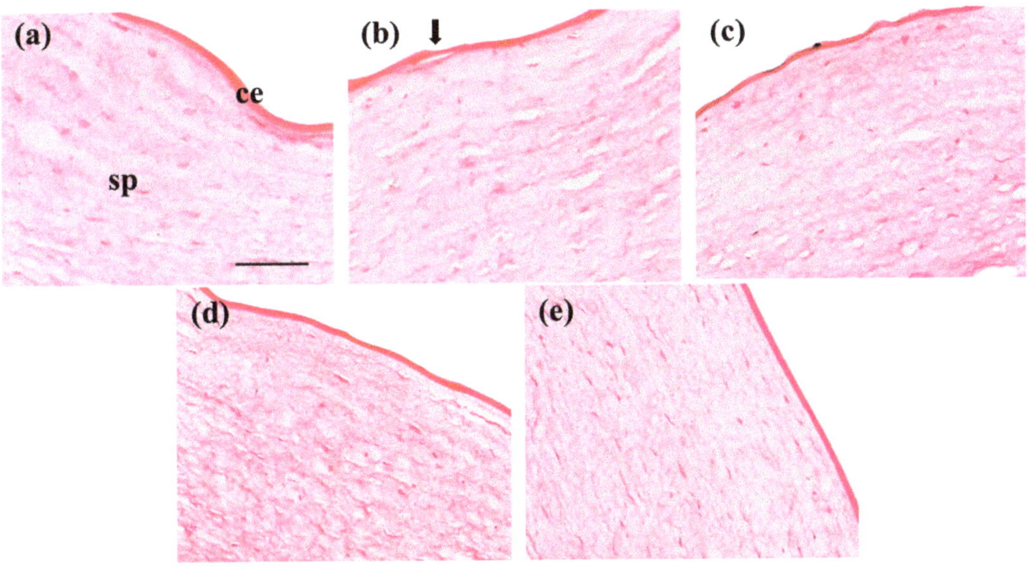

Figure 10. Cornea sections stained with hematoxylin and eosin. Cornea was treated with the following solutions: (**a**) serum (negative control conditions); (**b**) ethanol (positive control); (**c**) B-PLGA-95 NPs; (**d**) B-PLGA-75 NPs; and (**e**) B-PCL NPs. ce: corneal epithelium (non-keratinized stratified squamous epithelium); sp: substantia propria; arrow indicates disruption of ce. Magnification = 200×, scale bar = 100 µm.

4. Discussion

Three variations of NPs containing BTB were developed for use as immunosuppressive therapy for ocular diseases. NPs have great potential as drug carriers for eye treatments with improved drug absorption in the eye compared to eye drop solutions due to the slower elimination rate of the particles [42]. NPs are an effective solution for ocular drug delivery because of their small size, which causes minimal irritation, and their ability to provide a sustained release, reducing the need for frequent drug administration [40]. The NPs developed with both quantities of PLGA have a size smaller than 85 nm and a polydispersity ≤ 0.1, with B-PLGA-75 being the smallest. While the B-PCL NPs have the larger size, they are still suitable for the ocular administration [43]. The ability of NPs to transport drugs is dependent on their size, with smaller particles having a higher permeability [44]. Since smaller particles are better tolerated by patients than larger ones, the PLGA NPs may offer a more comfortable and long-lasting eye delivery system. A high ZP value increases the repulsive forces of NPs, resulting in better stability [45]. B-PLGA-95 is the formulation with the highest ZP, followed by B-PLC, while the lowest formulation is B-PLGA-75. The pH is similar for all NPs, with a value close to the ideal pH for ophthalmic formulations (that of tear fluid, around 7.4) being desired to prevent corneal injury [46]. The EEs of BTB on the different NPs are higher in B-PLGA-95 and B-PCL than in B-PLGA-75, with EE values higher than 60% for B-PLGA-95 and B-PCL and a value of about 27% for B-PLGA-75. A higher encapsulation efficiency with regard to a higher amount of polymer was also observed in the work of Lee and co-workers investigated the effects of polymer concentrations in relation to the drug when preparing NPs, and they observed that, when the polymer to drug ratios increased, maximum EEs were achieved [47].

The morphological study realized by TEM shows the distinctive shapes that the three NPs develop once formed, with B-PCL forming more aggregates, which is a drawback as it loses the homogeneity desired in these formulations. Human tears exhibit a non-Newtonian, shear-thinning viscoelastic profile that extends the contact time with the open eye and safeguards the eye's surface by reducing viscosity throughout the blinking process. Blinking creates significant shear rates within the tear film; as a result, a reduced viscosity is required to prevent harm to the epithelial areas. Viscosity and contact time may be associated with both the comfort experienced and the sharpness of vision. Consequently, these elements could affect patient compliance with their treatment plan and overall quality of life [48–50]. With this in mind, we evaluated the viscosity as an essential characteristic of the NPs developed for ocular delivery. Rheology influences the physical stability, ease of product application, spreadability, and residence time of an ophthalmic product. Ideally, an eye drop formulation that is not a solution should possess adequate viscosity to extend contact time on the eye, while also being easily flowable when subjected to shear conditions during blinking, to ensure an even distribution across the surface of the eye [51]. Extensibility is the evaluation of the formulation focused on its ability to spread; in other words, how easily the formulation spreads by blinking. The extensibility studies indicate that the formulations are appropriate for use as eye drops.

The stability study results showed that the three formulations, which were stored at room temperature and at 4 °C, had no significant differences between the two temperatures after a week. Differences can be observed between the values on the first day and those obtained after a month, and these differences are less significant for the formulations stored at room temperature. Despite the differences observed, these do not cause a negative impact at a critical level since the values are still within the acceptance criteria for stability [43].

Ophthalmic products are required to meet strict sterility standards [52,53]. Ensuring sterility is crucial since microbial contamination can lead to conditions such as conjunctivitis, keratitis, and even infectious endophthalmitis, which may result in vision loss [52]. Our NPs were sterilized via filtration through a syringe filter of 0.22 μm, and a Challenge Test was performed to evaluate the product's resistance—"robustness"—to microbial contamination during its use. Among the three NPs tested in microbiology studies, B-PLGA-75 NP is the nanoparticle that offers the best results against the harmful effects that contamination or microbial proliferation can cause during storage and use. For the rest, it would be necessary to add a suitable preservative to limit microbial contamination during storage and use or to store them in unidose packaging. In the treatment of pathological conditions, ocular drug delivery systems should ensure that drug molecules reach the site of action at determined concentrations and within the effective therapeutic dosage range. For this reason, a permeation study was performed. The concentration of Transcutol® was optimized to ensure sink conditions while maintaining the integrity of the corneal tissue. This was evaluated upon completion of the permeation study by determining the corneal hydration level. The corneal HL is a parameter frequently used to evaluate the damage to this tissue. Silva-Abreu et al. did an ex vivo cornea permeation with Transcutol®/water solution (60/40 v/v). They assessed the cornea integrity after the experiments by ocular HL, which showed values within the accepted range of 76–80% [28]. HL values 3–7% above the normal value denote damage of the epithelium or endothelium [54]. The HL for the assayed NP formulations was 78.24%, which corroborated the lack of damage on the corneal tissue.

Ex vivo corneas are often used to study drug permeation through the corneal tissue. The goal is to understand how the drug penetrates the cornea and reaches the anterior chamber of the eye after the ophthalmic application. The ex vivo permeation studies show that BTB reached the receptor fluid from all three NPs, suggesting that the NPs could be good carriers for transporting the drug to the aqueous humor. B-PLGA-95 is the formulation that promotes the most diffusion of BTB through the cornea, followed by B-PCL, and finally B-PLGA-75. The permeation enhancement ratio (ER) is a quantitative measure used to compare the permeation of a drug from a nanoparticle formulation to that from a plain solution. An ER greater than 1 indicates that the nanoparticle formulation enhances drug permeation compared to the plain solution. The higher the ER, the more effective the NPs are at enhancing permeation. NPs enhanced the permeation of BTB between 15 and 28 folds. The lag time (T_l) is the time it takes to observe BTB in the steady state, and this is similar for all three NPs. After a T_l of about 3 h, a linear relationship between the cumulative amounts of permeated drug versus time was observed for the assayed NPs, indicating that the permeation rate was constant. Regarding the partition and diffusion coefficients (P_1, P_2, respectively), P_1 describes the ability of the drug to leave the formulation and reach the cornea; the higher the value, the higher this capacity is. The only NP formulation with a significantly higher value is B-PLGA-95. On the other hand, P_2 explains the ability of the drug to diffuse through the cornea; this parameter does not have differences in regard to any of the studied NPs. The amount of BTB retained in the cornea (Q_{ret}) is statistically different between B-PLGA-95 NPs and B-PCL/B-PLGA-75. The highest drug retention was observed for B-PCL NPs, followed by B-PLGA-95 and, finally, B-PLGA-75. Therefore, the NPs would be in contact with the corneal surface, forming a drug depot from which BTB would be directly transferred.

The histological analysis of the corneas exposed to the NP formulations showed no alteration on the cornea surface. Although in the case of B-PLGA-95, a slight loosening of the substantia propria could be observed, the results of the HET-CAM study indicate that the NPs made from B-PLGA-95, B-PLGA-75, and B-PCL are well tolerated and do not have irritation potential in regard to the eye. This is in line with the biodegradable and biocompatible nature of these polymers, which are known to be well tolerated by ocular tissues [55,56]. The results of the cornea transparency test also support this conclusion, as the corneas treated with B-PCL or B-PLGA-75 had similar transparency to the negative control, further indicating that these NPs are safe for ocular administration. Conversely, corneas treated with B-PLGA-95 showed higher transmittance values compared to the negative control, indicating potential improvement in corneal tissue clarity.

The results obtained and analyzed in this section help us choose the B-PLGA-95 NPs as the best outlook for future studies. They are ophthalmologically compatible and appear to effectively increase the diffusion of the drug into the cornea, where the drug is retained and could have a local effect.

5. Conclusions

The results of this study aim to accomplish the demand for efficient ocular drug delivery systems. This study reports an approach on the use of a three-factor three-level central rotatable composite 2^{3+} star design in the optimization of PLGA and PCL NPs containing BTB using the solvent displacement method. Three different NP formulations were optimized for delivering BTB to the cornea. These NPs were stable for at least 30 days, stored at 25 and 5 °C, and enhanced the permeation of BTB. Among the three NPs, the encapsulation of BTB in PLGA-95 NPs showed the most promising results. In vivo studies should be addressed in the future to demonstrate the advantages of PLGA-based NPs loading BTB for treating ocular diseases.

Author Contributions: Conceptualization, M.M. and A.C.; methodology, N.B., N.G., R.M.-M., Ò.D., J.S.-C., M.J.R.-L., M.J.M., G.K. and M.M.; software, A.C. and P.B.-S.; validation, A.C., G.K., M.J.R.-L. and M.M.; formal analysis, N.B., N.G., R.M.-M., Ò.D., J.S.-C., G.K., M.J.R.-L., M.J.M. and M.M.; investigation, N.B., N.G., J.S.-C., R.M.-M., M.J.M. and G.K.; resources, A.C., M.J.R.-L. and M.M.; data curation, A.C., M.J.R.-L., P.B.-S. and M.M.; writing—original draft preparation, N.B., P.B.-S. and N.G. writing—review and editing, N.B., P.B.-S., N.G., Ò.D., J.S.-C., M.J.M., M.J.R.-L. and M.M.; visualization, A.C. and M.M.; supervision, A.C., P.B.-S., M.J.R.-L. and M.M.; project administration, A.C., M.J.R.-L. and M.M.; funding acquisition, A.C., M.J.R.-L. and M.M. All authors have read and agreed to the published version of the manuscript.

Funding: This research received no external funding.

Institutional Review Board Statement: The animal study protocol was approved by the Animal Experimentation Ethics Committee of the University of Barcelona (code 10619 approved on 10 January 2019).

Informed Consent Statement: Not applicable.

Data Availability Statement: The data are not publicly available due to this work is part of a doctoral thesis, and the data included in this work will be available once the project is completed and presented.

Acknowledgments: The authors would like to thank the Universitat de Barcelona for funding open access publishing. A special thanks to Lilian Sosa for her technical support and Chandler Quilchez for the English revision.

Conflicts of Interest: The authors declare no conflicts of interest.

References

1. Foster, C.S.; Vitale, A.T. *Diagnosis & Treatment of Uveitis; Jaypee Highlinghts*, 2nd ed.; Medical Publishers, Inc.: Philadelphia, PA, USA, 2002.
2. Kempen, J.H.; Gangaputra, S.; Daniel, E.; Levy-Clarke, G.A.; Nussenblatt, R.B.; Rosenbaum, J.T.; Suhler, E.B.; Thorne, J.E.; Foster, C.S.; Jabs, D.A.; et al. Long-Term Risk of Malignancy Among Patients Treated with Immunosuppressive Agents for Ocular Inflammation: A Critical Assessment of the Evidence. *Am. J. Ophthalmol.* **2008**, *146*, 802–812. [CrossRef] [PubMed]
3. Weber-Schöndorfer, C. 2.12—Immunomodulators. In *Drugs During Pregnancy and Lactation*, 2nd ed.; Schaefer, C., Peters, P., Miller, R.K., Eds.; Academic Press: Oxford, UK, 2007; pp. 321–334. ISBN 978-0-444-52072-2.
4. Yi, J.; Zhang, Y.; Liang, R.; Zhong, F. Beta-Carotene Chemical Stability in Nanoemulsions Was Improved by Stabilized with Beta-Lactoglobulin—Catechin Conjugates through Free Radical Method. *J. Agric. Food Chem.* **2015**, *63*, 297–303. [CrossRef]
5. Major, J.; Foroncewicz, B.; Paweł, J.; Mucha, K. Immunology and Donor- Specific Antibodies in Corneal Transplantation. *Arch. Immunol. Ther. Exp.* **2021**, *69*, 32. [CrossRef]
6. Coster, D.J.; Williams, K.A. The Impact of Corneal Allograft Rejection on the Long-Term Outcome of Corneal Transplantation. *Am. J. Ophthalmol.* **2005**, *140*, 1112–1122. [CrossRef] [PubMed]
7. Guilbert, E.; Laroche, L.; Borderie, V. Le Rejet d'allogreffe de Cornée. *J. Fr. Ophtalmol.* **2011**, *34*, 331–348. [CrossRef]
8. Garrós, N.; Mallandrich, M.; Beirampour, N.; Mohammadi, R.; Domènech, Ò.; Rodríguez-Lagunas, M.J.; Clares, B.; Colom, H. Baricitinib Liposomes as a New Approach for the Treatment of Sjögren's Syndrome. *Pharmaceutics* **2022**, *14*, 1895. [CrossRef] [PubMed]
9. Kaneko, Y.; Murakami, T.; Nishitsuka, K.; Takakubo, Y. Effectiveness of Baricitinib in Refractory Seronegative Rheumatoid Arthritis and Uveitis: A Case Report. *Front. Med.* **2022**, *8*, 764067. [CrossRef]

10. Gaudana, R.; Ananthula, H.K.; Parenky, A.; Mitra, A.K. Ocular Drug Delivery. *Am. Assoc. Pharm. Sci.* **2010**, *12*, 348–359. [CrossRef] [PubMed]
11. Li, S.; Chen, L.; Fu, Y. Nanotechnology-Based Ocular Drug Delivery Systems: Recent Advances and Future Prospects. *J. Nanobiotechnol.* **2023**, *21*, 232. [CrossRef]
12. Yang, C.; Yang, J.; Lu, A.; Gong, J.; Yang, Y.; Lin, X.; Li, M.; Xu, H. Nanoparticles in Ocular Applications and Their Potential Toxicity. *Front. Mol. Biosci.* **2022**, *9*, 931759. [CrossRef]
13. Wu, K.Y.; Tan, K.; Akbar, D.; Choulakian, M.Y.; Tran, S.D. A New Era in Ocular Therapeutics: Advanced Drug Delivery Systems for Uveitis and Neuro-Ophthalmologic Conditions. *Pharmaceutics* **2023**, *15*, 1952. [CrossRef]
14. Suri, R.; Nag, T.C.; Mehra, N.; Neupane, Y.R.; Shafi, S.; Sharma, D.; Sharma, K.; Sultana, Y.; Kohli, K. Sirolimus Loaded Chitosan Functionalized PLGA Nanoparticles Protect against Sodium Iodate-Induced Retinal Degeneration. *J. Drug Deliv. Sci. Technol.* **2023**, *82*, 104369. [CrossRef]
15. Khiev, D.; Mohamed, Z.A.; Vichare, R.; Paulson, R.; Bhatia, S.; Mohapatra, S.; Lobo, G.P.; Valapala, M.; Kerur, N.; Passaglia, C.L.; et al. Emerging Nano-Formulations and Nanomedicines Applications for Ocular Drug Delivery. *Nanomaterials* **2021**, *11*, 173. [CrossRef] [PubMed]
16. Janagam, D.R.; Wu, L.; Lowe, T.L. Nanoparticles for Drug Delivery to the Anterior Segment of the Eye. *Adv. Drug Deliv. Rev.* **2017**, *122*, 31–64. [CrossRef] [PubMed]
17. McMillan, J.; Batrakova, E.; Gendelman, H.E. *Cell Delivery of Therapeutic Nanoparticles*, 1st ed.; Elsevier: Amsterdam, The Netherlands, 2011; Volume 104, ISBN 9780124160200.
18. El Moussaoui, S.; Abo-horan, I.; Halbaut, L.; Alonso, C.; Coderch, L.; Garduño-ramírez, M.L.; Clares, B.; Soriano, J.L.; Calpena, A.C.; Fernández-campos, F.; et al. Polymeric Nanoparticles and Chitosan Gel Loading Ketorolac Tromethamine to Alleviate Pain Associated with Condyloma Acuminata during the Pre- and Post-ablation. *Pharmaceutics* **2021**, *13*, 1784. [CrossRef] [PubMed]
19. Anwer, K.; Muharram, M.M.; Mohammad, M. Development of Lipomer Nanoparticles for the Enhancement of Drug Release, Anti-Microbial Activity and Bioavailability of Delafloxacin. *Pharmaceutics* **2020**, *12*, 252. [CrossRef]
20. Andrade-Carrera, B.; Clares, B.; Noé, V.; Mallandrich, M.; Calpena, A.; García, M.; Garduño-Ramírez, M. Cytotoxic Evaluation of (2S)-5,7-Dihydroxy-6-Prenylflavanone Derivatives Loaded PLGA Nanoparticles against MiaPaCa-2 Cells. *Molecules* **2017**, *22*, 1553. [CrossRef]
21. Carvajal-Vidal, P.; González-Pizarro, R.; Araya, C.; Espina, M.; Halbaut, L.; Gómez de Aranda, I.; García, M.L.; Calpena, A.C. Nanostructured Lipid Carriers Loaded with Halobetasol Propionate for Topical Treatment of Inflammation: Development, Characterization, Biopharmaceutical Behavior and Therapeutic Efficacy of Gel Dosage Forms. *Int. J. Pharm.* **2020**, *585*, 119480. [CrossRef]
22. CPMP; ICH. *ICH Topic Q 1 A (R2) Stability Testing of New Drug Substances and Products*; European Medicines Agency: London, UK, 2003; Available online: https://www.ema.europa.eu/en/ich-q1a-r2-stability-testing-new-drug-substances-drug-products-scientific-guideline (accessed on 16 May 2022).
23. Bruna, T.; Maldonado-Bravo, F.; Jara, P.; Caro, N. Silver Nanoparticles and Their Antibacterial Applications. *Int. J. Mol. Sci.* **2021**, *22*, 7202. [CrossRef]
24. Documento de La Real Farmacopea Española Por Internet. Eficacia de La Conservación Antimicrobiana. Available online: https://extranet.boe.es/farmacopea/doc.php?id=50103# (accessed on 21 April 2022).
25. Mohammadi-Meyabadi, R.; Beirampour, N.; Garrós, N.; Alvarado, H.L.; Limón, D.; Silva-Abreu, M.; Calpena, A.C.; Mallandrich, M. Assessing the Solubility of Baricitinib and Drug Uptake in Different Tissues Using Absorption and Fluorescence Spectroscopies. *Pharmaceutics* **2022**, *14*, 2714. [CrossRef]
26. *ICCVAM-Recommended Test Method Protocol: Hen's Egg Test—Chorioallantoic Membrane (HET-CAM) Test Method*; NIH Publication: Bethesda, MD, USA, 2010; Available online: https://mdpi-res.com/data/mdpi_references_guide_v9.pdf (accessed on 2 June 2022).
27. Gopinathan, U.; Ramakrishna, T.; Willcox, M.; Rao, C.M.; Balasubramanian, D.; Kulkarni, A.; Vemuganti, G.K.; Rao, G.N. Enzymatic, Clinical and Histologic Evaluation of Corneal Tissues in Experimental Fungal Keratitis in Rabbits. *Exp. Eye Res.* **2001**, *72*, 433–442. [CrossRef]
28. Conde Penedo, A.; Díaz Tomé, V.; Fernández Ferreiro, A.; González Barcia, M.; Otero Espinar, F.J. Enhancement in Corneal Permeability of Riboflavin Using Cyclodextrin Derivates Complexes as a Previous Step to Transepithelial Cross-Linking. *Eur. J. Pharm. Biopharm.* **2021**, *162*, 12–22. [CrossRef] [PubMed]
29. Shaikh, M.I.; Jadhav, L.S.; Jadhav, R.K.; Kadam, J.V.; Pisal, S.S. Aceclofenac Organogels: In Vitro and In Vivo Characterization. *Curr. Drug Deliv.* **2009**, *6*, 1–7. [CrossRef] [PubMed]
30. Silva-abreu, M.; Calpena, A.C.; Espina, M.; Silva, A.M.; Gimeno, A.; Egea, M.A.; García, M.L. Optimization, Biopharmaceutical Profile and Therapeutic Eff Icacy of Pioglitazone-Loaded PLGA-PEG Nanospheres as a Novel Strategy for Ocular Inflammatory Disorders. *Pharm. Res.* **2018**, *35*, 11. [CrossRef] [PubMed]
31. Laignier Cazedey, E.; Carvalho, F.C.; Másquio Fiorentino, F.A.; Daflon Gremiao, M.P.; Nunes Salgado, H.R. Corrositex®, BCOP and HET-CAM as Alternative Methods to Animal Experimentation. *Braz. J. Pharm. Sci.* **2009**, *45*, 759–766. [CrossRef]
32. Danaei, M.; Dehghankhold, M.; Ataei, S.; Hasanzadeh Davarani, F.; Javanmard, R.; Dokhani, A.; Khorasani, S.; Mozafari, M.R. Impact of Particle Size and Polydispersity Index on the Clinical Applications of Lipidic Nanocarrier Systems. *Pharmaceutics* **2018**, *10*, 57. [CrossRef] [PubMed]

33. Teweldemedhin, M.; Gebreyesus, H.; Atsbaha, A.H.; Asgedom, S.W.; Saravanan, M. Bacterial Profile of Ocular Infections: A Systematic Review. *BMC Ophthalmol.* **2017**, *17*, 212. [CrossRef]
34. Wiacek, A.E.; Jurak, M.; Ładniak, A.; Szafran, K.; Przykaza, K. Cyclosporine CsA—The Physicochemical Characterization of Liposomal and Colloidal Systems. *Colloids Interfaces* **2020**, *4*, 46. [CrossRef]
35. Cheow, W.S.; Hadinoto, K. Colloids and Surfaces B: Biointerfaces Factors Affecting Drug Encapsulation and Stability of Lipid—Polymer Hybrid Nanoparticles. *Colloids Surf. B Biointerfaces* **2011**, *85*, 214–220. [CrossRef]
36. Trigo-Gutierrez, J.K.; Vega-Chacón, Y.; Brandao, A.; Garcia de Oliverira-Mima, E. Antimicrobial Activity of Curcumin in Nanoformulations: A Comprehensive Review. *Int. J. Mol. Sci.* **2021**, *22*, 7130. [CrossRef]
37. Onugwu, A.L.; Nwagwu, C.S.; Onugwu, O.S.; Echezona, A.C.; Agbo, C.P.; Ihim, S.A.; Emeh, P.; Nnamani, P.O.; Attama, A.A.; Khutoryanskiy, V.V. Nanotechnology Based Drug Delivery Systems for the Treatment of Anterior Segment Eye Diseases. *J. Control. Release* **2023**, *354*, 465–488. [CrossRef]
38. Han, H.; Li, S.; Xu, M.; Zhong, Y.; Fan, W.; Xu, J.; Zhou, T.; Ji, J.; Ye, J.; Yao, K. Polymer- and Lipid-Based Nanocarriers for Ocular Drug Delivery: Current Status and Future Perspectives. *Adv. Drug Deliv. Rev.* **2023**, *196*, 114770. [CrossRef] [PubMed]
39. Das, S.; Suresh, P.K. Drug Delivery to Eye: Special Reference to Nanoparticles. *Int. J. Drug Deliv.* **2010**, *2*, 12–21. [CrossRef]
40. Razavi, M.S.; Ebrahimnejad, P.; Fatahi, Y.; D'Emanuele, A.; Dinarvand, R. Recent Developments of Nanostructures for the Ocular Delivery of Natural Compounds. *Front. Chem.* **2022**, *10*, 850757. [CrossRef]
41. Rahdar, A.; Sargazi, S.; Barani, M.; Shahraki, S.; Sabir, F.; Aboudzadeh, M.A. Lignin-Stabilized Doxorubicin Microemulsions: Synthesis, Physical Characterization, and In Vitro Assessments. *Polymers* **2021**, *13*, 641. [CrossRef]
42. Rahdar, A.; Taboada, P.; Reza, M.; Barani, M.; Beyzaei, H. Effect of Tocopherol on the Properties of Pluronic F127 Microemulsions: Physico-Chemical Characterization and in Vivo Toxicity. *J. Mol. Liq.* **2019**, *277*, 624–630. [CrossRef]
43. Chauvet, A.; Lacoulonche, F.; Egea, M.A. PLGA Nanospheres for the Ocular Delivery of Flurbiprofen: Drug Release and Interactions. *J. Pharm. Sci.* **2008**, *97*, 5306–5317. [CrossRef]
44. Amrite, A.C.; Kompella, U.B. Size-Dependent Disposition of Nanoparticles and Microparticles Following Subconjunctival Administration. *J. Pharm. Pharmacol.* **2005**, *57*, 1555–1563. [CrossRef] [PubMed]
45. Samimi, S.; Maghsoudnia, N.; Eftekhari, R.B.; Dorkoosh, F. Chapter 3—Lipid-Based Nanoparticles for Drug Delivery Systems. In *Characterization and Biology of Nanomaterials for Drug Delivery*; Mohapatra, S.S., Ranjan, S., Dasgupta, N., Mishra, R.K., Thomas, S., Eds.; Micro and Nano Technologies; Elsevier: Amsterdam, The Netherlands, 2019; pp. 47–76. ISBN 978-0-12-814031-4.
46. Gökçe, E.H.; Sandri, G.; Eğrilmez, S.; Bonferoni, M.C.; Güneri, T.; Caramella, C. Cyclosporine A-Loaded Solid Lipid Nanoparticles: Ocular Tolerance and In Vivo Drug Release in Rabbit Eyes. *Curr. Eye Res.* **2016**, *34*, 996–1003. [CrossRef]
47. Lee, K.H.; Khan, F.N.; Cosby, L.; Yang, G.; Winter, J.O. Polymer Concentration Maximizes Encapsulation Efficiency in Electrohydrodynamic Mixing Nanoprecipitation. *Front. Nanotechnol.* **2021**, *3*, 719710. [CrossRef]
48. Arshinoff, S.A.; Hofmann, I.; Nae, H. Role of Rheology in Tears and Artificial Tears. *J. Cataract Refract. Surg.* **2021**, *47*, 655–661. [CrossRef] [PubMed]
49. Kapadia, W.; Qin, N.; Zhao, P.; Phan, C.M.; Haines, L.; Jones, L.; Ren, C.L. Shear-Thinning and Temperature-Dependent Viscosity Relationships of Contemporary Ocular Lubricants. *Transl. Vis. Sci. Technol.* **2022**, *11*, 1. [CrossRef]
50. Tiffany, J.M. The Viscosity of Human Tears. *Int. Ophthalmol.* **1991**, *15*, 371–376. [CrossRef] [PubMed]
51. Vo, A.; Feng, X.; Patel, D.; Mohammad, A.; Kozak, D.; Choi, S.; Ashraf, M.; Xu, X. Factors Affecting the Particle Size Distribution and Rheology of Brinzolamide Ophthalmic Suspensions. *Int. J. Pharm.* **2020**, *586*, 119495. [CrossRef] [PubMed]
52. Hasegawa, A.; Gulmezian-Sefer, M.; Cheng, Y.; Srikumar, R. Microbiological Considerations for Ophthalmic Products: Sterility, Endotoxin Limits, and Preservatives. In *Ophthalmic Product Development: From Bench to Bedside*; Neervannan, S., Kompella, U.B., Eds.; Springer International Publishing: Cham, Switzerland, 2021; pp. 199–227. ISBN 978-3-030-76367-1.
53. Wilson, C.G. Ophthalmic Formulation. In *Specialised Pharmaceutical Formulation: The Science and Technology of Dosage Forms*; Srivastava, V., Pitt, K., Tovey, G.D., Eds.; The Royal Society of Chemistry: Cambridge, UK, 2022; ISBN 978-1-83916-175-9.
54. Parra, A.; Mallandrich, M.; Clares, B.; Egea, M.A.; Espina, M.; García, M.L.; Calpena, A.C. Design and Elaboration of Freeze-Dried PLGA Nanoparticles for the Transcorneal Permeation of Carprofen: Ocular Anti-Inflammatory Applications. *Colloids Surf B Biointerfaces* **2015**, *136*, 935–943. [CrossRef]
55. Hirenkumar, M.; Steven, S. Poly Lactic-Co-Glycolic Acid (PLGA) as Biodegradable Controlled Drug Delivery Carrier. *Polymers* **2012**, *3*, 1377–1397. [CrossRef]
56. Diebold, Y.; Calonge, M. Applications of Nanoparticles in Ophthalmology. *Prog. Retin. Eye Res.* **2010**, *29*, 596–609. [CrossRef] [PubMed]

MDPI AG
Grosspeteranlage 5
4052 Basel
Switzerland
Tel.: +41 61 683 77 34

MDPI Books Editorial Office
E-mail: books@mdpi.com
www.mdpi.com/books

www.ingramcontent.com/pod-product-compliance
Lightning Source LLC
Chambersburg PA
CBHW040313270926
51850CB00007BC/1209
* 9 7 8 3 7 2 5 8 5 5 2 3 0 *